B. Bradshaw

B. Bradshaw's Dictionary of Mineral Waters, Climatic Health Resorts, Sea Baths and Hydropathic Establishments

B. Bradshaw

B. Bradshaw's Dictionary of Mineral Waters, Climatic Health Resorts, Sea Baths and Hydropathic Establishments

ISBN/EAN: 9783337258498

Printed in Europe, USA, Canada, Australia, Japan

Cover: Foto ©berggeist007 / pixelio.de

More available books at **www.hansebooks.com**

B. BRADSHAW'S
DICTIONARY

OF

MINERAL WATERS, CLIMATIC HEALTH RESORTS,
SEA BATHS,
AND HYDROPATHIC ESTABLISHMENTS.

Giving the Summer and Winter Residences of Doctors; Hotels which can be recommended with confidence; and other useful information.

WITH A MAP

Printed in 11 Colours,

SHOWING THE STATIONS NAMED, AND AN ITINERARY OF THE QUICKEST AND CHEAPEST ROUTES BY RAIL, BOATS, CARRIAGES, ETC.,

AND SEVEN SMALLER MAPS AND PLANS.

1884.

LONDON:

TRÜBNER & CO., 57 & 59, LUDGATE HILL.

BERLIN.   LEIPZIG.   VIENNA.
F. A. BROCKHAUS.

MILAN.
DUMOLARD FRÈRES.

PARIS.   NICE.
B. BRADSHAW & CO.

NEW YORK.   PHILADELPHIA.   CHICAGO.
THE AMERICAN EXCHANGE IN EUROPE.

OPINIONS OF THE PRESS.

A handy, well-arranged and practical guide; will prove useful to a very large class of tourists.—*The Times*, July 31st, 1884.

An excellent work. Can be thoroughly recommended as a trustworthy and convenient guide.—*The British Medical Journal*, October 27th, 1883.

The most practical and complete balneal guide we have seen.—*Der Cursalon Vienno*, July 2nd, 1884.

An admirably complete guide book.—*The Graphic*, October 27th, 1883.

Very cheap at the price charged, and we can add nothing to the praise given to its first issue.—*The Morning Post*, January 7th, 1884.

We can recommend it.—*The Daily News*, December 17th, 1883.

An invaluable book.—*Vanity Fair*, November 10th, 1883.

One of the most desirable manuals for travel. Our readers should make note of it.—*The Queen*, November 3rd, 1883.

It has no English rival. Most useful to medical men. Of great assistance to travellers and invalids.—*The Field*, October 27th, 1883.

A valuable guide, full of interest and variety. Every conceivable item of information for the benefit of the traveller.—*The Edinburgh Review*, October 19th, 1883.

Almost perfectly complete in information. The last contribution to comfort.—*The Court Journal*, November 3rd, 1883.

Imperatively needed; written by some one thoroughly conversant with the subject.—*The Dublin Daily Express*, October 12th, 1883.

An excellent guide of incalculable value to invalids.—*The Sportsman*, October 17th, 1883.

A work which may be consulted with considerable advantage.—*The Bristol Times*, November 1st, 1883.

Très utile et très pratique.—*Le Monde Elégant*, March, 1884.

A vast amount of information.—*The Birmingham Daily Post*, October 24th, 1883.

The most correct and authentic information.—*The Cambridge Express*, March 8th, 1884.

Of utmost interest to invalids and travellers. An immense amount of reliable information.—*The Freeman's Journal*, October 19th, 1883.

A capital guide. A perfect *vade mecum* for travellers and invalids.—*The Leamington Spa Courier*, October 30th, 1883.

A mass of general information not to be found elsewhere.—*The Liverpool Daily Post*, November 7th, 1883.

The author has understood to avoid the error of being too explicit at the expense of comprehensibility.—*The Scotsman*, December 25th, 1883.

Is as comprehensive and accurate as it well can be.—*The Yorkshire Post*, November 7th, 1883.

PREFACE.

In this issue, some of the most important English, Spanish, and Carpathian health resorts have had special attention.

In every instance the information has been brought up to to-day, thanks again to those friends of the work who have kindly co-operated with their experience.

A list of Specialists has, along with some further Plans and Maps, been added.

LONDON, *October 1st*, 1884.

CONTENTS.

	PAGE.
Advice to bathers	ix
Hints to visitors at watering places	xiii
Abbreviations	xxiii
Table of diseases, showing where they are most effectually treated	xxv
Comparative table of the more important climatic and winter stations	xliv, xlv
Table showing the quickest and cheapest routes to watering places and general health resorts	xlvii
Classification of stations according to their therapeutic indications	lxiii
General description of places (alphabetically arranged)	1
List of Doctors (with their residences in summer and winter)	357
Do. do. (Specialists)	392
Explanation of some technical terms	373
Classification of stations named according to their respective countries	377
Advertisements	393

ADVICE TO BATHERS.

In selecting a bathing-place, not only should the chemical composition of the waters be taken into account, but also the climate, altitude, resources of the station, habits of the patient, and his moral proclivities.

The family medical adviser should select the station most suitable for his patient, should furnish him with an introduction to the resident doctor, and should indicate the proper moment for departure. The warm months, which suit some persons, are not equally advantageous to others. Englishmen, owing to the high temperatures prevailing in July and August, often prefer to make their visits to watering-places in May and September. There are, however, many patients,— for instance, the lymphatic, the scrofulous and the rheumatic,— who require the direct heat of the sun quite as much as the thermal action of the waters. Such persons will, of course, prefer the warmest season for the mineral waters. Our ancestors would appear to have made some systematic preparations before going to the baths. *Non venite mai al bagno se non siete purgato!* We are a little too indifferent now, perhaps, to such preliminaries. We content ourselves generally with complementary treatment. The Germans, on the other hand, employ both. They have the preparatory treatment (Vorkur), and the after treatment (Nachkur), and deem the one not less important than the other.

To regain health, it is not sufficient merely to take a course of waters; it is also necessary to observe certain hygienic precautions. As regards treatment, there are perhaps no fixed rules. The indications must vary according to each individual case, for where one requires tonic, another may require alterative treatment.

Medical men are naturally the best qualified to decide what is good in each individual case. Whether, for example, an animal or a vegetable diet is in any given instance the more suitable of the two.

For some patients Bordeaux, for others a lighter wine is most suitable; the table water will also vary, from Carlsbad to Apollinaris, Oriol, Condillac, Vals or Royat, Vichy or Bourboule, as the case may be.

Many invalids find in exercise, and, if necessary, in massage and gymnastics, a complementary hygienic treatment. The want of exercise engenders disease, and is one of the most potent factors in the production of gout, gravel, obesity, anæmia, dyspepsia, and many other affections. Digestion is aided almost as much by the legs as by the stomach. It is by proper exercise that the appetite is sharpened, the digestion strengthened, nervous energy developed, and sleep induced.

Concerning wearing apparel, it may be remarked that, as a rule, mineral-water stations are situated in mountainous countries, where the temperature and hygrometric conditions of the atmosphere are subject to change every hour. For a patient who has experienced the depurative action of the waters, and has thereby been rendered more liable to take cold, a light woollen overcoat, worn as required, is always advisable.

The stimulating effects of the thermal treatment, the life led, the exercise taken, the obligation of early rising, all conduce to make sleep even more imperatively necessary than under ordinary circumstances.

It is not assuredly by leading in watering-places the same life which a man has led elsewhere—not, for example, by turning night into day—that he can expect to regain strength and nervous energy. Early rising and early retiring to rest must be the rule for all who wish to benefit by the waters.

Amusement is not absolutely necessary to the good effects of the mineral-water cure, nor is it true that it is its principal element. Bordell is said to have treated horses successfully with mineral waters, and as "amusement" or "imagination" are unknown to such animals, the most incredulous must allow that mineral waters have a curative power, due to their own specific principles. On the other hand, it is not less true, as Zimmermann remarks, that " Certain diseases can only be relieved by medicines acting on the mind, and, therefore, amusement and recreation play an important rôle in hydro-mineral medication." Persons, therefore, interested in the success of any particular spring, should bear this fact in mind, and remember that as there are patients to whom quietude, tranquility, the contemplation of nature, promenades on hills, in meadows, and so forth, may suffice, so there are others, with less bucolic tastes, who may prefer a good orchestra to the song of birds, and to whom the varied amusements of society, enjoyed in just measure, are alike useful and profitable.

The selection of an hotel is also an important matter. If an invalid is thrown among those whose conversation and habits are agreeable, and whose tastes and feelings are in harmony with his own, his sojourn will become a pleasant one. He may be induced to prolong it, and may separate with regret from his new acquaintances. How often has a friendship formed at a watering-place exercised a happy influence over the whole of after life!

HINTS TO VISITORS AT WATERING-PLACES.

DRINKING THE WATER.—It used at one time to be the custom to drink mineral waters by the gallon. Of late years, however, this practice has more or less declined, and it is now the rule to administer waters in graduated doses of one glass, half-a-glass, quarter-of-a-glass, or even, as at Eaux-bonnes, by the table-spoonful. The quantity should be diminished or increased in accordance with the indications of the disease, and the degree of tolerance shown by the patient. The waters are generally taken best in the morning, and their absorption is often promoted by a short walk. To go directly from the spring to the bath, or to drink the waters while in the bath itself, frequently disagrees, causing a sensation of weight in the stomach, or even nausea. Strictly speaking, all mineral waters might be taken internally. In practice, however, the hypothermal are reserved exclusively for external use.

INHALATION.—The object of inhalation is to cause certain medicinal substances to enter the lungs directly. These substances may be in the form of gases, vapours or spray. It has long been known that the pulmonary mucous membrane can be acted on by substances inhaled. It was this knowledge which led GALEN to send his patients to the sulphurous air of Sicily, and LAEMER to hang seaweed round the walls of the rooms occupied by sufferers from consumption. Pulmonary absorption often takes place with great rapidity, as is shown by the fact that merely passing through a malarial swamp will often induce an attack of fever. There is a double advantage in inhalation; it brings the medicament in direct contact with the affected part, and also relieves the stomach, already perhaps over-stimulated by excessive and long-continued dosing with drugs.

BATHS.—In the present state of our knowledge it is scarcely possible to say how mineral baths exactly act. It is reasonable to attribute their action to absorption through the skin. This however is denied by some physiologists. M. DURRIEN, for instance, maintains that the weight of the body is unchanged after a bath of moderate temperature, but that it gains or loses weight in proportion as the bath is under or over a mean temperature. SCONTETTEN maintains that a mineral water bath acts less by the mineral principles contained in it than by the electricity liberated. He suggests a study of the magnetic needle as a means of judging of the efficacy of a spring. A bath of from 50 to 76° F. is considered tepid, from 76 to 92° F. warm, from 92 to 112° F. hot. A mean between the two latter is the temperature of the reservoirs in the thermal establishments. It is of course impossible to give any one temperature which shall be applicable to all cases. Some individuals will be affected by a temperature of 76° more than others will by one of 82°. When, however, the temperature of the bath is unsuited to a particular case less benefit is derived from the waters. A patient should therefore note accurately the thermometer before entering his bath, and not, as is so often done, rely on his own tactile sensibility, or on that of his attendant. As a rule, to which, however, there are exceptions, the stay in the bath should not exceed one hour; nor should it be continued after the temperature begins to fall to luke warm, for a bath should either be hot or cold. If a warm bath be continued too long it is apt to produce determination of blood to the skin; if a cold bath be so continued, it is apt to produce a revulsion of blood towards the internal organs.

SHOWER BATHS AND DOUCHES.— The douche bath has only been prescribed systematically of recent years. Its chief advantages are that it can be used so as to impart a shock or stimulus, that it requires less time, and that it can be administered at what are comparatively extreme temperatures. The douche consists essentially of a jet of water, which as it were encloses the whole or any given portion of the body. Its effects can be modified by altering the temperature, direction, power, or form of the *jet d'eau*.

MASSAGE, or "Kneading," is a method of treatment which has prevailed in the East from time immemorial. It is now common enough in most balneal establishments in Europe. Long abandoned to quacks and empirics, who owed to it much of their success in treating joint affections, it has at length taken its place as a recognised method of treatment. Massage consists essentially of a series of acts of pressure and traction alternately on the skin, muscles and sinews. This manipulation can be combined, as desired, with the douche, the sweating bath, the simple bath (wet massage), or without water at all (dry massage). From a physiological point of view it may be said that the douche and massage act by augmenting the quantity of blood passing through a tissue or organ in a given time. The process of organic exchange thereby becomes more active, and nutrition is stimulated.

GYMNASTICS form a useful addition to hydrotherapeutics. The gymnastics here spoken of have of course nothing to do with professional gymnastics, but are simply and solely medical exercises. Their object is to restore the body to its normal form by suitable and continuous motion, and to compel feeble or undeveloped organs to undertake their fair share of physiological work. To this system of treatment the illustrious DELPECH gave the name of "orthomorphism."

MINERAL WATERS FROM A THERAPEUTIC POINT OF VIEW. The term "mineral waters" is applied to those waters which hold in solution a variety of mineral substances at various temperatures. Some writers have denied the therapeutic efficacy of mineral waters, basing their opinions on the minute quantities of mineral matter held in solution. In many cases however the waters act with so much energy that no scepticism as to their power is possible. Thus at Eaux-bonnes a season rarely passes without some sufferer inducing hæmoptysis by an indiscreet use of the waters. At Ax, Dr. GARRIGON has known death to occur from the injudicious and excessive use of the Vignerie spring. The *Carlsbad* waters, if taken by persons of a plethoric and sanguine habit of body, may occasion dangerous symptoms. Many of these mineral waters are of a much more complex character than the older chemists, with their less perfect methods of analysis, supposed. Quite recently Dr. THENARD has discovered arsenic in the waters of Mont-Dore, while in other springs the existence of rare elements, such as rubidium, can now be demonstrated. The human body indeed, as every physiologist knows, is itself sufficiently rich in minerals. Iron, lithium, manganese, lead and copper are found in it, besides phosphorus, sulphur, silica, chlorine, fluorine, potassium, sodium, calcium and magnesia in more or less intimate union with the common substances of all organic life, viz., oxygen, hydrogen, carbon and nitrogen—a sufficiently long list it will be allowed.

Modern pathology teaches that many diseased conditions of the body depend on deficiency or undue excess of mineral constituents. Thus a deficiency of iron shows itself in the blood by clorosis and anæmia, both conditions being soon removed when iron is supplied. Goitre, it has been suggested, depends on the absence of iodine in drinking water, just as stone in the bladder is often induced by drinking excessively hard water. Osteomalachia, or softening of the bones, as also rickets, depend primarily on a deficiency of the salts of lime. Hence, to restore these absent principles would seem at once the most efficacious and common-sense plan of treatment. This is exactly what mineral waters do, but they do it much more surely and quickly than any mixtures compounded artificially in the chemist's laboratory. The similarity existing between the constituents of the human body and of certain mineral waters did not escape the illustrious SYDENHAM, who remarked that chalybeate waters will sometimes restore patients to health more rapidly and thoroughly than the most lauded pharmaceutical preparations. Similarly, Dr. BAZIN, Professor of Dermatology at the Hospital St. Louis, Paris, holds that natural sulphurous waters are much more powerful therapeutically than any which can be produced by art. The mere fact that mineral waters have so long held their own in medicine, and are to-day more used than ever, is a strong argument in favour of their value. When rightly used they constitute a complex, but a very real remedy, the more valuable because prepared by Nature herself. One proof is as good as a thousand. The physiological effects of only 0·25 gram. of iron in spa water will show themselves quicker and more pleasantly than many times the amount of the same metal in an artificial compound. The reason is that from some cause or another natural iron water is more quickly absorbed and assimilated. It has been suggested that the presence of other minerals and chemical compounds render it more digestible. Whatever the reason may be the fact is indisputable, as many have learned from personal trial.

Medical experience goes to show that almost every affection which subsequently becomes chronic might at one period of its history have been relieved. All, at a certain stage are susceptible of relief by a suitable course of waters. There exists scarcely a diathesis, or "state of constitution," which may not be modified and improved by these natural remedies. Especially is this true of those faulty states to which the name "syphilitic," "scrofulous or strumous," "herpetic," and "gouty" have been given. A judicious and early use of mineral waters will often check the first manifestations of these unhealthy conditions, which, in the majority of cases, are hereditary rather than acquired. The action of mineral waters is well-nigh as varied and subtle as their composition. The great majority, however, of all medicinal water act essentially as tonics and alteratives. Hence the fact that a certain unhealthy condition of body may often be modified or changed by springs apparently widely different in composition. The important fact to bear in mind is that mineral waters exercise an influence above and beyond that which their merely chemical composition would seem to indicate. This is well-known to medical men in all the scientific centres of Europe. Hence it is that BAZIN sends sufferers from psoriasis to Royat, that HARDY sends herpetics to St. Gervais, that GUENEAU DE MUSSY sends chest cases to Mont-Dore, that RICORD and CALVO send syphilitics to Aix-les-Bains, that Sir HENRY THOMPSON of London sends calculous cases to Marienbad, Pullna, and Vals, that GARROD sends gouty and rheumatic sufferers to Aix-les-Bains, and CAVELOT of Naples lymphatic and strumous patients to Ischia.

b

CLASSIFICATION OF MINERAL WATERS.—It is by no means easy to classify mineral waters. Indeed, a simply arranged but scientific catalogue of the various springs of Europe is still a desideratum, for balneology has not as yet found its Linnæus or Jussien. Chemists are scarcely agreed as to what should form the basis of classification, whether it should be the mineral elements themselves, or the acids with which they are united in the form of various salts. The chief mineral salt present may be taken roughly as characteristic of a particular spring, and as imparting to it certain distinctive characters. It should, however, be borne in mind that the element of quantity is not always identical with that of quality, and that often chemical analysis is a most uncertain guide to therapeutic value.

The following classification may be adopted in the absence of a better, viz. :—

1. Indifferent, thermal, or cold.
2. Alkaline waters.
3. Alkali-saline waters.
4. Acidulated (bitter) waters.
5. Saline waters.
   - (a.) simple saline.
   - (b.) sea water.
   - (c.) iodo-bromide.
6. Sulphurous waters.
7. Ferruginous waters.
8. Lime and earthy waters

However defective this classification may be, it has the great advantage of indicating the mineral constituents of most importance from a therapeutic point of view. Thus, speaking in general terms, it may be said that lung and skin affections are benefited chiefly by sulphurous waters, the uric acid diathesis by the soda bicarbonate, the scrofulous by the alkaline, anæmia and chlorosis by the chalybeate, syphilis in its various manifestations by the bromiodides, and herpes by the arsenical.

MILK AND BUTTERMILK CURE.—The milks used for the former are those of the cow, goat, sheep, donkey, and mare. Of these five, that of the donkey contains most water and least organic matter, i.e., nearly 9 %; next comes goat's milk, with $13\frac{1}{2}$ %;

cow's, with 1¼ %; sheep's, with 16 %; and mare's, with nearly 17 %. Sheep's milk is, therefore that most generally used. Mare's milk is employed only in Russia, Hungary, and at a very few places in Germany. Sheep's milk is the most nutritive, but at the same time the most difficult of digestion.

Casein is contained in the various milks in the following proportions :—Cow's, 5 %; sheep's, 4¾ %; goat's, 4½ %; donkey's 2 %; and mare's, 1½ %.

Milk as a nutritive element has been used from time immemorial. With regard to their respective nutritive values, the various milks are classified as follows: -

(1) *Cow's.*
(2) *Goat's.*—Rich in albumen, and especially valuable in intestinal catarrh.
(3) *Sheep's.*—Very rich, and a valuable remedy in cases of reduced alimentation.
(4) *Donkey's.*—Relatively the poorest in nutritive constituents, but still one of the most important on account of its ready digestability and mild purgative action. It is especially valuable in chronic chest complaints, with frequent acute exacerbations and pyrexia, and where there is no intestinal catarrh.
(5) *Mare's milk,* though as yet but little employed as a curative agent, is actually the richest of all the five. It is very poor in protein, but is the richest in butter, as also in milksugar, and salts. Its chemical composition suggests that in chronic tuberculosis of the respiratory organs it might prove to be a very valuable therapeutic agent.

The special treatment advisable in each case differs considerably, and can best be decided by the medical practitioner. Some physicians forbid other nourishment whilst patients are undergoing the milk cure ; others admit a very small quantity, while others again continue the usual diet. Some mix the milk with water, others with coffee, and others prescribe it after the cream has been removed. The milk cure has given good results in dropsy, in difficult respiration depending on emphysema and catarrh, in chronic diarrhœa, in liver affections, in dyspepsia, in diabetes, in morbus Brightii, in diseases of the heart, in articular rheumatism, and in nervous and other affections.

BUTTERMILK CURE.—The ordinary residue of the milk after churning is sometimes used as a cure, and has given good results in cases of plethora abdominalis, chronic constipation, chronic heart disease, and especially in ulcers of the stomach.

WHEY AND KOUMISS CURE.—Whey is milk deprived of its casein. It has a pale green colour, is transparent, has a sweetish taste, and a flavour of goat's milk. It contains in 100 parts:—

	Sheep's.	Cow's.	Goat's.
Water	91·960	93·264	93·380
Albumen	2·130	1·080	1·140
Milksugar ...	5·070	5·100	4·530
Fat	0·252	0·116	0·372
Salt and other substances	0·588	0·416	0·578

The whey cure is generally prescribed in cases of scrofula, incipient tuberculosis, rheumatism, and hereditary gout, plethora abdominalis, chronic phthisis, pneumonia, and bronchitis, laryngitis, abdominal disorders, liver and kidney affections.

The cure is usually carried out in mountainous and woody places, as the air, climate, &c., work in conjunction with the whey. The daily quantities prescribed, vary according to individual habits, and range between ¼ to 1 litre a day. The latter amount, however, can be supported by very few, and is rarely given. The taste of whey is subacid and not unpleasant. Medical authorities are still undecided as to whether it is the whey or the mountain air, or the emanations of the forest which has so often done good in many of the above complaints.

Koumiss is fermented mare's milk and is held in high favour as a remedy against consumption with the nomad tribes of the Russian and Siberian Steppes. There are but few establishments in Germany, Austria or Switzerland, which include Koumiss in the list of their therapeutic agents. Taken in large quantities Koumiss allays the desire for solid food. Invalids can live for weeks, or even months, without any other nourishment but Koumiss itself.

The mode of preparation is too long to describe, besides which, it varies in almost every station. Koumiss creates a feeling of comfortable warmth in the stomach. Its taste is agreeable. When newly made it has a purgative, but when old an astringent action. As already remarked it is a valuable remedy in phthisis, chronic bronchial and intestinal catarrh, and diarrhœa, anæmia, chlorosis and scrofula. In Germany, as also in Switzerland, Koumiss is prepared from cow's, donkey's and goat's milk, but the results are not the same as those obtained with mare's milk. The Koumiss cure to do any good must be thorough. Taken in small doses it has no effect. Generally one bottle a day is used at first, and this quantity is augmented to five bottles later on. Very little solid food is taken, and that only in the evening when the treatment of the day has been finished. The temperature of the Koumiss should not be higher than 90° F. Nervous invalids and those who have an aversion to it, may take it at from 45° F. to 50° F.

GRAPE AND FRUIT CURES IN GENERAL:—Grapes contain in the mean in 1,000 parts—

Water	760	810
Sugar	106	330
Acid	3·5	10·2
Albumen	5·0	20·0
Pectin, &c.	2·5	30·0
Salts	2·0	4·0

The most important constituent is the grape sugar. This varies in quantity very considerably, according to the greater or less warmth of the climate. In Hungary—Ruszt and Tokay—301 parts of grape sugar are found; in Southern France, 240; Styira, 215; Moselle, 214; Worms, 201; Adlersberg, Hungary, 192; Stuttgart, 190; Bohemia, 185; Heidelberg, 180; on the Neckar, 159. Spain, Italy and Madeira, are not included here, as the grape in these countries is not employed therapeutically.

The quantity of grapes prescribed during the cure varies from 4 to 10, and sometimes even to as much as 12 pounds daily. It is but natural to suppose that the human body, subjected to such treatment for three and four weeks continuously must feel the effects of a diminished food supply, and of the stimulus given to secretion. The material absorbed operates chiefly on the kidneys and intestines.

The grape cure is indicated in cases of scrofula, anæmia, indigestion, convalescence after fevers, chlorosis, and menstrual disorders.

Favourable conditions or otherwise of climate play an important part in the results of treatment. Taken in very large quantities, 8 or 10 lbs. a day for instance, the grape cure is very effective in cases of plethora-abdominalis, haemorrhoids, constipation, hypochondriasis, and chronic catarrh of the respiratory organs.

In phthisis and gravel the results are somewhat uncertain.

Instead of grapes, other fruits may be similarly employed, as, for example, cherries, **strawberries**, the various **varieties of currants**, &c.

Raw ham, and other raw meats are likewise, though very rarely, prescribed as cures.

ABBREVIATIONS.

P.L.M. Paris, Lyons, and Mediterranean Railway.
S. & F. South of France (Midi) Railway.
N.F. North ,, ,,
E.F. East ,, ,,
O.F. West ,, ,,
Orl. Orléans Railway.
P.S.R. Prussian State Railways.
B.R. Belgian Railways.
D.R. Dutch Railways.
S.A.R. South Austrian Railway.
C.G.T. Compagnie Générale Transatlantic.
T. Temperature.
F. Fahrenheit.
F. Franc.
C. Centime.
M. Marks.
Pf. Pfennigs.
K. Kilomètres.

DISEASES:

AND PLACES WHERE THEY ARE MOST EFFECTUALLY TREATED.

Name of Disease.	Medium.	Places.
Abdominal swellings.	Iodo-bromide waters	Goczalkowitz, Königsdorff-Jastrzembs, **Ravone in Casaglia**, Sulzbath, Zajzon, **Bex**.
	Sea baths ...	North Sea, Atlantic and Mediterranean.
	Saline waters	**Acquae-Albulae, Ischl**, Amélie-les-Bains, **Baden-Baden**, Castro-Caro, **Cheltenham**, Gerace, Kissingen, Harrowgate, Leamington, Kreuznach, Mehadia, **Riolo, Uriage,** La Bourboule.
	Sulphurous waters	**Acireale, Acquae-Albulae, Aix-les-Bains**, Amélie-les-Bains, **Baden** (Austria), **Brussa, Castellamare**, Chianciano, Castel St. Pietro, Eaux bonnes, Gerace, Harkanyi, **Hélouan**, Ischl, Leamington, Parád, Strathpeffer, Telese, **Alveneu, Bagnères de Bigorre, Cauterets**.
Acne	Various waters ...	Ems, Enghien, Luchon, Schlangenbad, Uriage, Vals, Vichy, Dax, Gastein, **Ledesma**, Nanclares de Oca, **Schinznach**.
Adenitis	Saline iodurated waters	**Ischl,** Castro-Caro, Airthrey, Dürkheim, Gazost, Krankenheil, Lipik, Radein, Jvonicz, **Baden-Baden, Uriage,** Riolo.
	Alum waters ...	Erdöbenye, **Parád**.
	Sea baths	Brittany and French Atlantic coast.

Name of Disease.	Medium.	Places.
Albuminuria	Alkaline ...	Giesshübl, Fachingen, **Preblau**, Teinach, Salzbrunn.
	Ferruginous alkaline	Contrexéville, **Schwalbach**, Schlesisa-Wodsk, Harrowgate, Franzensbad.
	Alkaline earthy ...	**Chianciano, Bath**, Loëche, Contrexéville, Szkleno, Wildungen.
Amaurosis ..	Purgative waters...	Marienbad, **Castellamare**, Kissingen, Tardou, St. Alban, Brides.
	Hydropathy	**Divonne - les - Bains**, Baden-Baden, **Ischl, Vöslau**, Laubbach, Bushey.
Amenorrhoea ...	Sulphurous waters	**Aix-les-Bains, Baden** (Austria), **Castellamare, Acireale, Uriage, Eaux Bonnes.**
	Chalybeate waters	Schwalbach, **Malahá**, Contrexéville, **Pyrmont**, Cleve, Gleichenberg, **Chianciano, St. Moritz.**
	Indifferent waters and hydropathy.	**Laubbach**, Bushey.
Anæmia	Ferruginous waters	**La Bourboule**, Gleichenberg, **Chianciano**, Harrowgate, Szliacs, Schwalbach, **St. Moritz**, Alveneu, Rippoldsau, **Bagnères de Bigorre.**
	Hydropathy ...	**Divonne, Baden-Baden**, Godesberg, Mondorf, Strathpeffer, Bushey, **Droitwich,** Laubbach.
	Indifferent waters	Badenweiler **Vöslau**, Römerbath (Austria), Laubbach, Bushey, Plombières.
	Winter stations ...	**Acireale, Castellamare**, Madeira, the Riviera, Biarritz, Corfu, **Ajaccio.**
	Climatic air stations	**Ischl, Schmécks, Clifton.** Alveneu, **Loëche**, Campfer, Flims, **St. Moritz**, Pontresina, Mürren, Samaden, **Maloja**.
	Sea baths	North Sea, Atlantic, Mediterranean.

DISEASES.

Name of Disease.	Medium.	Places.
Anchylosis	Thermal sulphurous	Aix-les-Bains, Chianciano, Amélie-les-Bains, Dax, **Loëche**.
	Sulphurous saline	Acireale, Ischl, Baden (Switzerland), Mehadia, Balaruc, Bourbonne, **Tiermas**.
	Mud and sand baths	Acqui, St. Amand, Dax, Balaton-Füred.
	Thermal alkali-saline	**Teplitz-Schönau**, Karlsbad.
	Saline bromo iodurated	**Bex, Malvern**.
	Indifferent waters	Wildbad, Laubbach, Bushey, Plombières.
Angina	Sulphurous waters	Eaux-Bonnes, **Ischl**, Amélie, Allevard.
Arthritis	Hydropathy	Divonne-les-Bains, **Laubbach**, Bushey, Droitwich.
	Alkaline saline waters	Wiesbaden.
Asthma	Mountain air-cure	Ischl, Gleichenberg, Chamounix, Veldes, Grindelwald, La Bourboule. **Mürren**, Pontresina, **St. Moritz**, Rippoldsau.
	Sea baths	Biarritz, Houlgate, Doberan, Brighton, Scarborough.
	Brine baths	Droitwich, **Malvern**.
	Sulphurous waters	Eaux-bonnes, Acireale, Chianciano, Cauterets, **Schinznach**.
	Alkaline waters	Ems, Gleichenberg, **Recoaro**
	Mixed and ferruginous	Mont-Dore, **Alhama de Aragon**.
	Indifferent waters and hydropathy	**Laubbach**, Bushey, Droitwich.
Ataxy (Locomotor)	Ferruginous alkaline waters	Schwalbach, Pyrmont, Chianciano, Gleichenberg, Contrexéville.
	Sulphurous waters	Aix-les-Bains, Ischl, Challes, Baden (Austria).
	Hydropathy	Laubbach, Bushey.
Atrophy (Muscular)	Various waters	**Baden-Baden, Aix-les-Bains**, Loëche, Chianciano, **Acireale, Schwalbach**.

Name of Disease.	Medium.	Places.
Blenorrhoea	Various waters	Cervera del Rio Alhama, Anaélic, Pyrmont, **Teplitz-Schönau**, Recoaro, Castellamare, Schwalbach.
	Sea baths	Baltic and Mediterranean.
Boils and Abcesses	Sulphurous	Aix-la-Chapelle, Gazost, Guagno, Caldaniccia.
	Saline waters	Bourbon Lancy, Bourbon l'Archambault, Baruc, **La Bourboule, La Garriga**.
	Saline sulphurous	Baden (Austria), Mehadia, **Uriage**, Acireale, **Chianciano, Schinznach**.
Bones, diseases of the	Sea baths	North Sea and Atlantic.
	Saline waters	Ischia, **Ischl**, Nauheim, **Baden-Baden**, La Bourboule.
	Thermal-sulphurous waters.	**Aix-les-Bains**, Uriage, Alhama de Aragon, **Baden** (Austria), Mehadia, **Chianciano, Zaldivar**, Fitero, **Ledesma**.
	Sulphurous saline waters	Alveneu.
	Thermal hydropathy	Laubbach, Bushey.
Bright's disease	Hydropathy	**Malvern**, Bushey, **Droitwich, Laubbach**.
Calculus (Biliary)	Alkaline ferruginous waters	Gleichenberg, Pyrmont, Schwalbach, **Contrexéville**.
	Mild alkaline waters	Ems, Neyrac.
	Mixed	Castellamare.
Calculus (Urinary and Vesical)	Sulphurous waters	Acireale, Castellamare, **Salvadora**.
	Alkaline waters	Mont-Dore, Evinn, Gleichenberg, **Ibero**.
	Alkaline ferruginous waters	Pyrmont, **Contrexéville**, Belascoin, Sousas.

Name of Disease.	Medium.	Places.
Catarrhs (Bronchial)	Ferruginous waters	Gleichenberg, Mont-Dore.
	Hydropathy	Divonne-les-Bains, Veldes, **Malvern**, Laubbach.
	Sulphurous waters	Acireale, Eaux-Bonnes, Chianciano, Ledesma, **Zujar**.
	Saline arsenical waters	La Bourboule.
	Sulphurous saline waters	Alveneu, Acireale, Schinznach, Amélie-les-Bains, **Cambo, Cauterets**.
	Climatic air and winter stations	Corfu, **Ajaccio**, Flims, Campfer, **Wildbad**, Cambo, **Clifton**.
Catarrhs (Pulmonary)	Thermal-sulphurous waters	Eaux-Bonnes, Chianciano, Acireale, Ischl, Baden (Austria), Amélie-les-Bains, **Wildbad, Cauterets**.
	Various waters	Ems, Baden-Baden, Badenweiler, Gleichenberg, **La Bourboule**.
	Winter stations	Acireale, Hyères, Biarritz, Abazzia, Grasse, Cannes, **Mentone, San Remo, Cambo**.
Chlorosis	Ferruginous waters	Loëche, Plombières, Gleichenberg, Schwalbach, Mont-Dore, **St. Moritz**, Alveneu, Rippoldsau, Bagns. **de Bigorre**.
	Sulphurous waters	Harrowgate, Acireale, Ischl.
	Sea baths	North Sea and Atlantic.
	Hydropathy	**Divonne-les-Bains**, Godesberg, Laubbach, Bushey, Droitwich.
	Climatic air station	Campfer, Mürren, Corfu, Ajaccio, Pontresina, St. **Moritz**, Samaden, Maloja, **Clifton**.
Chorea or St. Vitus Dance	Thermal hydrotherapy	Aix-les-Bains, Baden (Austria), Bushey, Mehadia, **Baden-Baden**, Loëche, Divonne, **Laubbach**, Brighton, Plombières.
	Lake baths	Neusiedel, Balaton, Neuchatel, Geneva, Constanze.
	Bitter waters	**Frailes y la Rivera**.

DISEASES.

Name of Disease.	Medium.	Places.
Congestions (Cephalic)	Purgative and laxative waters	Castellamare, Püllna, Bilin, Ems, Niederbronn, Vöslau, **Chianciano**, Gerace.
	Thermal hydropathy	Laubbach, Bushey, Droitwich, Plombières.
Contractions (Tendinous or articular)	Various waters	**Aix-les-Bains**, Loëche, Baden (Austria), Uriage, **Baden-Baden**, **Dax**, Acireale, Bex, **Benimarfull, Malvern**, Plombières.
	Alkaline waters	Sierra Alhamilla, **Zujar**.
Coryza	Alkaline ferruginous waters	Mont-Dore, Contrexéville, Schwalbach.
Convalescence	Indifferent waters	Vöslau, Badenweiler, Veldes, **Wildbad**, Laubbach.
	Climatic air stations	Ischl, Axenstein, Bushey, Sonnenberg, Grindelwald, Schlangenbath, **La Bourboule, Loëche**.
	Climatic air and winter stations	Flims, Corfu, **Ajaccio**, Algiers, Pontresina, **Cambo, Maloja**.
	Ferruginous waters	Chianciano, Schwalbach, Pyrmont, Contrexéville, **St. Moritz**.
	Sulphurous and Winter station	**Acireale**, Corfu, Ajaccio.
	Saline waters	**Baden-Baden, La Bourboule**.
Debility (Nervous and chronic)	Chalybeate	**Alveneu**, Rippoldsau, **Maloja**, Tunbridge Wells, Gleichenberg, Schwalbach, Chianciano, Plombières.
	Climatic air station	**La Bourboule, Ischl**, Badenweiler, Wildbad, Axenstein, **Sonnenberg**, Clifton, Grindelwald, Villeneuve, Mentone, San Remo, Alveneu, **St. Moritz**, Corfu, **Ajaccio**, **Flims**, Samaden, Maloja.
	Indifferent waters	Wildbad, Laubbach, Bushey, Plombières.
	Brine baths	**Droitwich**.

DISEASES.

Name of Disease.	Medium.	Places.
Diabetes ...	Alkaline waters, and alkaline saline waters	Clifton, Royat, **Teplitz-Schönau**, Marienbad, Contrexéville, Vichy, Mont-Dore, **La Bourboule, Villatoya, Wiesbaden.**
	Alkaline ferruginous waters	Tunbridge Wells, Gleichenberg, Schwalbach, Contrexéville.
	Brine baths ...	Droitwich.
Dysenteria ...	Various waters ...	Niederbronn, Vals, Weissenburg, **Teplitz-Schönau.**
Dysmenorrhœa	Thermal hydrotherapy.	**Baden-Baden, Ischl,** Bushey, Gastein, Badenweiler, Topuszko, Laubbach.
	Chalybeate waters	**St. Moritz**, Alhama de Aragon, Plombières.
Dyspepsia ...	Gaseous waters ...	Condillac, Selters, Apollinaris, St. Alban.
	Alkaline waters ...	**Riva los Baños**, Gleichenberg, Chianciano, **Teplitz-Schönau.**
	Effervescent ferruginous waters.	Oriol, Pougues, **Schwalbach**, Pyrmont, Cheltenham, Rippoldsau, St. Moritz, **Maloja.**
	Effervescent saline waters.	Neiderbronn, Homburg, **Baden-Baden**, Wildungen, **Pyrmont.**
	Saline waters ...	Cestona-guezalaga, **Frailes y la Rivera.**
	Hydropathy ...	**Divonne**, Gräfenberg, Vöslau, Laubbach, Bushey, Plombières.
Emphysema ...	Ferruginous waters	Gleichenberg, Chianciano, Pyrmont, Recoaro.
	Climatic air and winter stations.	**Acireale**, Malaga, **Madeira, Abazzia,** Ajaccio, Schmécks, Zermatt, Chamounix, Cannes, **Mentone**, San Remo, **Laubbach.** Hyères, Bushey, Cambo.
	Sulphurous saline waters	Schinznach, **Cambo.**

DISEASES.

Name of Disease.	Medium.	Places.
Female disorders.	Hydropathy	Divonne-les-Bains, Vöslau, Droitwich.
	Ferruginous waters	Schwalbach, Loëche, Mont-Dore, Chianciano, Forges-les-Eaux, Contrexéville, Alveneu.
	Sulphurous waters	Aix-les-Bains, Bagnères de Bigorre.
	Indifferent waters	Wildbad, Schlangenbad.
Fractures	Saline thermal waters.	Baden-Baden
	Sulphurous waters	Baden (Austria), Aix-les-Bains, Chianciano, Uriàge, Acireale.
	Thermal alkaline saline waters	Teplitz-Schönau, Bertrich, Capvern.
	Indifferent waters	Wildbad, Laubbach, Bushey.
Ganglionic tumours	Sulphurous iodo-bromurated waters.	Saxon, Challes, Wildegg, Harrowgate.
	Saline iodo-bromurated waters	Ischl, Nauheim, Salins, Hall (Austria), La Garriga.
Gastralgia	Hydropathy	Divonne, Gräfenberg, Laubbach, Bushey, Ben-Rhydding.
	Effervescing mild mineral waters.	Wildbad, Rippoldsau, Badenweiler, Vöslau, Teplitz-Schönau, Schmécks, Sousas, Amphion, Alveneu.
	Saline sulphurous waters	Borines, Quinto, Cauterets.
Glands (swellings of)	Sulphurous and alkaline waters	Paracuellos de Jilloca, Cauterets,
	Bitter and ferruginous waters	Vilo or Rosas, Villaharta, Rippoldsau.
Goitre	Saline sulphurous iodurated waters	St. Genis, Challes, Marlioz, Chianciano, Alveneu.

DISEASES.

Name of Disease.	Medium.	Places.
Gout	Alkaline waters	Marienbad, **Wiesbaden, Teplitz-Schönau, Contrexéville.**
	Sulphate of lime waters	Aulus, Vittel, Capvern, Bath, Loëche.
	Various	**Uriage,** Gleichenberg, **Wildbad, Castellamare,** Tiermas, Schlangenbad.
	Sulphurous waters	**Aix-les-Bains,** Escaldas, Harrogate, Acireale, **Baden (Austria), Schinznach.**
	Thermotherapy	**Laubbach,** Bushey, **Droitwich, Malvern.**
	Ferruginous and lime waters	Alveneu, Rippoldsau, Plombières.
Gravel (Uric)	Alkaline waters	Carlsbad, **Contrexéville, Teplitz-Schönau,** Gleichenberg, Chianciano, Ibero.
	Ferruginous waters	**Amphion.**
Gravel (Phosphatic.)	Sulphate of lime waters	Vittel, Ischl, La Preste, Molitg.
	Mild mineral	Vöslau, Badenweiler, Schlangenbad, **Contrexéville,** La Salvadora.
Hæmorrhoids	Alkaline waters	Carlsbad, Le Boulou, Hovingham.
	Hydropathy	Divonne, **Laubbach,** Bushey.
	Acidulous	Castellamare.
	Sulphurous and chalybeate waters	Alveneu.
	Ferruginous	**Rippoldsau.**
Herpes	Saline sulphurous waters	**Aix-les-Bains, Castellamare, Baden (Austria), Schinznach,** Cambo, Barèges, Cauterets, **Uriage,** Chianciano, **Ischl,** Toleze, Amélie-les-Bains.
	Arsenical waters	**Baden-Baden, La Bourboule,** Cheltenham, St. Honoré, **Bagnères de Bigorre.**

Name of Disease.	Medium.	Places.
Hydropsia	Acidulous ferruginous sulphurous waters Hydropathy	Castellamare, Alveneu, **Bagnères de Bigorre**. Laubbach, **Bushey**.
Hypochondriasis	Purgative waters... Saline ferruginous waters Hydropathy Winter stations	Vacqueiras, Montecatini, Hunyadi-Yános, Epsom, Matlock, **Castellamare**. Rippoldsau, **Amphion**. Divonne, Coombe Wood, Bushey, Gleichenberg, **Laubbach**. The Riviera and Spanish Coast towns, **Corfu**.
Hysteria ...	Sedative Sulphurous bitter. Ferruginous and magnesian waters Climatic air and winter stations Sea baths ...	Schwalbach, Badenweiler, **Ischl, Vöslau**, Neris, Schlangenbad, Bormio, Plombières. Jabalcuz. Rippoldsau, **Luxeuil**, **Orezza, Römerbad** (Switzerland) Navalpino. Axenstein,Schmécks,Veldes, Pisa, Pau, Dax, Leghorn, Bushey, Abazzia, **Hyères, Laubbach**. **Livorno**, Hastings, Scarborough, Ventnor, Blankenbergh, Zandvoort, The **Riviera**.
Impotence ...	Various waters ... Chalybeate waters Saline waters Sea baths ... Sulphurous waters	Wildbad, **Gastein**, La Battaglia, Bushey, Vöslau, Spa, **Laubbach**. Pyrmont, Schwalbach,**Tunbridge-Wells**. Balaruc, **Cornetto**, Montecatini, **Droitwich**. North **Sea** and Baltic. Aix, Harrowgate, **Uriage, Baden (Austria)**.
Intermittens ...	Saline arsenical waters Indifferent waters	**La Bourboule**. Plombières.
Intestinal ... ... (Catarrhs, &c.)	Various waters ...	**Bagnères de Bigorre, Baden**-Baden, Cheltenham, Leamington, Chinciunou, Gleichenberg, Casstellamare, Vinadio, Vöslau, Alhama de Granada, Prelo, Rippoldsau, **Wiesbaden, Cambo**.

DISEASES.

Name of Disease.	Medium.	Places.
Intumescence (of liver and spleen).	Alkali saline waters Alkaline ferruginous waters Sulphurous earthy	**Teplitz-Schönau**, Bertrich, Sousas. Gleichenberg, Contrexéville. Salvadora.
Jaundice... ...	Muriatic saline and acidulous waters	Castellamare, **La Bourboule**, Prelo.
Kidney complaints.	Hydropathy ... Saline waters ... Alkaline waters ...	Laubbach, Bushey. **Baden-Baden, La Bourboule.** Sacedon.
Laryngitis ...	Climatic air and winter stations Sulphurous saline waters Alkaline-saline waters	Corfu, **Bex**, Ajaccio, Campfer, **Múrren**, Maloja, **Biarritz**, **Cambo, Clifton.** Betelú, Alveneu, Ormaiztegui, **Schinznach, Aix-les-Bains**, Cambo. **Wiesbaden.**
Leucorrhœa ...	Ferruginous waters Saline waters Sea baths ... Bitter waters	**Castellamare**, Schwalbach, Chianciano, Orezza Plombières. **Bigorre**, Solares, **Ischl**, Paterna. The North Sea and Atlantic. Buzot, Gravalos.
Liver obstructions.	Alkaline-ferruginous waters. Purgative waters... Hydropathy ...	Gleichenberg, Tepl **Chianciano**, Pyrmont, **Contrexéville, Rippoldsau. Cauterets,** Na. Sra. de Abellá, Sobron, Cambo, **Clifton.** Leamington, Brides, **Castellamare,** Cheltenham, Carlsbad, **Panticosa**, Escorinza. Divonne, Mondorf, Laubbach.
Lumbago ...	Hydropathy	Divonne, **Laubbach** Bushey, **Droitwich,** Malvern.

Name of Disease.	Medium.	Places.
Paralysis (Rheumatic)	Mud and sand baths	Acqui, Dax, Abano, St. Dalmas, Valdieri.
	Sulphurous waters	Aix-les-Bains, Uriage, Loëche, Baden (Austria), Mehadia, Harrowgate, Bigorre, Acireale, Chianciano.
	Sulphate of lime waters.	Dax, Bath, Hammam R'Irha, Tiermas.
	Thermal saline waters.	Baden-Baden, Nauheim, Ischia, Malvern.
	Thermal alkali-saline.	Teplitz-Schönau.
	Indifferent waters	Wildbad, Laubbach, Bushey, Plombières.
Paraplegia	Various waters	La Malou, Wildbad, Gastein, Chianciano, Graena, Loujo, Plombières.
Pharyngitis (Granular)	Sulphurous waters	Schinznach, Allevard, Eaux-Bonnes, Ischl, Marlioz, Ormaiztegui.
	Saline waters	Nauheim, Ischl, Bex.
	Mixed waters	Mont-Dore, Gleichenberg, Contrexéville, La Bourboule, Villaro, Wiesbaden, Clifton.
	Winter stations	Cannes, Biarritz, Acireale, Madeira, Malaga, Castellamare, Murcia, Abazzia, Mentone, San Remo, Ajaccio, Corfu, Maloja, Bagnères de Bigorre, Hyères.
Phlebitis (Abdominal)	Saline waters	Homburg, Kissingen, Ischl, Bigorre, La Bourboule.
Phthisis (Laryngial)	Alkaline waters	Gleichenberg, Mont-Dore.
	Sulphurous, saline and earthy waters	Acireale, Amélie, Eaux-Bonnes, Eilsen, Neudorf, Cauterets, Cambo.
	Climatic air and winter stations	Cambo, Hyères, Ischl, Mentone, Cannes, Schmecks, San Remo, Corfu, Ajaccio, Mürren, Maloja, Clifton.

DISEASES.   xxxix

Name of Disease.	Medium.	Places.
Phthisis ... ... (Pulmonary with lymphatism and atony)	Sulphurous waters Alkaline waters ... Sea air ... ... Climatic air and winter stations	**Cauterets, Eaux-Bonnes**, Amélie, Ischl, Acireale, Challes. Gleichenberg, Mont-Dore. Ventnor, Torquay, Atlantic and North Sea baths. Schmecks, Acireale, Madeira, **Hyères, Abazzia**, Biarritz, Malaga, Murcia, Elche, San Remo, **Le Cannet**, Cannes, Campfer, Pontresina, **Clifton**, Cambo, St. Moritz, Samaden, **Maloja, Laubbach,** Bushey.
Phthisis (With erethism)	Mild waters Mixed and arsenical waters Climatic air and winter stations	Ems, **Schlangenbad,** Römerbad (Austria), Schmecks. Mont-Dore, **Baden-Baden**, La Bourboule, Cheltenham. Abazzia, Mentone, Alassio, St. Moritz, Schmecks, Davos, **Mentone**, San Remo, **Hyères**.
Pleurodynia ...	Hydropathy	Divonne, Gräfenberg, Mondorf, Laubbach, Bushey, **Droitwich**.
Pleuritis	Sulphurous saline waters Mixed waters	Eaux-Bonnes, Acireale. Mont-Dore.
Pneumonia ...	Sulphurous waters and winter station Indifferent waters	Acireale, Castellamare, Madeira, Abazzia, Elche. **Wildbad.** Laubbach, Bushey, **Malvern**.
Rhachitis ...	Chalybeate waters Saline bromo-iodurated waters. Climatic mountain station.	Alveneu, Schwalbach, Spa, Pyrmont, St. Moritz, **Loëche**, Plombières. Bex, **Kreuznach**, Droitwich. **Wildbad**, Pontresina, Mürren, Schmecks, **Rippoldsau**, St. Moritz, Samaden, **Maloja**.

DISEASES.

Name of Disease.	Medium.	Places.
Rheumatism (Articular)	Thermal and hyperthermal, sulphurous and sulphurous-saline waters, mud and sand baths	Aix-les-Bains, Dax, Baden (Austria), Loëche, Abano, Acqui, Bath, Mehadia, Acque Albulae, Amélie-les-Bains.
	Thermal alkali-saline waters	Teplitz-Schönau, **Wiesbaden.**
	Saline waters	**Baden-Baden, La Bourboule, Droitwich.**
	Sulphurous saline waters	Alveneu, **Cauterets,** Ledesma, Tiermas, **Schinznach, Acque-Albulae, Bagnères-de-Bigorre.**
	Saline bromo-iodurated waters	Bex, **Malvern.**
	Chalybeate waters	Rippoldsau.
	Natural steam baths	Abano, Monsummano, Stufa di Nerone, La Battaglia, Plombières.
Rheumatism (With Neuralgia)	Thermal sedative waters	Neris, Olette, Acireale, **Droitwich,** Laubbach, **Wiesbaden, Wildbad,** Schlangenbad.
Rheumatism (With Dyspepsia.)	Thermal mixed effervescent waters	Teplitz-Trencsin, **Laubbach, Chianciano,** Gleichenberg, La Bourboule, **Wiesbaden.**
	Saline	**Baden-Baden,** Droitwich, **Malvern.**
Sciatica	Thermal sulphate of lime waters	Pistyan, Loëche, Bigorre, Ledesma.
	Thermal sulphurous waters	**Aix-les-Bains,** Harrowgate, Dax, Guagno, Aix.
	Thermal saline	**Baden-Baden,** Buzost, **La Bourboule, Malvern.**
	Mixed alkaline Various waters	Gleichenberg Gastein, **Wiesbaden,** Römerbad (Austria), Plombières, Aix.
	Hydropathy	Divonne-les-Bains, Gräfenberg, Laubbach, Busbey, **Droitwich.**
Scrofula	Saline waters	Ischl, Baden-Baden, Niederbronn, Kissingen, **Kreuznach,** Dumbane, Ashby-de-la-Zouche, La **Bourboule,** Bex, **Malvern.**

DISEASES.

Name of Disease.	Medium.	Places.
Scrofula (continued).	Cold ferruginous waters	Aberbrothwick, Gleichenberg, Plombières, Schwalbach.
	Sea baths	North Sea and Atlantic.
	Various waters	Amelie, **Wildbad**, **Eaux-Bonnes**, Schinznach, Abano, **Loëche**, Uriage, La Puda, Marlioz, **Castellamare**, Teplitz-Schönau, Baden (Austria), Chianciano, **Acireale**, Schwalbach, **Divonne**.
	Ferruginous waters	Alveneu, Bagnères de Bigorre, **Cambo**.
	Winter stations	Corfu, Ajaccio.
	Climatic air stations	Pontresina, St. **Moritz**, Samaden, Cambo.
Scurvy	Saline waters	Ischl, **Baden-Baden**, La **Bourboule**.
	Sulphurous waters	Caldaniccia, Baden (Austria), Dax.
	Ferruginous waters	Pyrmont, Castellamare, Chianciano Schwalbach.
Skin diseases	Saline waters	Ischl, **Baden-Baden**, **Malvern**, La **Bourboule**.
	Alkaline mixed waters	Mont-Dore, **Wiesbaden**.
	Sulphurous waters	Chianciano, Castellamare, Acireale, Schinznach.
	Sulphurous saline waters	Acque "Albulae", Alveneu, **Cauterets**, Aix-les-Bains.
	Earthy waters	Loeche.
	Natural steam baths	La Battaglia, Stufa di Nerone, Monsummano, Plombières.
Sores (Atonic)	Ferruginous waters	Chianciano, Recoaro.
	Sulphurous saline waters	Acque **Albulae**.
Sinews (Affections of the)	Thermal alkali-saline waters	Teplitz-Schönau.
Spinal affections	Saline waters	Ischl, Uriage, Nauheim, La **Bourboule**.
	Alkali-saline waters	Teplitz-Schönau.
	Hydropathy	Divonne, **Laubbach**, Ben Rhydding, Gräfenberg, Bushey.
	Indifferent waters	**Wildbad**.

Name of Disease.	Medium.	Places.
Splenic obstructions	Saline waters ...	Ischl, Castellamare, Kreuznach, **La Bourboule**, Sobrón.
	Alkaline ferruginous waters	Gleichenberg, **Recoaro**, Chianciano, Pyrmont, **Mont-Dore**, **Clifton**.
	Sulphurous ...	Escoriaza, **Cauterets**.
Sterility ...	Saline waters ...	Ischl, Droitwich.
	Ferruginous waters	**Forges-les-Eaux**, **Schwalbach**, **Gleichenberg**, Plombières.
Syphilis	Thermal sulphurous waters	**Aix-les-Bains**, Baden, **Schinznach** (Austria), Mehadia, Luchon, Abano, **Aulus**, Loëche, Amélie-les-Bains.
	Sulphurous earthy waters	**St. Filomena de Gomillaz**.
	Thermal saline waters	Baden-Baden.
	Iodurated waters	Krankenheil, Kreuth, Marlioz, **Quinto**, Alveneu.
Ulcers	Sulphurous waters	Uriage, Amélie, Eaux-Chaudes, Hartfel, Baden (Austria), Zaldivar, Schinznach, **Acque Albulae**.
	Silicated and mixed sulphurous waters	Valdieri, San Juan de Campos, **Ledesma**, Paterna.
Uric (diathesis) ...	Saline waters ...	Baden-Baden, Ischl, **La Bourboule**, Jabalcutz, Ibero.
	Indifferent waters.	Wildbad.
Urinary organs (catarrhs of the)	Indifferent waters	**Wildbad**, Vöslau.
	Alkali-saline waters	Teplitz Schönau, Contrexeville, Baden-Baden, **Wiesbaden**.
	Various waters ...	Bagnères de Bigorre, Cambo, Clifton.

DISEASES.

Name of Disease.	Medium.	Places.
Uterus ... ... (Diseases and affections)	Sulphurous waters	Marlioz, Luchon, **Baden (Austria)**, Chianciano, Ischl, Castellamare, **Alveneu,** Villaro, Cauterets.
	Clorinated waters	Lamotte, Niederbronn, **Droitwich**,
	Sea baths	Atlantic and North Sea.
	Mildly mineralised waters	Bains, Luxeuil, Schlangenbad, Badenweiler, **Vöslau**, Plombières.
	Alkaline effervescent waters	Franzensbad, Wiesenbad, **Contrexéville**.
	Effervescent ferruginous waters	Gleichenberg, Recoaro, Puerto-lano, Schwalbach.
	Saline iodo-bromurated waters	**Bex**, Malvern.
	Climatic air station	**St. Moritz**.
Vesical ... ... (disorders)	Various waters ...	Castellamare, **Contrexéville**, Wildungen, Fachingen, Gleichenberg, Teplitz, Eisenberg, Beschtau-baths, **Fonté**, Salvadora, Sobron, Villaharta.
	Hydropathy ...	Divonne, Laubbach, Bushey.
Wounds, incised or by fire arms	Thermal sulphurous or saline waters	Loujo, **Amélie**, Ischia, Ischl, Baden Baden, Aix-les-Bains, Mehadia, Chianciano, **La Bourboule**, Malvern.
	Indifferent waters.	Wildbad.
	Mud baths	Franzensbad, Acqui, St. Amand, Teplitz-Schönau.

TABLE OF COMPARISON

Death Rate °/₀₀	°/₀₀	Northern Latitude.	° '	Rainy days, Dec., Jan. and Feb.	No.
Genoa	37	Cairo	29·59	Palermo	40
Nice and Cannes	35	Funchal (Madeira)	32·28	Corfu	38
Corfu	32	Malaga	36·42	Pau	33
Naples	31	Algiers	36·47	Rome	32
Florence	30	Catania	37·30	Funchal (Madeira)	31
Catania	30	Palermo	38·06	Pisa	31
Palermo	29	Corfu	39·37	Baden-Baden	30
Madeira	29	Madrid	40·24	Naples	29
Görz	27	Naples	40·51	Wiesbaden	28
Ajaccio	26	Rome	41·53	Florence	27
Venice	25	Ajaccio	41·55	Nervi	25
San-Remo	24	Pau	43·20	Algiers	21
Meran	23	Mentone	43·47	Cannes	18
Geneva	23	San Remo	43·48	Nice	17
Rome	23	Nervi	44·22	Pallanza	17
Mentone	22	Venice	45·27	Venice	16
Pau	22	Arco	45·52	Mentone	16
Acireale	20	Görz	45·56	St. Moritz	16
Davos	20	Lugano	46·00	Remo	15
Wiesbaden	19	Montreux	46·26	Lugano	14
Pegli	19	Meran	46·41	Ajaccio	14
Bordighera	19	Baden-Baden	48·46	Cantania	13
Montreux	19	Wiesbaden	50·04	Malaga	12
Upper Engadine	19	Görbersdorf	50·40	Meran	11
Malvern	8	Berlin	52·30	Cairo	9

OF DIFFERENT STATIONS.

Relative dampness of air, from 1st Oct. to 30th Apr.	%	Height above sea level.	Feet.	Mean winter temperature, Dec., Jan. and Feb.	F. °
Wiesbaden	80	St. Moritz	5,800	Funchal (Maderia)	68·0
Baden-Baden	79	Samaden	5,500	Algiers	63·5
Venice	79	Davos	5,200	Cairo	62·0
Pisa	79	St. Beatenberg	3,500	Málaga	59·5
Pau	79	Schmécks	3,100	Palermo	57·5
Montreux	79	Reiboldsgrün	2,100	Catania	57·5
Corfu	77	Görbersdorf	1,776	Ajaccio	57·0
Ajaccio	76	Bex	1,380	Corfu	57·0
Palermo	75	Veldes	1,505	San Remo	54·0
Florence	73	Obermais	1,172	Mentone	52·5
Naples	73	Meran	1,100	Cannes	52·0
Lugano	72	Lugano	870	Naples	52·0
Catania	72	Gries	832	Nice	51·5
St. Moritz	72	Pallanza	611	Nervi	51·0
Madeira	72	Malvern	502	Rome	50·0
Görz	72	Wiesbaden	371	Pisa	48·0
Rome	71	Görz	302	Pau	45·0
Mentone	71	Arco	299	Venice	40·4
Lesina	69	Florence	231	Görz	40·2
Pallanza	68	Palermo	228	Arco	40·2
Meran	68	Rome	199	Pallanza	40·0
Cairo	67	Naples (Osservat.)	180	Lugano	39·5
San Remo	? 66	Catania	100	Montreux	38·0
Cannes	? 65	Cairo	70	Meran	37·0
Nice	? 62	Nice	63	Wiesbaden	33·0

TABLE

SHOWING THE QUICKEST ROUTE, MODE OF CONVEYANCE, TIME AND FARES TO VARIOUS HEALTH RESORTS.

Time and prices are calculated from Paris.

Fares for places in Germany and Austria are calculated—
1st Class in France.
2nd „ in Germany and Austria.

Name of Station.	Page.	Mode of Travelling.	Time. H. M.	Price. Fr. cts.
Abano (Italy)	1	P.L.M., *viâ* Mont Cenis, Turin, Milan, Padua	31 0	144 25
Acireale (Sicily)	2	P.L.M., *viâ* Mont Cenis or Ventimiglia, Genoa, Rome, Naples, Reggio, Messina	09 0	239 75
Acqui (Italy)	5	P.L.M., *viâ* Mont Cenis, Turin and Alessandria	24 45	114 30
Agueda (Sta.) Spain.	237	O. & S. F., *viâ* Bordeaux, Irun, Tolosa, Zummarraga (20 h. 11 m.), diligence or carriage, *viâ* Mondragon (4 h.)	24 10	126 35
Aix (Provence)	7	P.L.M., *viâ* Lyons and Marseilles, branching off at Rognac	18 36	106 18
Aix-les-Bains (Savoy).	7	P.L.M., *viâ* Macon and Culos	14 30	71 60
Aix-la-Chapelle (Prussia).	7	N. of F., *viâ* Erquelines and Namur-Verviers	9 30	48 0
Ajaccio (Corsica)	10	P.L.M. to Marseilles (16h.), thence by C.G.T. steamer (12 h.)	28 0	125 0
Aled (Aude)	12	P.L.M. and S. of F., *viâ* Toulouse, Montpellier and Cette	19 50	104 80
Alexisbad (Thuringia).	13	N. of F., *viâ* Cologne-Minden, Brunswick, Ballenstedt (29 h.), diligence (3½ h.)	32 15	118 95
Algiers	13	P.L.M. to Marseilles (16h. 25 m.), thence by C.G.T. steamer to Algiers (40 h.)	56 25	177 0
Alhama de Aragon (Spain)	13	O. and S. of F., *viâ* Bordeaux, Irun, Burgos, Madrid, and Guadalajara	43 38	238 75
Alhama de Murcia (Spain)	14	O. and S. of F., *viâ* Bordeaux, Irun, Burgos, Madrid, Carthagena, Murcia (52 h. 20 m.), carriage (5 h.)	57 20	306 40

Name of Station.	Page.	Mode of Travelling.	Time.	Price.
			H. M.	Fr. cts.
Allevard (Isère)	15	P.L.M., *viâ* Lyons and Chambery to Goneelin (15 h. 20 m.), thence by carriage to Allevard (1¼ h.)	16 50	85 0
Altwasser (Silesia), Alvenu.	17	N. of F., *viâ* Aix-la-Chapelle, Cologne, Berlin, and Kohlfurt	33 0	160 50
Amélie-les-Bains (Pyrenées).	17	S. of F., *viâ* Toulouse to Perpignan (21 h.), carriage from Perpignan (3 h.)	24 30	120 25
Amphion (Savoy).	18	P.L.M., *viâ* Lyons, Culoz, Geneva	13 0	81 20
Amsterdam	18	*viâ* Harwich by boat	15 30	42 50
		Or by N. of F., *viâ* Brussels and Antwerp, Rotterdam	15 40	69 40
Andabre (Aveyron).	19	S. of F., *viâ* Toulouse, Beziers, Bedarieux to St. Afrique (27 h. 30 m.), carriage from St. Afrique (2 h.)	29 30	141 0
Antogast (Baden).	20	E. of F., *viâ* Strasburg and Oppenau (14 h. 30 m.), carriage (½ h.)	15 0	66 50
Arcachon (Gironde).	21	S. of F., *viâ* Bordeaux, branching off at Lamothe	10 0	78 70
Archena (Spain)	21	O. and S. of France, *viâ* Bordeaux, Irun, Burgos, Madrid, Carthagena	51 0	292 65
Arechevaleta (Spain).	21	O. and S. F., *viâ* Bordeaux, Irun, Tolosa, Zumarraga (20 h. 11 m.), carriage (2 h.)	22 11	121 35
Arnedillo (Spain).	23	O. and S. of F., *viâ* Bordeaux, Irun, Tolosa, Vitoria, Miranda, Logroño, and Calahorra (32 h. 30 m.), carriage (4 h.)	36 30	159 15
Audinac (Pyrenées).	25	O. and S. of F., *viâ* Bordeaux, Toulouse and St. Girons (19 h. 25 m.), carriage (30 m.)	19 55	117 85
Aulus (Ariège)	25	S. of F., *viâ* Bordeaux, Toulouse, Montrejeau, Boussens to St. Giron (21 h. 30 m.), carriage from St. Giron (3 h.)	24 30	119 0
Avesne (Pyrenées).	26	P.L.M. and S. of F., *viâ* Lyons, Tarascon, Narbonne, and Bousquet d'Orb (19 h. 30 m.), carriage (50 m.)	20 20	133 60
Ax (Ariège).	26	S. of F. up to Foix (19 h.), carriage from Foix to Ax (4 h.)	23 0	118 0
Baden-Baden	28	E. of F., *viâ* Strasburg and Kehl	16 0	68 0
Baden (Austria)	27	E. of F., *viâ* Strasburg, Kehl, Munich, and Passau	27 0	152 50
Baden (Switzerland).	28	E. of F. *viâ* Basle and Aarau	13 30	71 15
Badenweiler (Baden).	30	E. of F. Railway, *viâ* Belfort and Basle (5 h.) carriage (¾ h.)	15 30	73 95
Bagnères de Bigorre (Pyrenées).	31	S. of F., *viâ* Bordeaux and Monceux, branching at Tarbes	21 45	105 0

Name of Station.	Page.	Mode of Travelling.	Time.	Price.
			H. M.	Fr. cts.
Bagnéres de Luchon (Pyrénées).		S. of F., viâ Bordeaux and Toulouse, branching at Montrejeau	23 50	106 35
Bagnoles de l'Orne.	33	W. of F. Railway, viâ Granville, Briacze, and La Fert-macé	7 0	29 50
Bagnoli, near Naples.	33	P.L.M., viâ Mont Cenis, Genoa, Rome and Naples	43 0	190 0
Bagnols ...	33	P.L.M. viâ Brioude and Villefort (15 h. 36 m.), carriage (3 h. 30 m.)	19 6	67 65
Bains (Vosges)...	33	E. of F., viâ Nancy and Epinal...	10 30	56 30
Balaruc (Hérault)	34	P.L.M., viâ Lyons and Cette	19 30	98 90
Balaton-Furëd (Hungary)	124	E. of F., viâ Avricourt, Strasbury, Munich, Vienna, Pesth and Sis-Fók (47 h. 15 m.), steamboat (1 hour)	48 15	223 90
Barbazan (Pyrenees).	35	O. and S. of F., viâ Bordeaux, Tarbes, Montreyeau and Lourdes (18 h. 40 m.), carriage (½ h.)	19 10	109 90
Barbotan ...	35	O. and S. of F., viâ Bordeaux and Mont à Marsan (15 h. 20 m.), carriage (3 h. 30 m.)	18 50	97 30
Barcelona ...	36	P.L.M. to Marseilles by boat, or S. of F., viâ Toulouse and Cette	28 25	151 25
Barèges ...	36	S. of F., viâ Bordeaux, branching at Lourdes for Pierrefitte (19 h. 30 m). Carriage from Pierrefitte (2 h.)	21 30	113 0
Basle ...	37	E. of F., viâ Belfort	10 50	63 20
Battaglia (La) Italy	38	P.L.M., viâ Macon, Culoz, Turin, Milan, Brescia and Padua	31 9	145 15
Bauche (La)	38	P.L.M., viâ Dijon, Macon, Chambéry (13 h. 45 m.), carriage (3½ h.)	17 15	79 35
Berlin	41	M. of F., viâ Venlo, Lehrte, Hannover or Cologne-Magdeburg	22 0	119 40
Berne ...	42	E. of F., viâ Belfort-Basle, Olten	14 25	74 50
Bertrich (Rhenish Prussia)	42	E. of F., viâ Metz, Treves and Bullay (12 h. 30 m.), diligence (50 m.)	13 20	68 85
Biarritz (Pyrenees).	44	S. of F., viâ Bordeaux and Bayonne	19 30	96 80
Bilin (Bohemia)...	45	N. of F., viâ Cologne, Hanover, Magdeburg, Halle, Leipzig, Dresden and Aussig	32 30	146 25
Bocklet (Bavaria)	47	E. of F., viâ Metz, Frankfort-on-Maine, Würzburg, Kissingen (20 h. 10 m.), carriage (1 h.)...	30 10	122 25
Bondonneau ...	49	P. L. M. viâ Lyons, Valence and Montélimart (14 h. 30 m.), carriage (30 m.)	15 0	83 55
Bordeaux ...	49	O. Railway	10 50	72 5
Bormio (Upper Val Tellino)	50	E. of F., viâ Basle, Lucerne, Bellaggio Colico (29 h. 40 m.), carriage (15 h.) ...	44 49	142 20
Boulogne ...	51	N. of F.	4 35	31 35

d

Name of Station.	Page.	Mode of Travelling.	Time.	Price.
			H. M.	Fr. cts.
Boulou, Le	51	O. and S. of F., *viâ* Bordeaux and Toulouse to Perpignan. (23 h.) carriage thence to Le Boulou. (1½ h.)	24 30	126 50
Bourbon-l'Archambault (Allier).	52	Orleans Railway to Sauvigny (5 h. 35 m.), carriage from Sauvigny (1 h.)	6 35	40 75
Bourbon Lancy (Saône and Loire).	52	Orléans and P.L.M. to Moulins and Gilly (7 h. 55 m.), thence by carriage (1¼ h.)	9 10	46 5
Bourbonne les Bains.	52	E. of F. Railway, branch at Vitrey (8½ h.)	8 30	48 75
Bourboule, La (Puy de Dôme).	52	P.L.M. Bourbonnais line *viâ* Clermont-Ferrand and to Laqueille (12 hours). carriage (1½ hours)	13 30	60 0
Brides les Bains (Savoy).	55	P.L.M., *viâ* Culoz and Chambery to Chamousset (15 h.); thence by carriage (8 h.)	20 50	93 0
Brückenau (Bavaria).	57	E. of F., *viâ* Metz, Frankfort-on-Maine, Würzberg and Kissingen (28 h. 40 m.), carriage (5 h.)	33 40	125 50
Brussels ...	58	Dover, Ostend	8 30	58 75
		Or from Paris by N. of F. Railway, *viâ* Maubeuge	9 20	39 70
Buda-Pesth (Ofen, Hungary)	230	E. of F., *viâ* Avricourt, Strasburg, Munich, Salzburg and Vienna	36 30	180 50
Buzot (Spain)	62	P.L.M. and S. of F., *viâ* Tarascon, Cette, Port-Vendres, Barcelona, Valencia, Eucina and Alicante (52 hours), carriage (2 h.)	54 0	289 15
Bussang (Vosges).	61	E. of F. Railway, line of Epinal	11 15	60 35
Cadéac (Pyrenees)	63	O. and S. of F., *viâ* Bordeaux, Montrejeau and Lannemezan (21 h. 45 m.), carriage, *viâ* Arreau, (2 h. 50 m.)	24 35	122 85
Caille, La (Savoy)	63	P.L.M., *viâ* Macon and Geneva (11 h. 50 m.), carriage, (3 h. 30 m.)	15 20	83 05
Cairo (Egypt)	63	P.L.M., *viâ* Mont Cenis and Brindisi (70 h.); thence by C.G.T. steamer (79½ h.)	149 30	551 25
Caldaniccia (Corsica).	63	P.L.M. and C.G.T. boats, *viâ* Lyons, Marseilles, and Ajaccio, train (15 h. 45 m.), boat (12 to 15 h.), carriage (1 h. 30 m.)	32 15	144 30
Caldas de Cuntis (Spain).	64	O. and S. of F., *viâ* Bordeaux, Irun, Palencia, La Corunna, Santjago, and El Carril (53 h.), carriage (5 h.)	58 0	241 25
Caldas de Mombuy (Spain).	64	P.L.M and S. of F., *viâ* Tarascon, Carcassonne, Port Bou and Mollet (28 h. 52 m.), carriage (2 h.)	30 52	176 70

li

Name of Station.	Page.	Mode of Travelling.	Time.	Price.
			H. M.	Fr. cts.
Caldas de Oviedo (Spain).	64	O. and S. of F., *viâ* Bordeaux, Irun, Vitoria, Burgos, Venta de Baños, Palencia, Busdongo and Pola de Leña (37 h. 46 m.), carriage (5 h.)	42 46	252 15
Caldas de Reyes (Spain).	65	Same itinerary as Caldas de Cuntis.	—	—
Cambo (France)	66	O. and S. of F., *viâ* Bordeaux, Dax and Bayonne (16 h. 10 m.), carriage (2 h.)	18 10	98 45
Campagne (France).	66	O. and S. of F., *viâ* Bordeaux, Toulouse, Carcasonne and Esperanza (17 h. 35 m.), carriage (½ h.)	18 5	121 60
Cannes ...	67	P.L.M., *viâ* Lyons, Marseilles, and Toulon	20 25	130 0
Cannstadt (Würtemburg).	69	E. of F., *viâ* Avricourt, Strasburg, Carlsruhe and Stuttgart	17 10	80 30
Capri (Island)	69	P.L.M., *viâ* Mont Cenis, Genoa, Rome, Naples, thence by boat (3 h.)	16 0	198 0
Capvern (Pyrenees).	70	S. of F., *viâ* Bordeaux and Tarbes	17 51	106 70
Carcanières (France).	70	P. L. M., *viâ* Lyons, Tarascon, Carcassone, and Quillan (24 h. 30 m.), carriage, *viâ* Axat, (4 h. 30 m.)	29 0	131 70
Carlsbad (Bohemia).	162	E. of F., *viâ* Strasburg, Heidelberg, Würzburg, and Eger	32 0	119 80
Carlsruhe	71	E. of F., *viâ* Forbach, Mayence, Mannheim	14 30	79 65
Carratraca (Spain).	71	O. and S. of F., *viâ* Bordeaux, Irun, Burgos, Madrid, Cordoba, and Gobantes (59h. 25m.), carriage (2½ h.)	62 0	336 60
Casciano	72	P.L.M. *viâ* Lyons and Genoa and Leghorn to Pontedera (44 h.); thence carriage (3 h.)	47 0	210 0
Cassel ...	72	N. of F., *viâ* Cologne-Giessen	19 25	90 12
Castellamare (near Naples.)	73	P.L.M., *viâ* Mont Cenis, Genoa, Rome, and Naples	59 20	218 0
Castéra-Verduzan (France).	74	O. and S. of F., *viâ* Bordeaux, Agen, and Auch (16 h. 10 m.), carriage (3 h.)	19 10	104 40
Catania (Sicily)	75	P.L.M., *viâ* Mont Cenis, Genoa, Rome, and Naples (45 h.); by steamer (24½ h.)	69 30	240 65
Cauterets (Pyrenees).	76	S. of F., *viâ* Bordeaux, branching from Lourdes to Pierrefitte (19 h. 30 m.), carriage from thence (2 h.)	21 30	111 90
Cauvalat-lez-le-Vigan (France).	77	P. L. M., *viâ* Lyons, Tarascon, Lunel, and Vigan (22 h.), carriage (½ h.)	22 30	111 45
Celles	77	P.L.M. branch at Livron, and stop at Voulte (14½ h.), thence by carriage (½ h.)	15 0	82 0

d 2

Name of Station.	Page.	Mode of Travelling.	Time.	Price.
			H. M.	Fr cts.
Cestona-Guesalaya (Spain).	78	O. and S. of F., viâ Bordeaux, Irun, Tolosa, and Zummaraga (20 h. 11 m.), carriage (2 h.)	22 11	122 35
Challes (Savoy)	78	P.L.M., viâ Culoz to Chambery (13½ h.) thence by carriage (¼ h.)	14 0	75 0
Charlottenbrunn (Silesia).	79	N. of F., viâ Cologne, Berlin, Liegnitz, and Altwasser (31 h. 54 m.), diligence (1 h. 20m.)	33 15	163 0
Chateauneuf	80	P.L.M., Bourbonnais line to Riome (8 h. 40 m.), carriage (3 h.)	11 40	57 35
Châteldon (France).	80	P. L. M. viâ Nevers, Moulins, and Vichy (8 h. 26 m.), carriage (2½ h.)	11 3	51 95
Chatel-Guyon	80	P. L. M., viâ Nevers and Riome, (8 h. 30 m.), carriage (1 h.)	9 30	53 10
Chaudes-Aigues (France).	80	P. L. M. and O., viâ Nevers, Clermont-Ferrand, Arvant, and Neussargues (12 h. 25 m.), carriage (9 h. 30 m.)	21 55	70 65
Chianciano (Italy).	81	P. L. M., viâ Modane, Turin, Genoa, Florence and Asciano (38 h. 30 m.), carriage (½ h.)	39 0	191 85
Chiclana (Spain).	82	O. and S. of F., viâ Bordeaux, Irun, Burgos, Madrid, Cordoba, Sevilla and Cadiz (62 h.), carriage (2½ h.)	64 30	372 70
Chitignano (Italy).	82	P. L. M., viâ Mont Cenis, Florence and Arezzo (34 h. 30 m.), carriage (2 h.)	36 30	186 70
Coblence	86	E. of F., viâ Forbach-Treves, and N. of F., viâ Cologne	12 30	67 50
Coise (Savoy)	87	P. L. M., viâ Culoz, Chambéry, and Cruet (14 h. 25 m.), diligence (1½ h.)	15 55	76 45
Cologne	87	N. of F., viâ Erquelines-Verviers Aix la Chapelle	10 0	60 0
Contrexéville (Vosges.)	88	E. of F., viâ Langres and Chalendrey	8 0	51 25
Cordoba	89	O. and S. of F., viâ Bordeaux, from Madrid, Toledo, and Ciudad Real	53 10	292 45
Courmayeur & La Saxe (Italy).	91	P. L. M., viâ Lyons, Culoz, Mont Cenis, and Aosta (19 h. 10 m.), carriage (5 h.)	24 10	128 25
Cransac	92	S. of F. Railway to Cransac	15 0	73 85
Cudowa (Silesia)	93	N. of F., viâ Cologne, Hanover, Madeburg, Leipzig, Dresden, Chotzen, Nachod (38 h. 30 m.), diligence (2½ h.)	41 0	173 70
Cusset	94	Orleans Railway to Cusset	10 30	45 0
Dax (Landes)	95	S. of F., viâ Bordeaux	15 0	90 80
Deauville (Calvados).	95	W. of F.	7 0	30 0

Name of Station.	Page.	Mode of Travelling.	Time.	Price.
			H. M.	Fr. cts.
Dieppe (Seine Inf.).	96	W. of F.	4 0	20 65
Divonne les Bains.	98	P.L.M. to Geneva, thence by rail to Nyon, carriage from Nyon (½ h.)	15 30	81 0
Dresden ...	101	N. of F., *viâ* Cologne, Berlin, Leipzig, or Cologne, Frankfurt, Leipzig	28 25	137 50
Driburg	101	N. of F., *viâ* Cologne, Hamm, and Altenbecken	17 40	82 50
Dürkheim an der Hardt.	103	E. of F., *viâ* Forbach, Saarbrücken and Neustadt ad. Haardt	15 0	68 30
Eaux-Bonnes (Pyrénées).	104	O. and S. of F., *viâ* Bordeaux, Pau and Laruns (17½ h.), carriage (½ h.)	18 0	105 0
Eaux-Chaudes (Pyrénées)	105	Same as above, ½ h. more by carriage	18 30	112 0
Elster (Bohemia).	108	E. of F., *viâ* Avricourt, Strasburg, Munich, Regensburg, and Eger	30 50	134 50
Ems (Nassau) ...	108	N. of F., *viâ* Erquelines, Namur, Cologne, Coblence, Oberlahnstein	16 15	71 15
Encausse (Pyrénées).	109	O. and S. of F., *viâ* Limoges, Toulouse, and St. Gaudens (19 h. 10 m.), carriage (1 h.) ...	20 10	116 85
Enghien (Seine and Oise).	109	N. of F.	0 20	1 35
Escaldas (Las) (Pyrénées).	110	P.L.M. and S. of F., *viâ* Tarascon, Cette, Perpignan, and Pradès (22 h. 30 m.), carriage (10 h.)	32 30	143 85
Étretat (Seine Inf.).	111	W. of F., to Ifs (4 h. 50 m.), carriage to Etretat (1 h.) ...	5 50	28 0
Euzet (France).	111	P.L.M., *viâ* Nevers, Langogne, and Alais (17h. 27 m.), carriage (1½ h.) ...	19 0	86 15
Evian (Savoy) ...	111	P.L.M., *viâ* Lyons, Geneva, and Bellegarde	13 0	82 20
Fécamp (Seine Inf.	113	W. of F.	5 20	7 30
Fitero (Spain) ...	115	O. and S. of F., *viâ* Bordeaux, Irun, Alsásua, Castejon (26 h. 15 m.), carriage (3½ h.) ...	29 45	164 75
Florence ...	117	P.L.M., *viâ* Geneva and Pisa ...	32 30	160 75
Foucaude (France)	120	P.L.M., *viâ* Lyons, Tarascon and Montpellier (19 h. 16 m.), carriage (½ h.) ...	19 46	104 85
Forges (Seine Inf.)	119	W. of F. to Fécamp, stop at Forges	3 15	14 0
Frankfurt ...	120	N. of F., *viâ* Cologne, Mayence, or E. of F., *viâ* Forbach, Mayence,...	18 30	77 15

Name of Station.	Page.	Mode of Travelling.	Time.	Price.
			H. M.	Fr. cts.
Franzensbad (Bohemia)	121	E. of F., viâ Strasburg, Ulm, Regensburg, and Eger	28 0	114 35
Friedrichshall (Bavaria)	122	E. of F., viâ Forbach, Mayence, Frankfort and Coburg (20 h. 45 m.), carriage (4 h.)	24 25	118 75
Fuencaliente (Spain)	123	O. and S. of F., viâ Bordeaux, Irun, Madrid, Ciudad, Real, Veredas (47 h. 55 m.), carriage (3½ h.)	51 25	267 40
Gastein (Tyrol)	125	E. of F., viâ Strasburg. Munich, Salzburg, to Lend (33 h.), carriage from Lend to Gastein (3 hr.)	33 20	143 85
Gazost (France)	126	O. and S. of F., viâ Bordeaux, Tarbes, Lourdes (19 h. 30 m.), carriage (1½ h.)	21 0	107 75
Geneva	127	P.L.M, viâ Macon, Culoz.	13 53	77 0
Genoa	127	P.L.M., viâ Mont Cenis Turin or Marseilles Nice	25 0	120 0
Gleichenberg (Austria)	130	E. of F., viâ Strasburg, Munich, Vienna, Gratz, Feldbach (40 h. 50 m.), carriage (1 h. 30 m.)	42 20	213 30
Görbersdorf (Silesia)	132	N. of France, viâ Cologne, Berlin to Dittersbach (38 h. 30 m.), carriage (½ h.)	39 0	167 50
Granada	133	O. and S. of F., viâ Bordeaux, Irun, Madrid, Toledo, Cordoba	61 40	364 85
Graválos (Spain)	135	O. and S. of F., viâ Bordeaux, Irun, Alsásua, Castejon (26 h. 15 m.), carriage (3½ h.)	29 45	167 75
Gréoulx (France)	135	P.L.M., viâ Lyons, Avignon, Pertuis, Manorgue (20 h. 30 m.), carriage (2½ h.)	23 0	109 20
Griesbach (Baden)	136	E. of F., viâ Strasburg and Oppenau (10 h.), diligence (1 h. 45 m.)	11 45	67 95
Guaguo (Corsica)	137	P.L.M. and C.G.T. boats, viâ Lyons, Marseilles, Ajaccio (30 h. 25 m.), carriage (8 h.)	38 25	156 25
Guillon (Doubs)	138	P.L.M. to Dijon, branch to Beaux-les-Dames (6 h.), carriage to Guillon (⅔ h.)	6 45	55 0
Hague	139	N. of F., viâ Brussels, Antwerp, and Rotterdam	15 20	66 70
Hanover	141	N. of F., viâ Brussels—Venlo—Lehrte	15 30	87 50
Heidelberg	144	E. of F., viâ Forbach—Mayence—Mannheim	18 0	76 05
Herculesbad Mehadia (Hungary)	146	E. of F., viâ Strasburg, Munich, Vienna, Pesth, Szegedin, and Temesvár	48 10	249 27

Name of Station.	Page.	Mode of Travelling.	Time.	Price.
			H. M.	Fr. cts.
Hof.-Ragatz (Switzerland).	255	E. of F., viâ Basle and Zürich ...	22 30	83 70
Homburg (Hessen)	149	N. of F., viâ Erquelines, Cologne, Mayence, Frankfurt	18 40	79 0
Huelva (Spain)...	151	O. and S. of F., viâ Bordeaux, Irun, Madrid, Cordoba and Sevilla	62 0	370 70
Hyères (Var)	152	P.L.M., viâ Lyons and Marseilles —Toulon	19 50	117 0
Interlaken ...	155	E. of F., viâ Basle and Berne ...	17 55	78 65
Ischia ...	156	Same as Naples (48 h.), boat (2½ h.)	50 30	288 25
Ischl	156	E. of F., viâ Strasburg, Kehl, Carlsruhe, Munich, and Salzburg	23 45	168 90
Kissingen (Bavaria)	166	E. of F., viâ Forbach, Bingen and Würzburg, branching at Schweinfurt	21 40	96 80
Kösen (Germany)	169	E. of F., viâ Metz, Saarbrücken, Mayence, Frankfurt, Bebra, Erfurt and Weimar	22 50	129 85
Krankenheil (Bavaria)	170	E. of F., viâ Strasburg, Carlsruhe, Stuttgart, Munich, Toelz (25 h. 20 m.), carriage (2 h.)...	27 20	119 80
Kreuth (Bavaria).	170	E. of F., viâ Strasburg, Carlsruhe, Munich, Schaffach (30 h.), diligence (3½ h.) ...	33 30	119 55
Kreuznach (Rhenish-Prussia)	170	E. of F., viâ Forbach	13 25	68 45
Labassère (Pyrenees)	172	S. of F., viâ Bordeaux and Monceux, branching at Forbes ...	21 45	105 0
Landeck (Slesia).	173	N. of F., viâ Cologne, Hanover, Berlin, Sagan, Breslau, Glatz (32 h. 40 m.), diligence (4 h.)	36 40	176 30
Laubbach	175	N. of F., viâ Cologne and Coblentz (12 h. 30 m.), thence carriage (¼ h.)	12 45	68 50
Lausanne (Switzerland)	176	E. of F. to Geneva and Lausanne	15 23	64 20
Lavey (Switzerland).	177	E. of F., viâ Pontarlier, Lausanne, St. Maurice (15 h. 30 m.), carriage (25 m.)	15 55	72 65
Ledesma (Spain).	178	O. and S. of F., viâ Bordeaux, Irun, Valladolid, Medina del Campo, Salaman a (52 h. 10 m.), carriage (2½ h.)	54 40	186 30
Leipzig	179	N. of F., viâ Erquelines, Cologne, or Forbach—Frankfurt ...	27 0	134 0
Liebenstein (Thuringia).	181	E. of F., viâ Saarbrücken, Francfort, a/m., Eisenach and Immelborn (21 h. 10 m.), diligence (1 h.)	22 10	132 75

Name of Station.	Page.	Mode of Travelling.	Time.	Price.
			H. M.	Fr. cts.
Liebenzell (Württemberg).	181	E. of F., *viâ* Strasburg, Carlsruhe, Pforzheim (15 h.), diligence (1½ h.)	16 30	76 70
Liebwerda (Silesia).	181	N. of F., *viâ* Cologne, Hanover, Magdeburg, Halle, Leipzig, Dresden, Görlitz, Seidenberg, Raspenau (32 h. 30 m.), carriage (30 m.)	33 0	150 25
Lippspringe (Westphalia).	184	N. of F., *viâ* Cologne, Hamm, Paderborn (16 h. 25 m.), diligence (1 h. 10 m.)	17 35	81 45
Louéche Valais (Switzerland)	186	P.L.M., *viâ* Geneva and Pontarlier to Susten (27 h.), carriage to Loëche (3 h.)	30 0	100 0
Lucca (Italy)	188	P.L.M., *viâ* Mont Cenis, Turin, Genoa, and Leghorn, to Lucca Station (31 h.), carriage to the baths (2½ h.)	33 30	116 0
Lucerne	188	E. of F., *viâ* Belfort, Basle	19 30	74 40
Luchon	189	O. and S. of F., *viâ* Bordeaux, Tarbes, and Montrejeau	19 30	103 50
Luxeuil (Vosges)	191	E. of F. Line to Mulhouse, St. Loup	11 40	60 0
Lyons	191	P.L.M.	9 0	63 50
Madeira (Canary Islands)	193	Steamers from Southampton, Bordeaux, or Lisbon; the whole journey from five to six days	90 / 100	500 0
Madrid	194	O. and S. of F., *viâ* Bordeaux, Irun, and Burgos	41 30	184 50
Maloja (Engadine)	195	E. of F., *viâ* Belfort, Basle, Zürich and Coire (23 h.), carriage or diligence (12 h.)	35 0	134 60
Malou (La) (Pyrenées)	173	P.L.M., S. of F., *viâ* Tarascon, Montpellier, Faugères, and Bedarieux (19 h. 15 m.), carriage (45 m.)	20 0	116 7
Malta (Island)	196	P.L.M., *viâ* Mont Cenis, Turin, Genoa, Rome, and Naples, (46½ h.); thence by steamer (34 h.); or direct from England by P. and O. steamers, nine days	80 15	315 0
Mannheim	198	E. of F., *viâ* Forbach and Bingen	17 30	70 80
Marcols (Ardèche).	198	P.L.M. line to Marseilles; branch at Livron to Lavoulte station (12½ h.); thence by carriage (4 h.)	16 30	84 0
Marienbad (Bohemia).	199	E. of F., *viâ* Strasburg, Heidelberg, Würzburg, Regensburg	30 12	126 60
Marlioz	199	Same as Aix-les-Bains.		
Mayence	201	N. of F., *viâ* Erqueliues-Cologne; or E. of F., *viâ* Forbach, Bingen	17 15	73 0
Mehadia or **Herculesbad** (Hungary).	202	E. of F., *viâ* Strasburg, Munich, Vienna, Pesth, Szegedin, and Temesvár	54 10	249 27
Mentone (France).	203	P.L.M., *viâ* Marseilles, Nice	24 35	137 0

lvii

Name of Station.	Page.	Mode of Travelling.	Time.	Price.
			H. M.	Fr. cts.
Mergentheim	205	E. of F., *viâ* Strasburg Heidelberg, and Königshofen	18 50	88 20
Milan	207	P.L.M., *viâ* Mont Cenis	27 0	116 75
Molar (El) (Spain).	209	O. and S. of F., *viâ* Bordeaux, Irun, Burgos, Valladolid, Madrid (46 h. 40 m.), carriage (4 h.)	50 40	188 40
Molinar de Carranza (Spain).	209	O. and S. of F., *viâ* Bordeaux, Irun, Vitoria, Miranda, Bilbao (27 h. 24 m.), carriage (6 h.)	33 24	156 25
Molitg (Pyrenees).	209	S. of F., *viâ* Perpignan to the station of Prades (22 h. 45 m.), carriage (1 h.)	23 50	126 0
Monaco and Monte Carlo	210	P.L.M., *viâ* Marseilles and Nice	24 30	135 0
Montbrun (Drôme).	213	P.L.M., *viâ* Lyons; branch at Sorges to Carpentras (16 h.), thence by carriage (5 h.)	21 0	95 0
Mont-Dore (Puy de Dôme).	213	P.L.M., Bourbonnais line to Clermont-Ferrand and Laqueille (9¼ h.), carriage (2 h.)	11 0	64 25
Monte-catini (Tuscany).	213	P.L.M., *viâ* Mont Cenis, Turin, Genoa, Pisa	33 0	145 0
Montemayor (Spain).	214	O. and S. of F., *viâ* Bordeaux, Irun, Burgos, Valladolid, Medina del Campo, and Salamanca (34 h.), carriage (10 h.)	44 0	202 30
Montmirail-Vacqueiras	214	P. L. M., *viâ* Lyons, Valence, and Orange (15½ h.), carriage (1½ h.)	17 0	89 40
Motte (La)	173	P. L. M., *viâ* Dijon, Macon, Grenoble, and St. George (14 h. 5 m.), carriage (2 h. 40 m.)	16 45	85 35
Munich	216	E. of F., *viâ* Strasburg, Kehl, Stuttgart	36 0	104 50
Nantes	219	W. of F., *viâ* Le Mans	17 30	48 75
Naples (Italy)	219	P.L.M., *viâ* Mont Cenis, Genoa, Rome	47 45	223 85
Nauheim (Hessen).	219	E. of F., *viâ* Forbach, Bingerbrück, Frankfurt	16 20	88 0
Neuenahr	221	N. of F., *viâ* Erquelines, Cologne, Remagen	12 30	64 75
Neris (Allier)	220	Orleans Railway, *viâ* Vierzon-Bourges to Chamblet (8 h. 45 m.), thence by carriage (45 m.)	9 30	43 0
Nice	224	P.L.M., *viâ* Lyons and Marseilles	22 0	134 20
Niederbronn (Alsatia).	226	E. of F., *viâ* Strasburg	12 40	60 85
Oeynhausen (Germany).	230	N. of F., *viâ* Erquelines, Cologne, and Düsseldorf	17 0	76 75

Name of Station.	Page.	Mode of Travelling.	Time.	Price.
			H. M.	Fr. cts.
Ofen (Buda-Pesth, Hungary).	239	E. of F., *via* Strasburg, Munich, and Vienna	60 0	244 0
Olette (Pyrenees)	231	P. L. M., S. of F., *via* Lyons, Tarascon, Narbonne, and Prades (24 h. 30 m.), carriage (2 h.)	23 30	122 30
Ontaneda y Alceda.	232	O. and S. of F., *via* Bordeaux, Irun, Venta de Baños and Penedo (36 h. 50 m.), carriage (1½ h.)	38 20	246 25
Palermo	235	P.L.M., *via* Mont Cenis, Genoa, Rome to Naples (46¾ h.), thence by steamer (10 h.)	66 15	239 25
Panticosa (Spain.)	236	O. and S. of F., *via* Bordeaux, Irun, Alsasua, Saragoza, Tardienta and Huesca (32 h. 10m. carriage, *via* Jaca (9 h.)	41 10	225 10
Pau (Pyrenées)	240	S. of F., *via* Bordeaux, branch at Dax	17 20	101 0
Pfaeffers (St. Gallen).	244	E. of F., *via* Basle and Ragatz (22 h. 30 m.), carriage to Pfaeffers (45 m.)	23 15	81 60
Pierrefonds (Oise)	244	N. of F. to Compiègne (1½ h.), thence by carriage (1 h.)	2 30	12 0
Pietrapola (Corsica).	244	P. L. M. and C. G. T. steamer, *via* Marseilles and Bastia (35 h.), carriages, *via* Miggliaccinro (13 h.)	48 0	164 30
Pisa	245	P.L.M., *via* Genoa	34 30	147 75
Plombières (Vosges).	246	E. of F. line to Mulhouse, branching at Port l'Atelier to Abbevillers (10 h.), thence carriage to Plombières (1 h.)	11 0	45 0
Porretta (La) (Italy).	249	P. L. M., *via* Mont Cenis, Turin, Parma, Modena, Bologna	31 0	163 55
Pougues (Nièvre).	250	P.L.M. Bourbonnais line	5 0	29 70
Prague	251	E. of F., *via* Strasburg, Kehl, Würzburg	38 0	130 80
Preste (La) (Pyrenees).	252	S. of F., *via* Perpignan (18h. 45m.), thence by carriage (5 h.)	23 45	145 0
Puda (La) (Spain).	252	P. L. M. and S. of F., *via* Lyons, Tarascon, Cette, Portbou, Barcelona, and Olesa (31 h. 52 m.), carriage (½ h.)	32 22	189 85
Pyrmont (Waldeck).	254	N. of F., *via* Erquelines, Cologne, Elberfeld, Soest, and Altenbecken	18 0	84 30
Pistyan or Pöstjén (Hungary).	245	E. of F., *via* Strasburg, Munich, Vienna, Pressburg and Tyrnau	45 10	190 90
Ragatz-Pfeffers (Switzerland).	255	E. of F., *via* Basle-Zürich, Rapperswyl and Sargans	1 31	83 70

Name of Station.	Page.	Mode of Travelling.	Time.	Price.
			H. M.	Fr. cts.
Ravone in Casaglia (Italy).	257	P.L.M., *viâ* Mont Cenis, Turin, Alessandria, Piacenza, Reggio, Modena	29 30	137 80
Recoaro (Venetia)	257	P.M.M., *viâ* Mont Cenis, Turin, Milan, to Vicenza (30 h. 45 m.), thence by carriage (4 h.)	34 45	139 75
Reinerz (Silesia)	258	N. of F., *viâ* Cologne, Hanover, Magdeburg, Leipzig, Dresden, Chotzen, Nachod (38 h. 10 m.), diligence (2½ h.)	40 40	179 70
Rheinfelden	259	E. of F., *viâ* Mulhouse and Basle to Rheinfelden	13 0	65 0
Rippoldsau (Baden).	261	E. of F., *viâ* Strasburg, Offenburg and Wolffach (13 h. 40 m.), diligence (2 h. 20 m.)	16 0	72 40
Rome	264	P.L.M., *viâ* Mont Cenis, or Marseilles, Nice, Genoa, Pisa	49 30	201 90
Rotterdam	268	N. of F., *viâ* Maubeuge, Brussels, Antwerp	14 15	64 40
Roucasblanc (France).	268	P.L.M. to Marseilles (16 h.), thence by carriage (30 m.)	16 30	107 0
Rouzat (Auvergne).	268	Orleans Railway to Riome (1 h. 45 m.) omnibus to Rouzat (30 m.)	2 15	9 35
Royat	269	P.L.M. Bourbonnais line to Clermont-Ferrand (9¼ h.), thence by carriage (15 m.)	9 30	51 75
St. Alban (France)	271	P. L. M., Bourbonnais line, *viâ* Nevers, Moulins, and Roanne (9 h. 30 m.), carriage (1 h.)	10 32	53 85
St. Amand (France)	271	N. of F., *viâ* Amiens and Valenciennes	6 53	31 25
St. Christau (France).	272	O. and S. of F., *viâ* Bordeaux, Pau, and Oléron (21 h.), carriage (45 m.)	21 45	105 75
St. Gervais	273	P.L.M. to Geneva (11 h.), thence by diligence (6 h.)	17 0	95 0
St. Honoré	273	P.L.M. Bourbonnais line to Nevers to Cercy-la-Tour (6 h. 50 m.) omnibus to St. Honoré (1 h. 30 m.)	8 20	40 0
St. Laurent (France).	274	P. L. M., *viâ* Nevers, and La Bastide (16 h. 42 m.), carriage (1½ h.)	18 13	77 75
St. Moritz (Engadine).	275	E. of F., *viâ* Belfort and Basle to Coire (23 h.), carriage to St. Moritz (11 h.)	34 0	130 00
St. Nectaire	276	P.L.M. Bourbonnais line to Coudes (10 h. 20 m.), thence by carriage (2 h.)	12 20	58 0
St. Raphael (Riviera).	277	P. L. M., *viâ* Lyons, Marseilles, and Toulon	19 30	130 20
St. Sauveur	277	S. of F., *viâ* Bordeaux, branching at Lourdes for Pierrefitte (19½ h.), carriage to St. Sauveur (1½ h.)	20 45	111 0
San Giuliano (Italy).	284	P. L. M., *viâ* Mont Cenis, Genoa, and Pisa	31 0	136 70

Name of Station.	Page.	Mode of Travelling.	Time.	Price.
			H. M.	Fr. cts.
San Remo	286	P.L.M., *viâ* Marseilles-Ventimiglia	25 20	140 0
Sacedon (Spain)	270	O. and S. of F., *viâ* Bordeaux, Irun, Burgos, Valladolid, Madrid and Guadalaxara (38 h. 45 m.), carriage (4½ h.)	43 15	200 65
Sail-les-Bains (France).	271	P. L. M., Bourbonnais line, *viâ* Nevers, Moulins, St. Germains les Fossés, St. Martin d'Estréaux (8. h. 50 m.), carriage (1 h.)	9 50	49 90
Sail-sous-Couzan (France).	271	P. L. M., Bourbonnais line, *viâ* Nevers and Clermont-Ferrand	12 10	61 95
Salies de Béarn (Pyrenées).	279	O. and S. of F., *viâ* Bordeaux, Dax, and Pujoo (15 h. 45 m.), carriage (15 m.)	16 30	96 25
Salins (France)	279	P. L. M., *viâ* Dijon, Dôle, and Mouchard	10 55	49 45
Salins-Moutiers (Savoy).	280	P. L. M., *viâ* Dijon, Mâcon, Chambery, and Chamont (14h. 36m.), carriage (5h. 45m.)	20 20	85 35
Salzbrunn (Silesia)	280	N. of F. *viâ* Cologne, Hanover, Berlin and Altwasser (33 h. 30 m.), diligence (30 m.)	34 0	160 37
Salzburg (Austria)	281	E. of F., *viâ* Strasburg, Bruchsal Ulm, Augsburg, Munich	25 55	108 30
Salzungen (Thuringia)	282	E. of F., *viâ* Saarbrücken, Frankfort, Bebra, Eisenach	23 40	111 35
Santa Agueda (Spain)	287	O. and S. of F. *viâ* Bordeaux, Irun, Tolosa, Zumarraga (20 h. 11 m.), diligence (4 hours)	24 11	126 35
Saragoza	289	O. and S. of F., *viâ* Bordeaux, Irun	35 5	184 30
Saxon (Switzerland).	291	P. L. M., *viâ* Dijon, Lausanne, St. Maurice, and Martigny	17 53	73 55
Scheveningen	292	N. of F. *viâ* Brussels and Antwerp to the Hague (15 h. 20 m.), thence by carriage (½ b.)	15 35	67 70
Schinznach	292	E. of F., *viâ* Belfort, Basle, and Aarau	17 0	71 30
Schlangenbad	293	E. of F., *viâ* Metz, Forbach, Bingerbrück, Eltville (11 b.), thence by carriage (1 h.)	19 10	75 20
Schwalbach	295	Same itinerary as above to Eltville, carriage (2 h.)	20 0	76 40
Schwalheim (Germany)	296	E. of F. *viâ* Saarbrücken, Mayence, Frankfort s/m, Nauheim (16 h. 30 m.), diligence (30 m.)	17 05	87 60
Sermaise	300	E. of F., *viâ* Epernay and Châlons-sur-Marne	5 12	28 40
Seville	300	O. and S. of F., *viâ* Bordeaux, Irun, Madrid, Cordoba	56 5	325 25
Siradan (Pyrenees).	303	O. and S. of F., *viâ* Bordeaux, Montréjeau Salichan (19 h. 45 m.), carriage (½ b.)	20 15	119 35
Soden	304	E. of F., *viâ* Saarbrücken, Bingerbrück, Frankfort and Soden	19 30	88 0

Name of Station.	Page.	Mode of Travelling.	Time.	Price.
			H. M.	Fr. cts.
Spa	307	N. of F., viâ Erquelines, **Namur**, Liège, to Pepinster and **Spa**	8 27	44 75
Spezzia (Italy)	307	P.L.M., viâ Genoa, stop at **Spezzia**	28 6	130 40
Strasburg	310	E. of F., viâ Nancy	14 8	70 80
Stuttgart	311	E. of F., viâ Strasburg-**Kehl**	17 15	80 0
Sulzbad	312	E. of F., viâ Nancy and **Saverne**	11 50	59 25
Sulzbach (Alsatia)	311	E. of T., viâ Strasburg, **Colmar** and Walbach (12 h. 40 m.), diligence (30 m.)	13 10	70 10
Sulzmatt (Alsatia)	312	E. of F., viâ Nancy, Strasburg, Saverne and Ruffach (13 h. 30 m.), diligence (50 m.)	14 20	65 90
Sylvanès (France)	65	O. and S. of F., viâ Limoges, Toulouse Beziers, Roqueredonde (26 h.), carriage (3 h.)	29 0	124 15
Szkleno (Hungary)	314	E. of F., viâ Strasburg, Munich, Vienna, Pesth, Garam-Berzencse (47 h. 10 m.), carriage (2½ h.)	49 40	243 85
Szliácz (Hungary)	314	E. of F., viâ Strasburg, Munich, Vienna, Pesth and Altsohl	46 20	135 37
Tabbiano (Italy)	315	P.L.M., viâ Mont-Cenis, Turin, Allessandria, Borgo Sau Domino (25 h. 48 m.) carriage (½ h.)	26 18	146 50
Tarasp (Schülz)	316	E. of F., viâ Belfort, Basle, Zürich to Coire (23 h.), thence by carriage (11½ h.)	34 30	112 0
Teplitz - Schönau	318	E. of F., viâ Strasburg, Stuttgart, Würzburg	32 0	145 50
Teplitz - Trenczin (Hungary)	325	E. of F. viâ Strasburg, Munich, Vienna, Pressburg, Trenczin (41 h. 40 m.), carriage (1 h. 20 m.)	43 0	195 60
Tercis (Pyrenees)	319	O. and S. of F., viâ Bordeaux, Dax (15 h. 5 m.), carriage (1 h.)	16 5	92 30
Trescore (Italy)	326	E of F., viâ Basle, Lucerne, Bergamo, Gorlago (26 h. 48 m.), carriage (½ h.)	27 18	128 15
Trillo (Spain)	327	O. and S. of F., viâ Bordeaux, Irun, Burgos, Madrid, Matilla (40 h. 21 m.), carriage (4 h.)	44 21	219 45
Trouville	327	W. of F. to Trouville	6 0	28 65
Turin	328	P.L.M., viâ Mont Cenis	21 30	100 20
Urberoaga de Alzola (Spain)	331	O. and S. of F., viâ Bordeaux, Irun, Zumarraga (20 h. 11 m.), carriage viâ Vergára (3½ h.)	23 41	129 35
Uriage	331	P.L.M., viâ Lyons to Giers, Uriage (14 h.), omnibus (40 m.)	14 40	79 0
Ussat	332	O. and S. of F. to Tarascon (19 h.), thence by carriage (3 h.)	22 0	115 0

Name of Station.	Page.	Mode of Travelling.	Time.	Price.
			h. m.	Fr. cts.
Valdieri (Italy)...	333	P.L.M., *viâ* Mont-Cenis, Turin, Coni (28 h. 30 m.), carriage (4½ h.)	33 0	116 05
Vals ...	334	P.L.M., branch at Livron to Voguc (16 h. 30 m.), carriage to Vals (1 h.)	17 30	87 0
Venice ...	336	P.L.M., *viâ* Mont Cenis, Turin, Milan	36 0	154 0
Vernet Le	337	S. of F. to Perpignan and Prades (18 h. 45 m.), thence by carriage (2 h. 20 m.)	21 5	136 90
Vevey	337	P.L.M. to Geneva, *viâ* Lausanne and Montreux	16 23	86 45
Vichy	338	P.L.M., Bourbonnais line	8 30	45 0
Vic sur Cère (France)	338	P.L.M. and O., *viâ* Nevers, Clermont-Ferrand, and Arvant	14 50	61 70
Vienna	339	E. of F., *viâ* Strasburg, Kehl, Munich, Salzburg	27 0	154 70
Vinadio (Italy)	342	P.L.M., *viâ* Lyons, Marseilles, Ventimiglia, Coni (24 h. 35 m.), carriage (6 h.)	30 35	120 05
Visos (Pyrenées)	342	O. and S. of F., *viâ* Bordeaux, Tarbes, Pierrefitte (20 h. 45 m.), carriage (2 h.)	22 45	109 30
Viterbo (Italy)...	343	P.L.M., *viâ* Mont-Cenis, Turin, Genoa, Florence, Orte (40 h. 46 m.), carriage (2½ h.)	43 16	203 25
Vittel ...	343	E. of F., *viâ* Langres and Chalendrey	8 30	48 60
Vöslau (Austria)	343	E. of F., *viâ* Strasburg, Kehl, Stuttgart, Munich, Salzburg, and Vienna	27 45	156 95
Warmbrunn (Silesia).	345	N. of F., *viâ* Cologne, Hanover, Sorau, Kohlfurt, Hirschberg (31 h. 30 m.), diligence (45 m.)	32 15	155 60
Weilbach	346	E. of F., *viâ* Metz, Forbach, Mayence, Frankfurt to Flörsheim (17 h. 30 m.), thence by carriage (10 m.)	15 0	79 80
Weissenburg (Switzerland).	347	P.L.M., *viâ* Pontarlier, Neuchatel, Berne, Thune (15 h. 35 m.), carriage (3 h. 30 m.)	19 5	81 05
Wiesbaden ...	348	E. of F., *viâ* Metz, Forbach, Mayence	15 0	73 05
Wildbad. ...	350	E. of F., *viâ* Strasburg, Carlsruhe, Pforzheim	16 0	76 20
Wildungen	351	N. of F., *viâ* Erquelines, Namur, and Marburg	19 10	97 12
Zürich ...	356	E. of F., *viâ* Belfort, Basle	18 8	73 55

CLASSIFICATION.

INDIFFERENT WATERS.

Abensberg
Alcala del Rey
Arapataka
Badenweiler
Barestrand-Syssel
Bejar
Berthemont
Bikszád
Bormio
Bray
Brouia
Bulgneville
Casamicciola
Caxamarca
Chambon
Como
Daruvar
Divonne
Empfing
Enu
Erlau
Foucaude
Fuente de Piedra
Gainfahrn
Gastein
Geisslingen
Gräfenberg
Guitera
Hammam el Enx
Hevitz
Iceland
Ilmenau

Ischia
Job
Jose
Kács
Kaissariani
Kirstenpils
Kis-Kalau
Kleinengstingen
Laubbach
Lauterberg
Leprese
Leutstetten
Limpach
Loka
Lund
Lutraki
Mallow
Monestier de Clermont
Moingt
Moselli
Nelefina
Niederweil
Nieratz
Nocera
Nook
Orebro
Oroslau
Pelagio
Perruches
Petersthal
Pfaeffers

Plombières
La Preste
Rajecz-Teplicz
Ragatz
Renaisson
Römerbad (Austria)
Röthelbad
Rothenburg
Rudolstadt
Säckingen
St. Blasien
St. Georgen
St. Roman le Puy
St. Yorre
Schlackenbad
Schlangenbad
Seraglio
Sutinsko
Sztubicza
Tannenbrunnen
Tapolcza
Tatenhausen
Teruel
Tonnstein
Topolschitz
Topuczko
Tüffer
Visibachbad
Vöslau
Wildbad
Yverdon

ALKALINE WATERS.

Abach
Adelholzen
Acqua-Bolle
Ajnacskő
Alfáro
Al-Györy
Alica
Allegrezza
Almeida
Alt-Haide
Alt-Turn

Almeida
Anagui
Aspio
Attisholz
Atya
Aulus
Baden-Baden
Bagnères de Bigorre
Bagni-a-Morba
Bagno d'Apollo

Bagnoli
Bagnolino dei Rhachitici
Bains, near Arles
Balsócz
Balzach
Baños
Barzun
Belascoin
Bellerieve
Benedekfalva

ALKALINE WATERS—*continued*.

Beuetutti	Gehringswalde	Nuestra Sa de Abellá
Benyus	Geilnau	Nuestra Sa de las
Beschtau baths	Giesshübl	Mercedes
Bilén	Gisi	Obertiefenbach
Bisztrá	Giuneo-Marino	Olette
Bodok	Goldbach	Oni
Boesing	Göppingen	Ormaiztegui
Bondouncau	Gythium	Osztrovsk
Borra	Hackelthal	Paracuellos de Gilloca
Borsa	Haldensten	Pergine
Boudes	Hambach	Pizzofalcone
Boulou, Le	Harkanyi	Planchamps
Bouquéron	Heiligekreuzbad	**Preblau**
Brunnenthal	Heilstein	Prelo
Budis	Helenskilde	Pystián
Caccio-Cotto	Heppingen	Rio-Meo
Caldas de Malavella	Holbeck	Riva-los Baños
Caldas de Oviedo	Houches, Les	Rodna
Caldini	Hovingham	Römerbad (Switzerland)
Calvello	**Ibero**	
Cambo	Jacintos	Röslibad
Caprenue	Jamnicza	Royat
Cascano (San)	**Jobsbad**	Sail sous-Couzan
Castelletto d'Orba	Johannisbad	St. André
Caz di Bagno	**Kanitz**	St. Galmier
Chaudes-Aigues	**Kastenloch**	St. Honoré
Cinciano	Keruly	St. Jean d'Aulph
Citára	Kis-Sáros	St. Laurent
Clifton	Klieningen	St. Nectaire
Contrexéville	Kochel	Salzbrunn
Deutsch-**Kreutz**	Krähenbad	San Georgen
Diós-Györ	Krapina-Töplitz	**Sau Vincens**
Disznopátak	**Krynica**	Schelesna-Wodsk
Doktorka	**Kugelbad**	Schimberg
Dolka	**Kuppis**	Schmécks
Dombhát	**La Caldare**	**Schwalbach**
Dorfbad	**Landskron**	Schuns
Douai	**Leccia**	Sesavnik
Draitschbrunnen	**Levana**	Solters
Dubowa	**Lienzmühl**	Shap
Egerdach	**Lipara**	Sid
Ems	**Lucan**	Sigliano
Ermetschwyl	Lucarnena	**Sinzig**
Escaldas, **Las**	Luxburg	**Sulz**
Essentuk	Macerrato	**Szaldobos**
Estadilla	Madonna á Papiano	Szinye-Lipócz
Evian	Malmédy	Tarasp
Pachingen	**Malvern**	Teinach
Fällorne	Micmo	Telgart
Falú-Szlatina	Montancjos	Tenos
Fellathal	**Montbrison**	**Teplitz-Schönau**
Felső-Visso	Mont-Dore	Thingoë-Syssel
Ferranehe	Nave dell' Inferno	Töplika
Fonga	Neris	Tresclaix
Fonfredo	Neuenahr	Tschawitz
Fontenello	Neu-Ragoczi	Tübingen
Ficoncella	Neuschmécks	Unterbad
Forges-les-Bains	Niederselters	Urberoaga de Ubilla
Fortyogó	Nocceto	Urnässchen

CLASSIFICATION.   lxv

ALKALINE WATERS—continued.

Vals
Vellebro
Vetzel
Vichy
Villa delle Caselle

Vittel
Walby-Brunnen
Weilbach
Wildungen
Wolfs

Wolfsegg
Zagwera
Zen
Zürich

ALKALI-SALINE WATERS.

Adolfsberg
Assmannshausen
Badstofuhver
Bagni (Acqua di)
Bagne à Baccanella
Bagno Fresco
Barèges
Bertrich
Bilin
Blumenstein
Boll
Borschom
Borszek
Brides
Brussa
Brüttalen
Bykowicz
Caldiero
Camarès
Caprifico di Valaspra
Catena
Chitignano
Chiusa di Monaci
Concepcion de Peralta
Cotto
Czigelka
Dombhát
Dreykirchen
Dubogrudsk
Ebriach
Elster
Enatbühl
Epidaurus
Felines
Feiso-Nereszincze
Fonfrede
Fläsch
Fossino

Fosso degli Ontani
Gaberneg
Godesberg
Gross-Schlagendorf
Gyüzy
Hanau
Heiligekreuzbad
Hofgeismar
Ischia
Irno (Valley of the)
Josza
Karlsbad
Kis-Czeg
Katharinenbad
Kirchberg
Kiszlawodsk
Klein-Chocholna
Kobersdorf
Kronthal
Kungara
La Malou
Landeck
Lanjarron
Langenberg
Liebenzell
Liebwerda
Lochbachbad
Luhatchowitz
Luhi
Marienbad
Mehadia
Monte-Ortone
Noceto
Ober-Rauschenbach
Olah Szt. György
Olenyova
Orel
Oni
Passugg

Peiden
Pötschnig
Prutzerbad
Querzola
Reinerz
Riedbad
Roisdorf
Rolle
Rita
Rocca San Felice
Royat
St. Laurent
St. Vincent
Salceti
Salice
Sangerberg
Schwarzenberg
Serapis-Temple
Sierra Alhamilla
Siete Aguas
Sisso
Sobron
Souzas y Caldellinas
Stoika
Suliguli
Szátou
Szczawnicza
Szolyva
Teplitz-Schönau
Ueberkingen
Vale-Vinului
Velejte
Vicarello
Vichy
Vidago
Weissenburg
Wenzelsbad
Wiesbaden

BITTER (ACIDULATED) WATERS.

Acqua-Santa-di-
  Buyhuto
Alap
Alhama de Aragon
Alicum
Almeria
Alsó-Sebes
Anguillara

Aranjuez
Argentona
Arenosillo
Ascoli
Bari
Bellus
Birmansdorf
Boras

Braque
Busk
Buzost
Cambo
Casale
Cefalú
Cheltenham
Chiclana

BITTER (ACIDULATED) WATERS—continued.

Dubograedsk
Elisabeth-Salz-
  baths
Epidaurus
Epsom
Felső-Alap
Foradade
Frailes y La Rivera
Friedrichshall
Galthof
Georgenbad
Gran
Glenn Sulphur Springs
Herg
Hermione
Huniady-Yános
Ivanda
Jabalcuz

Jaen
Jood
Kilburn
Kirchbrunnen
Kreuth
Lepanto
Ligurio
Loujo
Mala
Mergentheim
Montmirail
Nanclares
Nebouzat
Niedernau
Oelves
Ofen
Olliergues
Orel

Püllna
Quinto
Saidschütz
Santa Ana
Sarepta
Sasso di Maremma
Sedlitz
Soulieux
Spital
Stuben
Tűr
Valle de Rivas
Vescovo
Viallavieja
Villaharta
Villar del Pozo
Vilo or Rosas
Zwieselsalpe

SALINE WATERS.

Abas-Tumam
Aboukir
Absac
Acquæ-Albulæ
Acqua-Santa
Acquæ-Subreni-
  Homini
Adelheidsquelle
Adorf
Aegina
Agnano
Ahioli
Aigle
Aincelle
Ain-nonicy
Airthrey
Akná-Szlatina
Alange
Albano
Alcantud
Alicum
Almamező
Alveneu
Amelie-les-Bains
Andabre
Antrim Spa
Arborme
Archena
Arenosillo
Arnedillo
Arnstadt
Arpad
Artejo
Artern
Arzilhe
Ascoli
Ashby-de-la-Zouche

Astrakan
Augustusbad
Aussee
Availles
Baassen
Bacskó-Rahó
Baden-Baden
**Bagnères de Bi-
  gorre**
Bagnet, Le
Bagnos
Bains de la Reine
Balaruc
Baldini
Balyspellan Spa
Baran
Battaglia, La
Bebra
Bejar
Berg
Beringerbad
Betelú
Bex
Birkenfeld
Bleichbad
Bobbio
Bodenfelde
Bonar
Boriues
Borkút
Borla
Boulogne-sur-Mer
**Bourbon l'Archam-
  bault**
Bourbon-Laucy
Bourbonne-les-Bains
Bourboule, La

Bramstadt
Bridge of Earn
Buca dei Fori
Budimir
Builth
Burnham
Busk
Buzost
Caldanella di Cam-
  piglia
Caldas de Besaya
Caldas de Estrac
Caldas de Mombuy
Caldas da Rainha
Caldas de Tuy
Caldillas de San Miguel
Canillejas
Cannstadt
Carballo
Carballino
Casares
Casa Stronchino
CasinodelleCurrigliane
Casiola
Castellamare
Castiglione
Castel San Pietro
Castro-Caro
Cestona-guezalaya
Cave
Chaldette, La
Charlottenburg
Chatel-Guyon
Chatenois
Cheltenham
Chazam
Ciechocinek

CLASSIFICATION. lxvii

SALINE WATERS—continued.

Cipollo
Colberg
Corcoles
Cormus
Cos (Island)
Crieff
Daetlingen
Diemeringen
Diree
Dofana
Domène
Dovadola
Drennon Springs
Droitwich
Driburg
Drohobycz
Drumgoon
Drumlane
Dürrheim
Dürkheim **an der Haardt**
Eberbach
Eehaillon
Ecquevilly
Elmen
Ems
Erlaubad
Eufemia
Farkas-Mezö
Felso-Apsa
Felsö-Bajom
Feredschik
Filetta
Fitero
Fonsalada
Fontaceia
Fonté
Forbach
Fordignano
Fortuna
Frankenhausen
Frankfort-ou-Maine
Galaxidion
Garriga (La)
Gebangau
Gerace
Gleissen
Gloucester
Gmunden
Goczalkowitz
Grosskarben
Grull
Guarda-vieja
Guillon
Hall (Austria)
Hall (Tyrol)
Hall (Württemberg)
Halle
Hallein
Hammam-Melouane

Hammam-Meskutin
Harrowgate
Harzburg
Hellopia
Helmstädt
Helouan-les-Bains
Hermida, **La**
Horeajo de Lucena
Hubertusbrunnen
Hüttersbach
Hypate
Ikaria
Ildjak
Ilmenau
Inowrazlaw
Inola
Inninchen
Inselbad
Inverleithen
Ischl
Ivonicz
Jaxtfeld
Joanette
Jouche
Jungbrunn
Jurowla
Kalinaneste
Karlshafen
Katharsion
Kerö
Keuchreæ
Kimpalungi
Kissingen
Kondrau
Königsborn
Königsdorf-Jastrzembs
Korond
Korytinca
Kosia
Kösen
Köstritz
Krankenheil
Kreuznach
Kronberg
Kronthal
Kythmos
Laer
La-Malou
La-Motte
Langenbrücken
Lauchstadt
Laurion
Leamington
Lemnos
Lesbos
Les Salins
Letantus
Linmer
Lindenholzhausen

Lipik
Llandrindod
Lons-le-Saulnier
Loujo
Loreta
Lu
Lucas
Lüneburg
Luxeuil
Maekviller
Malahá
Malvern
Marsala
Masino
Mehadia
Meinberg
Melksham
Mezières
Moffatt
Molar, El
Molinar de Carranza
Monda
Mondorf
Monegrillo
Monfalcone
Monsão
Montalceto
Monte-Catini
Montegrotto
Morgins
Mortagone
Münster a/ Stein
Nauheim
Naples
Naxos
Neffiach
Nenndorf
Neuhaus
Neustadt a/ Saale
Newtondale
Niederbronn
Niederhall
Nohanend
Oberhergern
Oberladis
Oeynhausen
Offenau
Okarben
Oldesloe
Orb
Osterfingen
Palazzolo
Pannanich
Panticosa
Páros
Paterna de la Riviera
Paterna y Gigonza
Paterno
Peebles
Peiden

e 2

lxviii   CLASSIFICATION.

SALINE WATERS—continued.

Peleikitou	Roselle	Steinfurth
Penna, La	Roucas-Blanc	Steinheyde
Perrière, La	Rosswein	Stolypin
Pertino	Rothenfelde	Stuben
Petriolo	Russwyl	Suderode
Pillo	Saarguemines	Sulza
Pitcaithly	Sack	Sulzbad
Pitigliano	Sadshütz	Szalatnya
Plan de Phazy	St. Blasein	Talamonaccio
Plane	St. Genis	Temburg
Platimgan	St. Laurent	Tercis
Poggetti	Salces	Termopylae
Poggibonzi	Salies de **Béarn**	Thermas, Las
Poggio-Rosso	Salies	Thorpe Arch.
Poutamafrey	Salins	Thermia
Pouillon	Salins-Moutiers	Tiermas
Pozzuoli	Salonichi	Titus, **Bünos** de
Préchac	Salz	Torda
Pré St. Didier	Salzhausen	Torpa
Puda, La	Salzschlirf	Torres-Vedras
Puente-viej	Salzburg **(Hungary)**	Torretta, La
Pyrmont	Salzungen	Traishorloff
Quinto	Salzuffeln	Trescore
Rabka	San Fedele	Tritoli
Raddusa	San **Juan** de Campos	Truskowice
Radeberg	Santa-Gonda	Tusnád
Radein	Santa Restituta	Ugód
Rapolano	Santenay	**Uriage**
Rappenau	Saratoga	Ussat
Redruth	Sarepta	Venafro
Rehme	Saubuse	Vignale
Ravone in Casaglia	Saulce, La	Vignolles
Recklinghausen	Saxon	Vinadio
Reichenhall	Schieder	Viuça
Retorbido	Schmalkalden	Vippach-Edelhausen
Rheinfelden	Schmeckwitz	**V**olterra
Riando	Schönbeck	Vonitza
Rietenau	Seguara	Wielliczka
Riolo	Segura	Wiesbaden
Rio-Mayor	Semur	Wiesenbad
Rio-Sordo	Sierk	Wildegg
Rippoldsau	Soden	Wimpfen
Roggendorf	Soest	Wilhelmsbad
Rohitsch	Sotteville	Willoughby
Rodenberg	Spalato	Wittekind
Rothenfels	Staraja-Rossa	Woodhall
Rothesay	Staden	Zajzon
Rothwell		

BROMO-IODURATED WATERS.

Abano	**Bex**	Castelnuovo
Adelheidsquelle	Bondonneau	**Castro-Caro**
Airthrey	Borkut	Challes
Baassen	Bourbon l'Archambault	Czigelka
Bacskó-Rahò		Dürkheim an der Haardt
Bains de la Reine	Bourbonne-les-Bains	Felső-Alap
Balaruc	Bourboule, La	Gazost
Beringerbad	Casa Stronchino	

BROMO-IODURATED WATERS—continued.

Goczalkowicz
Grasville l'heure
Greifswalde
Hall (Austria)
Hechingen
Heilbrunn
Ivonicz
Knuitz
Konigsdorff-Jas-
  trzembs
Krankenheil
Kreuznach
Leamington
Lu
Luhaczowicz
Marlioz

Molinara de Carranza
Mondorf
Monfalcone
Montecatini
Münster am Stein
Pertino
Rabka
Radein
Ravone-in-Casag-
  lia
Ronneburg
Rothenfelde
St. Genis
St. Jean d'Aulph
Salies de Béarn

Salins
Salzburg (Hungary)
Saxon
Soest
Sotteville
Staraja-Rossa
Sulzbad
Tatzmannsdorf
Torda
Torpa
Valenza
Weilbach
Wildegg
Wimpfen
Zajzon

SULPHUROUS WATERS.

Abano
Abas-Tuman
Acireale
Acqua-Santa
Acqua-Acidola
Acqua-Puzzolente
Acqua-Raineriana
Acquæ Albulae
Acqui
Aghuloo
Agnano
Aguas Calientes
Aguas de Camangilas
Aias
Aidos
Aitora
Aix-la-Chapelle
AIX-LES-BAINS
Aláraz
Alcafache
Alcama
Alceda
Alexandersquelle
Alfaro
Alhamude Aragon
Alhama de Granada
Ali
Alica
Allevard
Allume
Almás
Almeida
Almeria
Also-Sebes
Alveneu
Amelie-les-Bains
Anagni
Andeer
Archena
Aregos

Arenosillo
Armajolo
Ascoli
Astroni
Athimonus
Ax
Baccanella
Baden (Austria)
Baden (Switzerland)
Bagnères de Bi-
  gorre
Bagui-à-Acqua
Bagni-di-Crana
Bagno
Bagno-in-Romagna
Bagnoles
Bagnols
Bains
Bains near Arles
Bajfalú
Balzach
Bania-Louka
Barambio
Baran
Barbazan
Barèges
Bari
Bath (Jamaica)
Batignolles
Battaglia, La
Baza
Bellus
Bernstein
Bertua
Beschtau or
  Maschukabaths
Betelú
Bivuto di Termini
Blue Sulphur Springs
Bobotsch

Bocklet
Boll
Bonn
Borgo-Maro
Borines
Bottaccio
Bourtscheidt
Braga
Bramstadt
Bräsa
Bréb
Bruca
Brussa
Buclesore
Budosko
Builth
Bujak
Buncome
Bundorran
Buschbad
Busk
Butterby
Buyères de Nava
Caccio-Cotto
Cadéac
Caldas de Geres
Caldas de Nossa Sen-
  hora
Calliarhoë
Calw
Cambo
Canal Grosso
Carballo
Carratraca
Casa Nuova
Casares
Casciani
Casiola
Castellamare
Castelleto Mascagni

SULPHUROUS WATERS—continued.

Castel San Pietro
Canquedes
Cauterets
Cecinella
Cervera
Challes
Cheltenham
Chianciano
Christenhofsbad
Chulilla
Citára
Civilliano
Civita Vecchia
Cocomiso
Colombajo
Cordéac
Cortegada
Crailsheim
Cuervo
Dagh Hamman
Dax
Deux-Lots
Dingolfing
Dirce
Disznopátak
Domène
Dorres
Dotit
Dragomerfalva
Dronis Namullock, or Dramsna
Drumgoon
Durenhof
Dürrwangen
Eaux-Bonnes
Eaux-Chaudes
Ebed
Ebningen
Echzell
Edenkoben
Egartbad
Eghell
Egelhof
Elorrio
Emutbühl
Enghien
Escaldas, Las
Escouloubre
Escoriaza
Eski-cherrer
Esparaguera
Essentuk
Falkenberg
Feldalfing
Feletekút
Finceschti
Fiumorbo
Fontaccin
Fonté
Forstegg

Fortyogó
Fuen-Alamo
Fuente del Toro
Fuente Amargosa
Fuente Sta. de Lorca
Fuente St. de Gayangos
Fumades
Gaieiras
Gadara
Gafete
Galera
Galleraje
Gaudesa
Garnyswyl
Gaviria
Gazost
Georgenbad
Gerace
Gisland
Gleisslibergerbaths
Gmünd
Golaise, La
Gori
Graena
Gravalos
Grödeck
Gross-Wunitz
Grüben
Guarda-vieja
Guesalivar
Guitiriz
Gythium
Gyüzy
Häberubad
Halsbrücke
Hardeck
Harkanyi
Harrogate
Hechingen
Heckinghausen
Heiden
Heiligekreuzbath
Hellopia
Hèlouan-les-Bains
Hermannsbad
Hermonville
Herrscha
Hervideros de Fuen Santa
Heselwangen
Höhenstadt
Holbeck
Homok
Horn
Hozumezó
Hradiszko
Hypathe
Iberg
Ilkeston
Ischaurto

Imola
Innichen
Ischl
Isola-Bonn
Jabalcuz
Jallova
Jamnieza
Janiszek
Jano
Jaraba
Jorullo
Joanette
Kaifa
Kahauria
Kalimaneste
Kammietz-Podolsk
Karithena
Karpfen
Kastanowka
Keked
Kemmern
Kerö
Killymard
Kimpalungi
Kiralyi
Kirchheim
Kirkilisoo
Klemntzion
Klosters
Klutschewsk
Kornwestheim
Kosia
Kostendil
Krevenish
Krzessow
Kunda
Kunzendorf
Labassère
Laemnoli
Lago d'Averno
Langassa
Langenbrücken
Laroche-Posay
La Salvadora
Landeck
Lavey
Leamington
Lebetzoba
Ledesma
Lemnos
Lepanto
Lintzi
Lisbon
Lisdoonvarna
Llandwryd
Loëches
Los Banos
Lu
Lucan
Lnucho

SULPHUROUS WATERS—continued.

Lugo
Luxburg
Macerrato
Mád
Magyar-Szent-Lazlo
Malahá
Malnas
Marlioz
Marching
Maria in Bagno
Martos
Mehadia
Meidling
Meinberg
Mercantale
Methana
Miemo
Moffatt
Moggiona
Molar, El
Moncada y Reitach
Montefiascone
Montemayor y Bejar
Morto-Ortone
Mscheno
Münsterberg
Müskau
Nenndorf
Neudorf
Neumarkt
Neusohl
Nevis
Nierstein
Nissyros
Nottington
Nuestra Señora de las Mercedes
Nydelbad
Obertiefenbach
Offenau
Okmé
Olette
Offenstein
Ohmenhausen
Oldesloe
Oll**mütz**
Olivera
Oloneschti
Ormaiztegui
Osthofen
Otschni
Parád
Pantalaria
Páramo de **Ruiz**
Paterna y **Gigonza**
Patmos
Penna, La
Peissenberg
Petriolo
Pierrefonds
Pietrapola
Piguieu
Pixigueiro
Poggetto-Theniers
Pomaret
Porretta, **La**
Portoria
Poschiavo
Pozo-Amargo
Pozzuoli
Pretriolo
Pseknps-springs
Puzzichello
Quez
Raddusa
Rapoláno
Ratzes
Rede de Corvaçeira
Retorbido
Reutlingen
Reyrieux
Riedbad
Riga
Riolo
Rio-Real
Rio-Vinagre
Rohnau
Romeyer
Ronneby
Rostona
Rothesay
St. **Cassien**
St. **Domingo**
St. **Genis**
St. **Gervais**
St. **Honoré**
St. **Jean d'Aulph**
St. **Martin Lantosque**
St. **Sauveur**
Salinetas
Salinas
Salonichi
Sandefjord
San Casciano
San Georgen
San Gregorio de Brozas
San Michele
San Martino
San Pedro do Sul
San Vincens
Santa Agueda
Santa Barbara
Santa Cesarea
Santa Filomena de Gomillaz
Sardara
Saturnia
Saudon
Schinznach
Schimberg
Schmeckwitz
Schmarden
Schwarzseebad
Schwefelbad
Schwefelbergerbad
Sciacca
Sesavnik
Sebruch
Segesta
Segura
Sennfeld
Serboneschte
Sergiewsk
Sibö
Sibitschudi-Sus
Sierra Elvira
Sironabad
Sitka
Skara-Chori
Skleno
Slepzoff-Michaeloff Springs
Smokobe
Sombor
Spag
Spalato
Sprofondo
Stachelberg
Starbeck
Stolypin
Strathpeffer
Sujo
Swanlibar
Szent-Ivan
Szinyak
Szmirdak
Szobrancz
Szombat-Falva
Talamonaccio
Talloires
Tabbiano
Telese
Tercis
Terme Luigiane
Termini Castroreale
Termini-Imerese
Teufen
Thalgut
Thera
Thermopylae
Thuez
Thusis
Tiflis
Ticrmas
Torre de San Miguel
Torres
Trenczin-Teplitz
Trescore
Trillo
Ullersdorff

CLASSIFICATION.

SULPHUROUS WATERS—continued.

Unterhallau
Urbalacone
Urberonga de Ubilla
Uriage
Ussat
Valdieri
Vale-Szkragye
Valeuza
Veruet, **Le**
Vialla
Vignale

Villaro
Villavieja de Nules
Vinadio
Vinça
Visone
Viterbo
Vizella
Voltaggio
Voltri
Warm Springs
Walby Brunnen

Weilbach
White Sulphur Springs
Wiesloch
Wildbausbad
Wilhelmsbad
Willoughby
Winterbach
Wittekind
Wolfs
Yeuzet
Zante

FERRUGINOUS WATERS.

Abbecourt
Aberayron
Aberbrothwick
Aberystwith
Acqua-acetosa
Acqua-acidola
Acqua-Bolle
Acqua-Subreni
Adolfsberg
Ahioli
Ajuacskö
Alais
Albano
Albens
Alexisbad
Aliseda
Allezani
Allume
Almamezö
Altsohl
Altwasser
Alveneu
Amphion
Andabre
Antognst
Arcs, Les
Arcidosso
Argentona
Arlam
Ascoli
Asinolunga
Aspio
Athlone water
Audinac
Auerbach
Augnat
Augustusbad
Auteuil
Baccanelli
Faczuch
Bagnaccio di Colombajo
Bagno Bossale

Bagnolino dei Rhaeli tici
Ballyspellan Spa
Balnea d'Avignone
Balsócz
Barbazan
Barberie
Barbotan
Bath
Bauche, La
Beaulieu
Beaupreau
Bela
Bellesme
Belloc
Benavente
Ben Haroun
Berg
Berencze
Berg-Gilfshueb
Bernos
Bernstein
Besenyöfalva
Bétaille
Bigorre
Birkenfeld
Birlenbach
Birresborn
Bisztra
Blanchemon
Bleichbad
Bleville
Blumenstein
Bochegginno
Boisse, La
Bonar
Bonnington
Borra
Borrone
Borsaros
Bottaccio
Boulogne-sur-Mer
Boursol

Bourbon-Lane
Bournemouth
Braga
Bramstadt
Brighton
Brownstown Spa
Brückenau
Brüttalen
Budis
Builth
Burnham
Burrone
Busignargues
Bussang
Buzias
Caldini
Calw
Camarès
Cambo
Campagne
Candin
Canena
Cannstadt
Caprifico de Valaspra
Caprenne
Carlsbrunn
Casnefouls
Castel-Connell
Castel-jaloux
Castellamare
Castroreale
Castel San Pietro
Catauin
Cattenaja
Cauquedes
Cayln, La
Cecinella
Celles
Cesalpino
Cetona
Chabetout
Chapelle-Godefroy, La
Charbonnières

FERRUGINOUS WATERS—*continued.*

Charlottenbrunn
Charlottenburg
Chateau-Gontier
Chateauneuf
Chateldon
Chatel-Guyon
Chaudefontaine
Cheltenham
Chemille
Chiatamone
Chiusa dei Monaci
Cinciano
Civilliano
Clermont
Cleve
Combe-Giràrd
Contrexéville
Corneille de la Rivière
Corticella
Couchon
Courmayor
Courpière
Cours
Courtomer
Craislheim
Crèche
Credo
Crol, Le
Cudowa
Cuervo
Czaeko
Czarskow
Daneverd
Desvres
Dinan
Dios-Györ
Dios-Jenö
Dirsdorf
Disznopatak
Dives
Dobberan
Dolka
Domeray
Dorfgeismar
Dorna
Donai
Drahowa
Draitschbrunnen
Driburg
Drise
Drumrastel
Durtal
Ebeaupin
Ebrinch
Eckartsbrunnen
Ecuillé
Eghegh
Einöd
Eisenbach
Eisenberg

Elba
Elizabethbath
Elöpathak
Elster
Engistein
Erdöbenye
Erlachbad
Ernabrunnen
Escoriaza
Evolena
Falciano
Falkenberg
Falü-Szlatina
Farette
Felsö Neresznicze
Felsö-Visso
Ferranche
Ferreira
Ficoneella
Fideris
Flinsberg
Folkestone
Fonsainte
Fonsrouilleuse
Fontagre or Sorède
Forceral
Forges-les-Bains
Forges-les-Eaux
Fossino
Franzensbad
Frasersburgh
Freienwalde
Freyersbach
Fuen-Caliente
Gaberneg
Gabian
Gagliana
Gais
Galway Sya
Garryhill Spa
Gava
Gavorano
Geilnau
Genestelle
Gerace
Geroldsgrün
Giglio
Giunco Marino
Glanagarin
Gleichenberg
Gleissen
Gloucester
Godelheim
Godesberg
Golaise, La
Goldbach
Goldberg
Gonten
Gortwa-Kisfalú
Gournay

Gracna
Grandeyro
Grasnawawoda
Grasville l'heure
Greifenberg
Griesbach
Griesbad
Grinnneaux
Grüben
Grundhofen
Gustafsberg
Halsbrücke
Hambach
Hamma
Hanau
Hardeck
Harrowgate
Harsfalra
Hartfelt
Harwich
Hassan-Pascha
Heckinghausen
Heiden
Heiligenstadt
Heinrichsbad
Heinrichbrunnen
Helenskilde
Helmstadt
Herse, La
Hermaunsbad
Hernösand
Hervideros de Fontillesca
Hervideros de Fuen Santa
Hervideros de Villar del Pozo
Heucheloup
Hinnewieder
Hofgeismar
Hohenberg
Hohenstein
Holywell
Holywood
Homburg
Homok
Homorod
Horley-Green
Horn
Houches, Les
Hozumezö
Humera
Hunstanton
Ilkeston
Imnau
Imola
Innichen
Inselbad
Islington
Ivanyi

FERRUGINOUS WATERS—continued.

Ivonicz
Jacintos, de los
Jacobfalva
Jahodnika
Jalleyrac
Jamnicza
Jaróslaw
Jelen
Jenatz
Joanette
Johannisbad, nr. Melnik
Johannisbad, nr. Pardubitz.
Johnstown
Jordansbad
Josefsbad
Kabolapolyána
Karlsruhe
Kaschin
Kaudenbach
Kellberg
Keruly
Kilkenny College
Kilrush
Kinsale
Kiralymező
Kis-Sarós
Klaussen
Klein-Chocholna
Klieniugen
Kobersdorf
Königswarth
Korsow
Korytinca
Kötschenowa
Kronberg
Krzessow
Kugelbad
Kunzendorf
Kuppis
Lacvillers
Lago d'Averno
Laifour
Lamscheid
Lanaskede
Langenberg
Larivière
Leamington
Lendershausen
Levana
Levico
Lichtenthal
Liebenstein
Liebwerda
Lindenholzhausen
Lipezk
Lippspringe
Llandrindod
Llandwrtyd
Löbenstein

Lochbachbad
Lodova
Loëche-les-Bains
Luchon
Lucsky
Luisenbad
Luxburg
Luxeuil
Macon
Mád
Madonna-a-Papiano
Maguac
Malmedy
Maloja
Marcols
Marienbad
Marmolejo
Martigné-Briant
Martinique
Martres de **Veyre**
Mastinecz
Mattigbad
Mauer
Mecina-Burbarou
Medewi
Melksham
Melos
Meltingen
Mercantale
Mina Nova
Mirandella
Mitterbad
Modum
Moha
Molla
Moncada y Reitach
Mondariz
Mondon
Montalceto
Mont Amiata
Moutanejos
Montchausou
Montégut Ségla
Montligon
Mont-Louis
Montmirail
Monte Rotondo
Montner
Morba
Morgins
Mscheno
Mula
Münchshofen
Münsterberg
Muskau
Naumburg
Naples
Navajais
Navalpino
Neudorf

Neskutschnoie
Neu-Lublau
Neumarkt
Neu-Ragoczi
Neuschwalheim
Neustadt-Eberswalde
Neyrac
Nieder-Langenau
Niedernau
Nitrolis
Nördlingen
Nowosseija
Ober-Brambach
Oberhergern
Oberunendig
Obernhaus
Ober-Rauscheubach
Obladis
Ochsenhausen
Olahfalú
Olliergues
Oppenau
Orezza
Origny
Oriol
Osterfingen
Osterspay
Otschin
Palazonia
Pandraux
Pannanich
Pantano
Panticosa
Parád
Parchim
Passy
Paterno
Penna, La
Pejo
Pergine
Pesolina
Petraglia
Pierrefonds
Pignol
Piguien
Pilsen
Pisciarelli
Pizzofalcone
Plaine, La
Plau, Le
Planchamps
Plombières
Po-Cseviczc
Poggetti
Poggetto Theniers
Poggio Curatale
Poggio Pinci
Pojan
Polzin

FERRUGINOUS WATERS—*continued.*

Pont à Mousson
Pontano
Popoh
Porla
Porte
Port Thareau
Portugos
Pötschnig
Potsdam
Pougues
Prenzlau
Prompsad
Provins
Prugues
Puertolano
Puzzichello
Puzzola
Pyrawarth
Pyrmont
Rabbi
Radeberg
Rakós
Randamel
Ranigsdorf
Rastenberg
Ratzes
Recoaro
Reiboldsgrün
Reinerz
Reulaigne
Rennes
Reyrieux
Riguardo
Rima-Brezó
Rio
Riolo
Rio Tinto
Rippoldsau
Rita
Rodna
Rohitsch
Roisdorf
Rolle
Roncegno
Roncevaux
Ronneburg
Ronneby
Ronya
Roselle
Rosenheim
Rosswein
Rosuna
Royat
Roye
Ruhla
Sail-les-Bains
Sail-sous-Couzan
Sadschütz
St. Alban
St. Denis

St. Diery
St. Martin de Fenouilla
St. Moritz
St. Myon
St. Nectaire
St. Ours
St. Pardoux
St. Peter
St. Pierre
St. Prieste
St. Prieste la Roche
St. Quentin
St. Remy
St. Sautin
St. Ubrich
St. Vallier
Ste. Heléne
Ste. Madeleine
Ste. Marie
Salins Moutiers
Salies
San Bartolomé de la Cuadra
San Bernhardino
San Casciano
San Filippo
Santa Filomena de Gomillaz
Sangerberg
San Giuseppe
San Leopoldo
Santa Cesarea
Santorin
Sarcey
Sasso di Maremma
Savergnolles
Saucats
Saundersfoot
Scarborough
Schandau
Schelesna-Wodsk
Schimberg
Schmerikon
Schooley Mountains
Schwalbach
Schums
Schüols
Schwallungen
Schwelm
Sciacca
Sesavnik
Seewen
Segrny
Segre
Sentein
Scravalle
Shotley
Sibitschudi-Sus
Sid
Sickeriki

Sigliano
Silvanés
Siradan
Sixt
Soden
Source des Cèdres
Sohl
Spa
Starbeck
Stavenhagen
Steben
Steinheyde
Sternberg
Sujo
Sulzbach
Suot-Sass
Szaldobos
Szczawnicza
Szliacz
Szombhat-Falva
Szulin
Talalmanaccio
Tapolcz-Bisztra
Tarna
Tarascon
Tarasp
Tatzmannsdorf
Telgart
Tenby
Teneke
Teplitz
Teplitz-Schönau
Terme Luigiane
Termini Castroreale
Ternaut
Terrau
Testa
Thalgut
Tharandt
Thera
Thermia
Thesbis Spring
Tivoli
Tolpa
Toropetz
Torres-Vedras
Trebas
Tremiseau
Trillo
Trois-Torrens
Trolliére, La
Truskowice
Tunbridge Wells
Tusnad
Twer
Tynemouth
Ueberlingen
Ugod
Undary
Unter-Miczinye

FERRUGINOUS WATERS—continued.

Utzera
Uzsók
Vale Szkragye
Valdeganga
Valle de Rivas
Vals
Vilo or Rosas
Villar del Pozo
Várgéde
Varennes
Vaugniéres
Veierbach
Velmont
Vicarsbridge
Vic-le-Comte
Vic-sur-Cer
Victoire
Vidago
Vihnye
Visk
Vitry
Vittel
Vizella
Waldstatt
Wassacherberg
Wattenweiler
White Suphur Springs
Wiesau
Wight, Isle of
Wildungen
Wilhelmsbad
Wolfach
Wörth
Zafarana
Zagwera
Zajzon
Zanyka
Zerbst
Zögg
Zovány

LIME AND EARTHY WATERS.

Acqua Santa
Acqui
Adelholzen
Aix (Provence)
Aled
Al-Gyógy
Alhama de Murcia
Allevard
Altsohl
Andeer
Aranzarre
Archevaleta
Atya
Baden (Austria)
Badenweiler
Bagni-à-Acqua
Baracza
Barambio
Baréges
Bath
Benimarfull
Bentheim
Bex
Bigorre
Brigg
Bubendorf
Caille, La
Caldas de Reyes
Cauvalat-le-Vigan
Cave
Cervera
Charlottenburg
Cheltenham
Chulilla
Clifton
Cransac
Cristo (Acqua di)
Derindaff
Digne
Driburg
Draitschbrunnen
Ehrenbreitstein
Euns
Encausse
Eschelloh
Escoriaza
Euzet
Faënza
Farkas-Mezö
Felsö-Ruszbach
Féron
Ferreira
Filetta
Fitero
Fuente del Rosal
Fumades
Füred
Gagliana
Gais
Gath
Gavorano
Gehringswalde
Gempelenbad
Giengen
Gonten
Gränichenbad
Gravalos
Griesbad
Grosskarben
Grosswardein
Gyrenbad
Hechingen
Heckinghausen
Heilstein
Heppingen
Hohenstein
Ibenmoos
Iberg
Ivanyi
Jobsbad
Kéméud
Kirchberg
Kirchlecnau
Kiszlnwodsk
Klieningen
Klokocs
Koroud
Krähenbad
Krapina-Töplitz
La Salvadora
Lama
Lauchstadt
Loëche-les-Bains
Losdorf
Lucca
Malvern
Mariabrunnenbad
Marmolejo
Mirabello
Molina de Aragon
Monbarri
Montafin
Montbrun
Montagnoue
Monte Rotondo
Münchshofen
Nave dell' Inferno
Neskutschnoie
Nenndorf
Neuschwalheim
Oberladis
Osztrovsk
Peiden
Piguien
Pitiglinuo
Po-Cseviczc
Poggetti
Pougyelok
Pontano
Pré St. Didier
Pystián
Quinto
Radeberg
Ransbad
Rennes
Rietenau
Römerbad (Switz.)
Röslibath
Rosswein

LIME AND EARTHY WATERS—continued.

Röthenbach
St. Amand
St. Boes
Salies
Salinetas
San Georgen
San Gregorio de Brozas
San Marziale
Santa Cesarea
Santa Gonda
Santa Filomena de Gomillaz
Schmeckwitz
Secon
Serdopol

Sierra Elvira
Sirona
Sisso
Staden
Stuben
Sulz
Szalatnya
Szklo
Szombát-Falva
Tatzmannsdorf
Teinach
Theissholz
Theussenbad
Trois-Torrens
Tschawitz

Tüebingen
Ugod
Unterlbad
Unter-Miczinyc
Uriage
Urnäschen
Vellebro
Vihnye
Villaro
Waldstatt
Wolfs
Zaisenhausen
Zaldivar
Zujar
Zurich

HYDROPATHIC ESTABLISHMENTS.

Acireale
Airthrey
Albisbrunn
Alexandersbad
Alexisbad
Amphion
Assmannshausen
Baden (Austria)
Baden-Baden
Bath
Bellaggio
Ben-Rhydding
Bishops-Teignton
Bushey (The Hall)
Berg
Bex
Brighton
Bournemouth
Buxton
Cannstadt
Champel
Charlottenburg
Cheltenham
Claremont Park
Coombe Wood
Crieff
DIVONNE
Droitwich
Eckerberg
Eichwald
Elgersburg
Empfing

Ernsdorf
Falkenstein
Feldberg
Gainfahrn
Geltschberg
Gleisweiler
Giesshubl
Görbersdorf
Grafenberg
Harsfalva
Hilversum
Herne Bay
Hohenstein
Homburg
Ilkley
Ilmenau
Ischia
ISCHL
Johannisberg
Kaltenleutgeben
Königsbrunn
Königstein
Kreuzen
Langenberg
Laubbach
Lauterberg
Leamington
Liebenstein
Maloja
Malvern
Michelstadt
Nassau

Nice
Ottenstein
Peebles
Pitlachy
Plombières
Recoaro
Reichenau
Riolo
Rippoldsau
Roznau
Ruhla
Ryde
St. Moritz
St. Radegund
Scarborough
Schmecks
Schöneck
Schöenbrunn
Schwarzenberg
Schweizermühle
Strathpeffer
Teplitz-Schönau
Troutbeck
Ullswater
Ustrom
Veldes
Villa d'Este
Vichy
Vöslau
Wartenberg
Wemyss Bay
Wiesbaden

NATURAL STEAM GROTTOS (Stufa).

Abano
Amoniac Gas Spring
Astroni
Battaglia, La
Budos
Caccinto
Castiglione
Citára

Civita Vecchia
Felsö-Ruszbach
Grotto del Cane
Gurgitello
Lemnos
Lipari Islands
Monsummano
Naphta

Nerone
Puntalaria
Petraglia
Pisciarelli
Plombières
San Germano
Testaccio
Torre del Greeco

SEA BATHS.

Abazzia
Abermyron
Aberbrothwick
Aberdaron
Aberystwith
Aboukir
Agon
Ajaccio
Alassio
Aldborough
Alicante
Allonby
Ambleteuse
Amlwich
Anacapri
Ancona
Appledore
Arcachon
Ardrossan
Arnside
Arromanches
Asnelles
Ballycotton
Bangor
Bantry
Bari
Barmouth
Beaumaris
Bernières
Berwick **(North)**
Beuzeval
Bexhill
Biarritz
Birchington
Blankenbergh
Bognor
Boltenhagen
Borbye
Borkum
Boulogne-sur-Mer
Bourg d'Ault
Bournemouth
Bray
Bridlington Quai
Bridport
Brighton
Broadstairs
Brindisi
Broughty **Ferry**
Bruneval
Bude
Budleigh **Salterton**
Cabourg
Cadiz
Calais
Campbelltown
Cannes
Cap d'Antibes
Carbagnal
Carteret

Carnarvon
Castellamare
Catania
Cayeux
Cherbourg
Clacton-on-Sea
Cleethorpes
Clevedon
Clynnog Vawr
Colberg
Colwyn Bay
Coney Island
Corfu
Courseulles
Courtmacsherry
Coutainville
Cowes
Criccieth
Criel
Croisic, **Le**
Cromer
Crosshaven
Crotoy, **Le**
Cullercoats
Cuxhaven
Dale
Dalkey
Dangast
Dartmou**th**
Deal or **Walmer Castle**
Deauville
Devonport
Diedenow
Dieppe
Dievenow
Dinard
Dives
Dobberan
Douarnez
Douglas
Douville
Drogheda
Dunbar
Dundrum
Dunkerque
Dunmore
Dunoon
Durness
Dusternbrook
Eastbourne
Eckernförde
Erqui
Exmouth
Falmouth
Fano
Fécamp
Felixstowe
Fleetwood
Flint
Folkestone

Fontarabbia
Fowey
Frasersburgh
Freshwater
Friedrich
 Wilhelmsbath
Glengariff
Gourock
Grandcamp
Grange-over-Sands
Granville
Grno
Gravesend
Great Yarmouth
Gustafsberg
Hafkreuz
Hapsal
Hartlepool
Harwich
Hastings
Hâvre, Le
Hayling Island
Helensburgh
Heligoland
Hennebon
Heringsdorf
Herne Bay
Heyst
Holkham
Holyhead
Holywood
Home-Varaville, Le
Honfleur
Hornsea
Houlgate
Hourdel
Hovehampton
Howth
Hunstanton
Hyères
Hythe
Ilfracombe
Instow
Ismailia
Junqueiro
Katwyk
Kiel
Kilkee
Kilrush
Kinsale
Klutz
Kopenhagen
Laugrune
Largs
Leo
Legué St. Brieul
Lepouliguen
Lion-Sur-Mer
Lisbon
Littlehampton

SEA BATHS—continued.

Littlehaven
Livorno
Llanduduo
Llanfairfechau
Llangranog
Llanstephen
Long**branch**
Lowestoft
Luc sur Mer
Lulworth, **West**
Lymington
Lyme Regis
Lynmouth
Lytham
Mablethorpe
Madeira
Maestrand
Malahide
Malta
Margate
Marienlust
Marsala
Marseilles
Massa
Melcombe-Regis
Mentone
Mers
Messina
Middle**kerke**
Millport
Minehead
Misdroy
Monaco
More**cambe**
Mumbles
Mum**by**-cum-Chapel
Muritz
Nahaud
Nairn
New Brighton
New Quai (Eng.)
New Quai (Wales)
Nice
Norderney
Nou**velle**, La
Odessa
Oese**l**
Oporto
Ostende
Paignton
Paim**pal**
Palermo
Param**é**
Passage
Peg**li**
Pembrey
Penarth
Pendine
Penmaenmawr
Penzance

Pesaro
Petites Dalles, Les
Pizo, Il
Plymouth
Poutaillac
Pontrieux
Pornic
Portland
Porto**bello**
Port Bail
Port en Bessin
Port Rush
Port Said
Port Steward
Pouroille
Putt**bus**
Pwllheli
Queen**stown**
Quinéville
Ramsay
Ramsgate
Redcar
Regneville
Rhyl
Rimini
Roscoff
Rosstrevor
Royan
Runcorn
Ryde
Ryhope
Sables d'Olonne
St. Andrew
St. Aune's-on-Sea
St. **Bee's**
St. **Briac**
St. **Davids**
St. **Jean de** Luz
St. La**wrence**
St. Leonards-on-Sea
St. **Lunaire**
St. **Malo**
St. **Marie**
St. **Mary's**
St. **Pair**
St. **Raphael**
St. V**alery en** Caux
St. V**alery sur** Somme
St. V**aast la** Hougue
Ste. Adresse
Salcoaths
Saltburn-by-the-Sea
Saltfleet-Haven
Sandefjord
Sandgate
Sandown
San Lorenzo
San Remo
San Sebastian
Sassnitz

Saundersfoot
Scarborough
Scheveningen
Seaford
Seascales
Seaton
Seaton-Carew
Sea-View
Shanklin
Shap
Sheerness
Sidmouth
Sillotth
Sinigallia
Skegness
Skinburness
Sorrento
Soulac-les-Bains
Southampton
South**bourne**
Southend
Southport
Southsea
South Shields
Southwold
Spezia, La
Stromstad
Sutton
Swanage
Swansea
Swinemünde
Sylt
Taormina
Teignmouth
Teste, La
Thurso Bay
Torquay
Totland Bay
Toulon
Towyn
Tramore
Trani
Travemünde
Tremblade, La
Trepot, **La**
Trieste
Trouville
Upton
Val André, Le
Valencia
Vaxholm
Venice
Ventnor
Vcules
Venlettes
Viareggio
Villefranche
Villers-sur-Mer
Villerville
Walton-on-the-Naze

SEA BATHS—*continued*.

Wangeroog
Warnemünde
Warrenspoint
Warnicken
Watchet
Wells
Wemys Bay
Westerland Sylt

Westgate-on-Sea
Weston-Super-Mare
Westward Ho
Weymouth
Whitburn
Whitby
Withernsea
Worthing

Wyk
Wyk aan Zee
Yarmouth
Yport
Zandvoort
Zoppot
Zwolle

CLIMATIC AIR STATIONS.

Abazzia
Adelsberg
Aigle
Ajaccio
Alassio
Algiers
Alicante
Almeria
Alton Towers
Alveneu
Amalfi
Ambleside
Amelie-les-Bains
Amphion
Anacapri
Annecy
Antibes
Arcachon
Arco
Arenzano
Auerbach
Aussee
Axenfels
Axenstein
Baden (Austria)
Baden-Baden
Badenweiler
Bagnères de **Bigorre**
Ballater
Beggenried
Belalp
Bellaggio
Beuren
Bex
Biarritz
Bodendorf
Bordighera
Bormio
Botzen
Bridge of Allan
Brig
Brixham
Bürgenstock
Bushey (The Hall)
Cadenabbia
Cairo
Cambo
Campfer

Cannes
Cannstadt
Capri
Carlsbrunn
Casamicciola
Castellamare
Catania
Cauterets
Chamounix
Chiatamone
Chiavari
Cliftou
Coire
Colico
Como
Corfu
Cornigliano Ligure
Courmayor
Davos
Divonne les Bains
Eaux-Bonnes
Eckerberg
Elche
Ems
Engelberg
Feldberg
Flims
Geltschberg
Gerace
Gersau
Gmunden
Godesberg
Görbersdorf
Göez
Grasse
Greifenberg
Gries
Grindelwald
Gurnigel
Harsfalva
Hastings
Helouan
Heidelberg
Heiden
Hyères
Ilfracombe
Ilmenau
Interlaken

Ischl
Johannisbad
Korytinea
Kreuth
Kreuzen
Laubbach
Lausanne
Leprese
Lesina
Liebwerda
Lippspringe
Lisbon
Livorno
Löbenstein
Loëche-les-Bains
Looe
Lucsky
Lugano
Luksor
Madeira
Malaga
Maloja
Malta
Malvern
Martigny
Mattigbad
Mentone
Meran
Merligen
Meta
Messina
Moffat
Mogador
Modum
Monaco
Monte-Carlo
Montreux
Morgins
Murcia
Mürren
Naples
Nervi
Neuschmecks
Nice
Oban
Obermais
Ober-Rauschenbach
Oberladis

CLIMATIC AIR STATIONS—continued.

Oberwinter
Ottenstein
Palermo
Palestrina
Pallanza
Parád
Pau
Pegli
Penzance
Pisa
Plombières
Pontresina
Preblau
Pyrawarth
Ramleh
Rapallo
Reiboldsgrün
Rippoldsau
Rome
Römerbad
Rosenheim
Roznau
St. Cergues
St. Moritz
St. Raphael
Salcombe

Salerno
Salzburg (Salzkammergut)
Samaden
San Remo
Santa Margherita
Schandau
Schleusingen
Schmécks
Schmerikon
Schönbrunn
Schweizermühle
Seelisberg
Sestri Ponente
Sorrento
Spezia, La
Steinabad
Streitberg
Stresa
Sutinsko
Tambach
Tarasp
Taormina
Tatzmannsdorf
Tegernsee

Teufen
Thoune
Tissington
Torquay
Tobelbad
Triberg
Trieste
Tüffer
Tuzsnád
Valencia
Veldes
Venice
Vevey
Veytaux
Viareggio
Villefranche
Villeneuve
Vihnye
Vöslau
Wartenberg
Weggis
Wiesbaden
Wildbad
Zermatt
Zug

WINTER STATIONS.

Abazzia
Acireale
Ajaccio
Alassio
Algiers
Alicante
Almeria
Amélie-les-Bains
Anacapri
Antibes
Arcachon
Arco
Arenzano
Beaulieu
Biarritz
Bigorre
Bordighera
Brussa
Cairo
Cambo
Cannes
Cap d'**Antibes**
Capri
Catania
Corfu
Cornigliano-Ligure

Dagh Haman
Elche
Görberstorf
Görz
Grasse
Hammam R'irha
Helouan-les-Bains
Huelva
Hyères
Laubbach
Losina
Lisbon
Livorno
Luksòr
Madeira
Malaga
Maloja
Malvern
Mentone
Meran
Messina
Monaco
Monte-Carlo
Murcia
Mustapha-Supérieure
Naples

Nervi
Neuschmécks
Nice
Palermo
Pau
Pegli
Pisa
Rome
St. Jean de Luz
St. Moritz
St. Raphael
Samaden
San Juan de Campos
San Remo
Santa Margherita
Schmecks
Sestri-Ponente
Sevilla
Smyrna
Spezia, La
Taormina
Valencia
Venice
Villefranche
Wiesbaden

GRAPE CURE.

Bex	Lausanne	Neustadt **an der**
Bodendorf	Lesina	**Haardt**
Botzen	Levico	Sinzig
Dürkheim an der	Locarno	Sulza
Haardt	Lugano	Vevey
Erdöbenye	Madeira	Vihnye
Gleisweiler	Maloja	Vöslau
Görz	Markammer	Wachenheim
Johannisberg	Meran	**Wiesbaden**
Laubbach	Montreux	Zug

WHEY CURE.

Aigle	Kierling	Rosenau
Aix-les-Bains	Kirchberg	Rothwell
Albisbrunn	Kreuth	Roznau
Andeer	Kreuznach	**St. Moritz**
Appenzell	Kronberg	**Salzbrunn**
Axenfels	**Laubbach**	Salzburg (Tyrol)
Axenstein	Liebenzell	Schinznach
Baden (Austria)	Liebwerda	**Schönbrunn**
Baden-Baden	**Loëche**	Schöneck
Badenweiler	**Maloja**	Seelisberg
Belalp	Marienbad	Seeon
Beuren	Montreux	Sinzig
Bex	Morgins	Steinabad
Bistritz	Münster am Stein	Sternberg
Ems	**Mürren**	Streitberg
Ernsdorf	Neundorf	Szczawnicza
Flims	Neuenahr	Tegernsee
Franzensbad	Neustadt an der Haardt	Teplitz-Schönau
Friberg	Nieder Langenau	Tüffer
Gais	Ober-Rauschenbach	Ullersdorf
Giesshübl	Oberladis	Unterseen
Gleissweiler	**Orenburg**	Ustrom
Gmunden	Ottenstein	Veldes
Görbersdorf	**Parád**	Vevey
Görz	Partenkirchen	**Vöslau**
Grindelwald	Piguien	Waidhaldenbad
Heiden	**Plombières**	Warmbrunn
Interlaken	Pontresina	Wattwyl
Ischl	Rabbi	Weggis
Jenatz	Reichenau	Weissenstein
Jungbrunn	Reichenhall	Wildhausbad
Kanitz	Reutlingen	**Wiesbaden**
Karlsbad	Rigi	**Wildbad**
Kammer	**Rippoldsau**	Zug

THE DICTIONARY.

ABACH.—Germany, Bavaria, near Ratisbon, on the right bank of the Danube.
Alkaline waters.
Special indication: Abdominal plethora.

Abano.—Italy, Venetia, near Padua. For routes see table.
Brom-iodide and sulphurous waters, 200° F.
Mud baths.
Special indications: Rheumatism, scrofula, herpes, gout, ovarian and uterine inflammations, secondary and tertiary syphilitic affections.
This place, known to the Romans as Acquae-Aponenses, contains now about 4,000 inhabitants. The water is used almost exclusively for baths. The mud is applied locally to affected parts and the vapour in steam-baths for inhalation and locally.

Abas-Tuman.—Russia, Caucasus, near Achalzich.
Sulphuro-saline waters, 110° to 120° F. Similar to Aix-les-Bains and Cauterets, but less rich in mineral constituents.
Special indications: Rheumatism, articular contractions, arthritis, atonic ulcers and mercurial dyscrasia.
The establishments offer but little comfort. The scenery is wild and grand, and game very abundant. There is a tradition that the waters were used by Alexander of Macedon.

Abazzia.—Austria, Istria, district of Fiume.
Winter station.
Season: 1st October to 1st May.
Special indications: Affections of the larynx, pharynx and chest.
Abazzia is a new and rising station, one hour's drive from Fiume. It is situated on a small bay of the Adriatic, and is well protected against north, east and west winds. The Bora is deprived of much of its evil influence by the adjacent island of Veglia. Statistics are still wanting.
Doctors: Dr. von Hansen.
Hotel: The Grand Hotel Abazzia.

A

Abbecourt.—France, Department of Seine and Oise, commune of Pesey.
Cold ferruginous bicarbonate waters.

Abensberg.—Germany, Bavaria, situated on the Abens, near Ratisbon.
Mixed carbonate waters.
Special indications: Herpes and arthritis.

Aberayron.—Wales, Cardiganshire, between Cardigan and Aberystwith.
Chalybeate waters and sea bathing. Sandy beach.
Doctor: Dr. Williams.

Aberbrothwick, or **Arbroath.**—Scotland, Forfarshire. 16 miles from Dundee.
Sea baths and cold ferruginous carbonate waters.
Special indications: Scrofula.
Season: July-September. The climate is mild, and the town sheltered from easterly winds.
Doctor: Dr. Monroe.
Hotel: The *Albion*.

Aberdaron.—Wales, Carnarvonshire, on the northern coast of Bay of Cardigan.
Sea bathing; sandy beach; very picturesque scenery.

Aberystwith.—Wales, Cardiganshire, by London and North-Western Railway, in nine hours.
Sea baths and ferruginous waters.
Season: May-October. Sandy sloping beach.
There is an establishment for hot and cold baths. The amusements are varied, and include concerts, theatre, archery, tennis, cricket, musical promenades, balls, horse racing, good fishing and boating, grouse and partridge shooting.
Doctor: Dr. Morgan.
Hotel: The *Royal*.

Aboukir.—Algiers, Province of Oran, near Mostaganem, and 79 k. from Oran.
Sea baths and saline spring.

Abzac.—France, Department of Charente, *via* Bordeaux, stop at Ruffec; thence by omnibus.
Cold saline waters.
Special indications: Atony, constipation, congestions, ague and intermittent fevers.

Acireale.—Italy, Sicily; 35,000 inhabitants; 550 feet above sea level; on the southern slope of Mount Etna; two hours from Messina. For routes see table.
A winter station of great importance.
Sulphuro-saline waters of St. Venere, 70° F. Milk and whey cure, Hydropathic establishment.
Season: All the year round.

Special indications: Pulmonary phthisis, **chronic bronchial catarrh**, emphysema, pneumonia, anæmia and **chlorosis are** benefited by the climate in winter; rheumatism, **gout, tho uric or oxalic acid** diathesis, anchylosis, obesity, latent **scrofula, paralysis and neuralgia,** mercury, lead and arsenical **poisoning are benefited by** the waters.

The climate of Acireale is very dry and **mild. The death-rate is 19 per thousand.** Mean **winter temperature 55° F. The variations are never more than 10°.** The number of rainy days in the year range between 30 and 50, but do not exceed the latter. Snow and fogs are unknown. Windy days are less in number than on the Riviera, and there is a complete absence of dust. The prevailing winds are the S.E., S.W., S.S.E, and E.S.E. The air is very rich in ozone, owing to the vicinity of the sea and the luxuriance of the vegetation.

Post and Telegraph in the Hôtel des Bains. Doctors are attached to the establishments.

Hotels: The *Grand Hôtel des* **Bains**, facing directly **south**, surrounded by gardens, 100 bed-rooms, music, reading and billiard rooms. Board, 300 fcs. per month, 80 fcs. per week, and 12 fcs. per day.

Acquaacetosa.—Italy, Rome, outside Porta del Vopolo. For routes see table.

Bicarbonate ferruginous waters, 33° F.

Special indications: Convalescence from **intermittent fever.**

Acqua-Acidola.—Italy, Tuscany, near Montalceto.

Sulphurous and ferruginous waters, 70° F.

Special indications: Atonic affections **of tho stomach and intestines, including various forms of dyspepsia and** nephritis.

Acquæ-Albulæ.—Italy, 35 minutes from Tivoli; 50 minutes from Rome by steam tramway.

Sulphurous-saline waters, 75° F.

Season: Open all the year round, but the principal season **is from April till N**ovember.

Special indications. Chronic ulcers, atonic and relaxed conditions of **the** mucous membrane generally, chronic skin diseases and rheumatism.

The establishment is one of the best of its kind in Europe as regards buildings and **appliances.** The supply of **water amounts to** 5,000 quarts a second. In this respect, **and also in the** quantity of **sulphurous acid they hold** in solution, **these waters are second to none in** Europe of **their kind. They were very famous in Roman times,** and much frequented. Owing to the copious supply, the water is being constantly **renovated in the** baths. The establishment contains four large swimming baths, without including tho commoner baths **for** soldiers **and the poor.**

There are also 200 separate baths, with dressing-rooms, pavilions for families, cold and sulphurous water douches, inhalation chambers and a large restaurant, at which the prices are moderate. All appliances are thoroughly efficient.

Three doctors are constantly in attendance at the baths.

For further information address the joint proprietor, Com. Ign. Faust Anderloni, Rome.

Acqua-Bolle.—Italy, Tuscany, near Grosseto, in a very lonely and cheerless valley.

Alkaline-ferruginous waters 60° F.

Special indications: Gravel and calculous affections of the urinary and digestive organs.

Acqua-Puzzolente-di-Livorno.—Italy, Tuscany, near Leghorn.

Sulphuro-saline waters, varying in temperature, according to the season, from 50° to 75° F.

Special indications: Scabies, herpes, eczema, and other skin diseases, scrofulous and rheumatic affections.

Acqua-Raineriana.—Italy, Venetia, near Venice, in the vicinity of the volcanic lake of Arqua.

Sulphurous waters, 70° F., containing a considerable quantity of free sulphuretted hydrogen.

Special indications: Skin diseases such as herpes and psoriasis, scrofulous affections, and debility of the digestive organs.

Notwithstanding its disagreeable taste, the water is chiefly used internally; if used externally, it is generally mixed with seawater.

Acqua-Santa (Terme di).—Italy, Province of Ascoli-Piceno. Railway to St. Benedetto del Tronto, thence by carriage, 4½ hours. Omnibus, 3 fcs. each person. Cariages, 15 fcs. one horse; 25 fcs. two horses. 1,200 feet above sea level.

Saline-sulphurous waters, with traces of bromine and iodine, 85° F.; steam, shower and mud baths. Inhalation.

Season: June–September.

Special indications: Chronic inflammation of the skin, muscular and articular rheumatism, syphilis, scrofula.

The establishment built in 1843 has been recently furnished with all modern appliances. Several hotels, and pensions ranging from 6 fcs. to 10 fcs. a day.

Acqua-Santa is a quiet spot, with delightful walks and excursions. Post and Telegraph Office.

Doctors: Drs. Bellini and Tucci.

Acqua-Santa-di-Buyhuto.—Sicily, near Palermo and close to the sea.

Acidulated and bitter waters.

Special indications: Constipation and obesity.

Acqua-Santa.—Italy, near Voltri, *viâ* Genoa.
Sulphate of lime waters, 57° F.
Special indications: Scrofula and herpes.
Doctor: Dr. Bagliello.

Acquæ-Subreni-Homini.—Italy, on the road from Naples to Pozzuoli.
Saline ferruginous waters, 95° F.
Special indications: Affections of the genital organs, amenorrhœa, anæmia.

Acqui.—Italy, Piedmont, near Alessandria.
Iodide and sulphate of lime waters, 178° F. The muds are more used than the waters.
Special indications: Indolent articular affections, local paralysis, rheumatism and sores of an atonic character and syphilis.
The sojourn at Acqui is not considered particularly agreeable, owing to the dampness of the climate. The thermal establishment is well arranged, and the accommodation comfortable.
The climate of Acqui is very changeable and humid in consequence of the vapours from the waters.
Doctor: Dr. Dom. de Alessandri.
Hotel: The Thermes.

Adelheidsquelle.—Germany, Bavaria, near Tölz.
in the village of Oberheilsbrunn.
Bromo-iodurated saline waters, 24° F. 5·93 per cent. mineral constituents.
Special indications: Scrofula, obesity, affections of the bones, goitre, inflammatory affections of the ovary, uterus and eyes.

Adelholzen.—Germany, Bavaria, near Frauenstein.
Bicarbonate of lime waters.
Special indications: Anti-arthritic.

Adelsberg.—Austria, Illyria, near Trieste.
A climatic station in spring and autumn, very charmingly situated.
The town is visited chiefly on account of the "Grotto of Adelsberg," which is without exception the grandest and most remarkable natural excavation in Central Europe. To visit it thoroughly requires about two hours.
Hotel: The *Adelsberger Hof*.

Adepsos.—Greece, Island of Eubea.
Thermal sulphurous waters renowned in antiquity, 205° F.
Special indications: Rheumatic gout, anchylosis, paralysis, plethora, &c.

Adolfsberg.—Sweden, Läns Örebro, near Axberg.
Alkali-saline and ferruginous waters.
Special indications: Chronic diarrhœa, gout, rheumatism, anæmia.
One of the most picturesquely situated towns in Sweden. The arrangements are all good.

Adorf.—Germany, Saxony, in the circuit of Zwickau.
Three springs; one contains sulphate of soda, and two are saline.

Ægina.—Greece, Morea, near the sea coast.
Thermal saline waters.

Aghaloo.—Ireland, county of Tyrone.
Saline sulphurous waters.
Special indications: Skin diseases. Only used by the residents.

Agnano.—Italy, near Naples, by tramway. On the lake of Agnano.
Sulpho-aluminous and saline springs, 122° F.
The lake of Agnano contains various springs, which owe their origin to the condensation of steam from the volcanic soil.
The Grotta del Cane with its emanations, proves, however, a greater attraction to visitors than the waters.

Agon.—France, Department of La Manche, from Paris via Coutances.
Seabaths; sandy beach; increasing in popularity.

Aguas-Calientes.—Mexico, State of Tacatecas.
Sulphurous waters, 214° F.
They give rise to the river of Aguas-Calientes.

Aguas de Camangilas.—Mexico, State of Guanaxuato.
Sulphurous water, 228° F., only slightly mineralised.

Ahiolo.—Turkey, about 55 miles from Constantinople, near Rumili.
Saline ferruginous waters, 100° F.
Special indications: Abdominal affections, amenorrhœa, scrofula, &c.

Aias.—Turkey, Asia Minor.
Sulphurous and cuprous waters.
The spring rises in a valley rich in silver and copper mines.

Aidos.—European Turkey, at the foot of the Balkans, near the Black Sea.
Saline sulphurous waters, 113° F.

Aigle.—Switzerland, Canton Vaud, between Montreux and Bex, 1,350 feet above sea level.

Saline waters.

The Kurhaus Hôtel is well fitted-up and thoroughly comfortable. The scenery is romantic.

Doctor: Dr. Vercy.

Aincille.—France, at the foot of the Pyrenees.

Saline water, containing 135 grammes of salt to the litre.

The mineral constituents are derived from a mass of rock salt.

Ain-nonicy.—Algiers, Province of Oran, arrondissement Mostaganem, 48 miles from Atteville.

Strong saline waters, which act as a purgative.

Airthrey.—Scotland, in the neighbourhood of Stirling.

Brine waters; 6 springs.

Special indications: Skin diseases, scrofula.

A fine picturesque town.

Doctor: Dr. Patterson.

Aitora.—Italy, Tuscany, near Monte-Catini.

Sulphurous waters, 55° to 60° F.

Special indications: Used externally in chronic rheumatism, in scrofula, gout, and skin diseases; internally in cases of gravel and vesical catarrh.

Aix.—France, Bouches du Rhone. For routes see table.

Earthy waters, 81° F., useful as a sedative and alterative.

Doctor: Dr. Louzet.

Aix-la-Chapelle (Aachen).—Germany, Rhenish-Prussia. For routes see table.

Sulphurous alkaline waters, 107° to 120° F.

Special indications: Skin diseases of all kinds, rheumatism, chronic diseases of the nervous centres, neuralgias, paralysis, and certain class of diseases in which mercurial treatment is generally indicated.

Bookseller: Rud. Barth, Holzgraben 8.

English church: Annastrasse.

Doctors: Drs. Beissel, Lersch, Mayer, Reumont, Brandes and Schuhmacher.

Hotels: The *Grand Monarque*, The *Union*.

Aix-les-Bains.—France, Savoy, 850 feet above the level of the sea; 8 hours from Turin, 4 hours from Lyons, 3 from Geneva, and 2 from Annecy. A steamboat plies on the Rhone between Aix and Lyons. For routes see table.

Thermal sulphurous waters, containing chiefly sulphuretted hydrogen. Temperature from 112° to 114° F.

AIX-LES-BAINS.

Season: April to November, but open all the year.

As many as 20,000 strangers annually visit the town.

Special indications: Rheumatism and gout are the diseases most successfully **treated** at Aix. Local affections dependent on or **connected with** these, such as sciatica, dyspepsia, cutaneous **diseases, are also, with few exceptions,** either cured or relieved

Chronic catarrh of the neck of the womb, amenorrhœa, metritis, syphilis, bronchitis, laryngeal and nasal catarrh, pharyngitis, wounds by fire-arms, the sequelæ of fractures or sprains, form the bulk of cases in which the value **of the Aix** waters **is most generally admitted by medical men.**

The world-wide reputation of the thermal establishment of Aix is easily accounted **for,** in **the** first instance, by the abundance **of the natural** hot-water springs, **the supply of which am**ounts **to a million** gallons daily secondly, by **the well-known attention and skill of** the shampooers: thirdly, by **the** close proximity of **Marlioz and Challes, the two** strongest cold sulphur springs **on the Continent. They are very** useful as therapeutic auxiliaries **to the mineral waters of Aix.**

The town of Aix is picturesquely situated at an altitude of **about 800 feet above the** sea **level, on** the lower slopes of the **hills which form the** base of **Mont** Revard. It overlooks the romantic lake of Le Bourget, and lies amidst verdant meadows, vineyards and gardens, in which figs, almonds, and even pomegranates grow luxuriantly.

The resident popu**lation of the** town amounts to **some** 3,500 **inhabitants.**

The mean temperature during the season, from April to October, is 70° **F.**

The thermal establishment **is open** the **whole** year round Dr. MACPHERSON says: "There **are probably no** sulphur baths **in which** the arrangements are so complete as in those of Aix."

"**The** douche," says Dr. Garrod, "forms a very important **specialty** in the treatment at Aix. It is used in conjunction with **hot vapour,** with ordinary bathing, and with the waters internally. When a **larger** amount of sulphur, in the form of sulphate of **sodium, is desirable, the cold** waters of Marlioz, situated **about one mile from** Aix, **or the Ch**alles water, near Chambery, which **contains a large** amount not only of sulphur, **but also** of **iodine** and bromine, **may** be **employed** internally. In **gouty** affections **the Aix treatment has proved most useful in these cases, (1) where** the gout **has manifested itself in cutaneous eru**ptions, **such** as psoriasis or eczema, whether combined **or** not with **joint** affections; (2) where much passive swelling and stiffness **have** remained after attacks of articular gout; (3) in that **form** of gout which is not connected with portal congestion, and in which other thermal waters, such as those of Wildbad and **Teplitz, are usually found** beneficial."

In speaking of the treatment of rheumatic gout (rheumatoid arthritis), Dr. Garrod adds:

"Waters which I have found very beneficial in rheumatoid arthritis are those of Aix-les-Bains, at which place the douche is a very important part of the treatment."

The Grand Cercle d'Aix-les Bains is situated in the centre of the town, in close proximity to the Hotels, the railway station, the parks, the bathing establishments and the post and telegraph offices. This club, which has recently been much improved by the addition of an elegant little theatre and large reception rooms, affords every facility for social amusement to its members. Every evening there is a *soirée musicale* given by the Septuor and Italian Orchestra, under the guidance of L. Olivieri. Twice daily concerts are given in the gardens (40 musicians) under the same conductor. Three times a week there are representations by the Opera Comique and Italian Opera troupes (Director, M. Santino Costa); also, three times weekly, a grand instrumental concert by the whole orchestra (60 musicians), directed by M. E. Clonne, of the *Châtelet*, Paris. Every Tuesday, *fête de nuit*, ball, illumination of the parks, and *feu d'artifice*. Visitors desiring to attend these *fêtes* must be properly introduced. In July and August some of the chief stars of the Paris theatres, as well as many foreign celebrities, appear. Balls for children, marionettes, music, billiard and card rooms, and pigeon matches, are also open to members. There is a telegraph office in the club café. The season is from the 1st May to 31st October. The *foyer* is a very handsome apartment, affording a magnificent view of the gardens and chain of mountains. For further information address the Director.

An excellent race-course has lately been purchased, and races will take place in June and July. The course is situated on the Avenue de Marlioz.

Walks and excursions are very varied. The environs being most picturesque, an ample field is given for outdoor amusements. The principal excursions are: Annecy and the lake, les gorges du Fier, la vallée du Fier, la grande chartreuse, la vallée des Beauges, the abbey of Haute-combe, la cascade de Gresy, le moulin de Primor, etc., etc.

Aix possesses a museum with lacustrine collections from the lake of Le Bourget.

Cabs and carriages: one horse, course, 1 fr.; two horses, course, 1½ fr.; by hour, one horse, 2½ fr.; two horses, 3½ fr.; by day and by month, as agreed upon.

Post and telegraph office.

English church.

Bookseller: H. Bolliet, Place du Marché.

Dentist: Dr. Harwood.

Bankers and Exchange Office: Domenge, Monestès & Co. Payment of letters of credit, circular notes, cheques, &c., &c.

Doctors: Drs. Blanc (Inspector of the Baths), Menzies, Wakefield, Macé, Cessens, Brachet and Guilland.

Hotels: The *Hotel de l'Europe* is a first-class and well-appointed house, suitable for English and American families. There are two large villas in the grounds of the hotel, which may be hired by families. Lift and lawn tennis ground. J. M. Bernascon, proprietor.

The *Splendid Hotel* is a first-class house, in a fine situation overlooking the park. G. Rossignoli, same proprietor as *Hotel Venat* and *Bristol*.

Ajaccio.—Corsica. 14 hours by P.L.M. to Marseilles; thence by steamer of the C. G. T. to Ajaccio 15 hours; direct boat from Nice in 10 hours every Saturday. 18,000 inhabitants.

A winter station of great importance. Sea and warm baths all the year round.

Season : For the former, April to November; Winter season, November to end of May.

Special indications : Chest diseases generally; anæmic and chlorotic affections, rheumatism, paralysis, diabetes.

Built at the foot of a sloping hill, the town extends in the form of an amphitheatre, and is well protected from cold northerly and westerly winds. Ajaccio, as a winter health resort, has a great future.

Meteorological observations are as yet deficient. Mean temperature October—April, 57° F. Rainy days, 34. Variations slight, and never rapid. Mean barometric pressure, 775 $^m/_m$; death-rate, 26 per thousand; relative humidity of the air, 76 %; fogs are scarcely known. There is almost complete absence of dust; atmospheric disturbances occur but seldom; the mean of temperature during the day and throughout the year is very constant, much more so than on the Riviera, from month to month.

Bankers and Exchange Office : Lanzi Frères.

Chemist : Cours Grandval.

Bookseller : Peretti, Place du Marché.

Public Library : Library Fesch (with 30,000 volumes), open to the public.

The drives and walks in the neighbourhood are beautiful, and the roads well kept. Flowers and shells are found in abundance.

Excellent shops and circulating library.

The water of the Gravona river, brought by aqueduct to the town, gives an abundant supply, the overflow from which cleanses the drains.

The railway to Bastia, *via* Corte, is progressing rapidly and will soon be open for traffic.

Guagno, with its celebrated sulphur springs, 128° F., is only 38 miles distant; the famous Orezza springs are 81 miles distant by an excellent road.

The scenery in Corsica is very beautiful and quite unique. From the nature of the soil dust is seldom an annoyance in Ajaccio; while the mountains on the opposite side of the gulf protect the eye from the glare, so often complained of on the Riviera.

Post and Telegraph Office, Cours Napoléon.

English church during winter season; the chaplain is licensed by the Bishop of Gibraltar. The church has been built by Miss Campbell.

Doctors : Drs. Keith, Valentiner, Colonna, Wagner and Hugemann.

Hotels : The *Continental Hotel.*

The *Hotel Belle-Vue.*

Ajnacskö.—Austria-Hungary, comitat of Gömör.

Alkaline ferruginous waters.

Special indications : Gout, rheumatism, hæmorrhoidal affections, disorders of the urinary and intestinal tracts.

Alais.—France, department of Gard, near Nismes.
Chalybeate waters.
Special indications: Nervous debility, migraine, **anæmia**, chlorosis.

Alange.—Spain, Estremadura, near Merida.
Saline waters, 80° F.
Special indications: One of the most powerful of Spanish waters in cases of rheumatism, paralysis, atonic sores and irregularities of the lower intestines.
Very primitive arrangements
Doctor: Dr. Hernandez.

Alap or Felső-Alap and Alsö-Alap.—Austria, Hungary, near Stuhlweissenburg.
Acidulous or bitter waters.
Special indications: Irregular functions of the intestinal organs, affections of the mucous membrane, chronic skin diseases and plethora.

Alaraz.—Spain, province of Avila, near the town of same name and Peñaranda.
Sulphurous waters.
Special indications: Amenorrhœa, hysteria, migraine, and nervous complaints.

Alassio.—Italy, between Ventimiglia and Genoa, 5,500 inhabitants.
Winter station. Charming and quiet locality.
Its position is very sheltered and romantic at the foot of wooded mountains. Very fine parks at some of the villas. Little wind and no dust. It has been visited frequently by members of the Italian Royal family, and General Garibaldi was here cured of his bronchial attacks. There is an English church and several primary schools.
Banker and Exchange Office: Gurgo, who also gives all information gratis on application.
Doctors: Drs. Dickinson and Schneer.
Hotels: The *Grand Hôtel d'Alassio*, well situated, full southern aspect; comfortable, with moderate prices; baths of every description in the hotel; omnibus meets all trains. Rafael Paggi, proprietor.

Albano.—Italy, in the Campagna of Rome. For routes see table.
Ferruginous saline waters, 71° F.
The muds and waters are principally used in cases of rheumatism and old wounds.

Albens.—France, Savoy, on the road from Aix to Annecy.
Effervescing ferruginous waters.

Albisbrunn.—Switzerland, canton of Zürich, near Hausen; 1,960 feet above sea level.
Hydropathic establishment, milk and whey cure.
A very comfortable and elegant establishment, picturesquely situated.
Doctor: Dr. Brunner.

Alcafache.—Portugal, district of Beira, in the Sierra Estrella.
Sulphurous waters, 100° F.
Special indications: Chronic syphilis and skin diseases.

Alcala del Rey.—Spain, province of La Mancha, Sierra Morena.
Indifferent waters.
Special indications: Debility of the stomach, nervous complaints.

Alcamo.—Sicily, near Palermo.
Sulphurous waters, 176° F.

Alcantuz.—Spain, province of Cuenca, near Alcarría.
Saline waters.
Special indications: Constipation and rheumatic paralysis.
These waters enjoy so high a reputation, that notwithstanding the fact that invalids have to find accommodation in huts made from branches of trees—no other accommodation existing—they are very much frequented.
Doctor: Dr. Sevillana.

Alceda or Ontaneda.—Spain, near Santander and Riviero. Thence by diligence.
Sulphurous waters, 65 to 70° F., and winter station.
Special indications: Skin diseases and affections of the mucous membranes.

Aldborough.—England, Suffolk.
Sea baths, shingly beach; much frequented.
Doctor: Dr. Hele.

Alderney.—Channel Islands, 15 miles from Guernsey.
Sea baths; bad beach.
Doctor: Dr. Ross.

Aled.—France, Department of Aude. For routes see table.
Bicarbonate of lime waters; sedative to the nervous system.
Doctor: Dr. Gourdon.

Alexandersbad.—Germany, Bavaria, circuit of Upper Main, near Wunsiedel, 1,800 feet above sea level.
Cold bicarbonate of lime water; hydropathic establishment and pine baths.
Doctor: Dr. Cordes.
Hotels: Weber, Kurhaus.

Alexandersquelle.—Russia, Caucasus, near Pjätigorsk.
Sulphurous waters, seven springs, varying in temperature from 110 to 165° F. Very abundant supply.
Special indications: Gout, rheumatism, old sores, anchylosis, syphilis.
The situation is extremely picturesque, and the number of visitors have increased considerably of late. Accommodation first-class, but expensive.

Alexisbad.—Germany, Thuringia, in the Duchy of Saxe-Anhalt.
Ferruginous waters, electro-therapy, hydropathy.
Doctor: Dr. Heusinger.
Hotels: *Golden Rose* and *Alexisbaths.*

Alfaro.—Spain, Province of Almeria, and near this town.
Sulphurous alkaline waters.
Special indications: Skin diseases and rheumatics.
The establishment is only partially finished, and the accommodation is scanty
Doctor: Dr. Serrano y Sanchez.

Algiers.—Africa, capital of French colony of Algiers. For routes see table.
Winter station.
Season from November to end of April.
Special indications: Arthritis, asthma, bronchitis, phthisis.
Owing to the proximity of the Atlas Mountains, the climate of Algiers is warmer and **more** humid **than** that of any other Mediterranean station. **The mean temperature** during winter is **61° F.** The mean **barometric** pressure **762** mm., and the number of rainy days in the year, 87.
Doctors: Drs. **Thompson, Savill** and Feuillet.
Hotel de la Régence.

Al-Gyógy.—Austria-Hungary, Transylvania, district of Karlsburg.
Alkaline-earthy waters.
Special indications: Chronic catarrh in its various forms.
The **establishment is in keeping with** the place.

Alhama de Aragon.—Spain, **Province** of Saragossa, district **of Atéca.**
For routes see table.
Sulphurous magnesian and ferruginous waters, 100° F.
Season: Spring and autumn, but open all the year round.
Special indications: Asthma, paralysis, debility of kidneys and bladder, skin diseases, dysmenorrhœa.
The **establishment** and accommodation can compare favourably **with those** of **many of** the **most**-frequented watering **places in Germany and France.**
Doctor: Dr. Tomas Parraverde.

Alhama de Granada.—Spain, near Granada and Malaga.
Sulphurous magnesian waters, 110° F. The spring is seven inches **in** diameter, and contains much sulphretted hydrogen.
Season: April 15th to October 31st; climate healthy and mild.
Special indications: Intestinal obstructions, **affections** of the **mucuous** membranes, **nervous** debility, **rheumatism,** atonic sores and skin diseases.
The baths were very much frequented under the Saracen rule, **the** existing old Moorish edifices **showing** the importance

the baths then had. The Caliph of Granada is said to have received in some years as much as 500,000 ducats of revenue from these baths.

Doctor: Dr. Perales Churt.

Alhama de Murcia.—Spain, 20 m. from Murcia.
Sulphate of lime waters, 84 to 107° F.

Special indications: Affections of the nervous centres, hemicrania, palsy, anæmia, chlorosis, constipation, difficult menstruation, pyrosis, uterine affections.

Notwithstanding the length of time these baths have existed and the efficacy of their waters, the accommodation is still of the worst kind.

Doctor: Dr. Fortuni.

Ali.—Sicily, between Messina and Taormina, on the coast.
Hydrosulphuric waters, 100° F.

Special indications: Sciatica, rheumatic affections and skin diseases.

Frequented chiefly by the inhabitants of Messina.

Alica.—Italy, Tuscany, valley of the Era, on the right bank of the Rigone.
Sulphurous and alkaline waters, 60 to 65° F.

Special indications: Kidney diseases, calculous affections of the urinary system in general, menorrhagia, blennorrhœa, for one spring; rheumatism gout, hysteria and anæmia for the other.

Alicante.—Spain, on the Mediterranean.
Winter station of first importance.

Special indications: Invalids requiring a stimulating climate; also in cases of latent scrofula, asthma and bronchorrhœa, pleuritis, rheumatism and albuminuria.

Mean Winter temperature, 60° F. Air, very dry; rainy days fewer than on the Riviera; mean rainfall, 177 $^{m}/_{m}$. The town is well sheltered and the bay open only to southerly winds. No mistral or dust.

Doctor: Dr. Roman.

Alicum.—Spain, on the Fardes, near Guadix.
Saline waters, 90° F.

Special indications: All diseases due to debility and atony, hemiplegia, palsy, neuroses, and scrofula.

Doctor: Dr. Ferrer.

Aliseda.—Spain, in the Sierra Morena, near Leguas de Carolina and the Castle of Las Navas de Tolosa.
Ferruginous waters.

Special indications: Chronic stomachic and kidney affections, chronic diarrhœa, intermittent fevers.

Alkna-Szlatina.—Austria-Hungary, near Szigeth.
Saline waters, 21 per cent. of salts.
Arrangements somewhat primitive.
If properly managed, this might become one of the most important brine and inhalation bathing places.

Allegrezza.—Italy, district of Montale.
Alkaline waters.
Special indications : Gravel, vesical catarrh, and skin diseases.

Allerheiligen.—Switzerland, Canton of Solothurn, 5 miles from this town; 4,000 feet above sea level.
Carbonate and sulphate of lime waters.
These waters are well suited to nervous individuals; they have a stringent, depurative and tonic action.

Allevard-les-Bains.—France, Department of Isère; 1,400 feet above sea level. For routes see table.
Sulphate of lime waters, 52° F. Inhalation.
Season, 15th May to 1st October.
Special indications : Very efficacious in chronic inflammation of the laryngeal and pharyngeal mucous membrane, in nasal catarrh, syphilis and skin diseases.
Post and Telegraph.
Doctors : Drs. Isoard, Kostens and Niepce, fils.
Hotels : The *Hôtels du Louvre* and *de la Planta*, very comfortable. Special omnibus meets all trains at Goncelin. Berthet, proprietor.

Allezani.—Corsica, arrondissement of Corte.
Ferruginous bicarbonate waters.

Allonby.—England, Cumberland, near Silloth.
Sea baths. The surf being rather heavy, these baths should not be taken by the weak or nervous. Sandy beach.

Allume.—Italy, on the island of Giglio.
Sulpho-ferruginous waters, 68° F.
Special indications : Constipation and chronic skin diseases.

Almamező.—Austria-Hungary, comitat of Bereg-Ujocza,; see Zányka.
Several saline-ferruginous springs not yet analysed.

Almas.—Austria-Hungary, comitat of Grau, on the Danube.
Tepid sulphurous waters.

Almeida.—Spain, province of Leon, near Bonar.
Sulphurous-alkaline waters, containing sulphate of iron, 190° F.
Special indications : Gout, Rheumatism, paralysis, skin diseases.

Almeria.—Spain, on the south coast.
Winter station, with sulphurous magnesian waters.
Special indications: Rheumatism and paralysis, phthisis, chlorosis, nervous affections.
Mean winter temperature 60° F. One of the most sheltered of winter stations; frost and snow are entirely unknown here, and the temperature rarely falls to 45° F. A clear, dry and calm air characterises Almeria.

Alnmouth.—England, Northumberland, between Newcastle and Berwick.
Sea baths and sandy beach.

Alpnacht.—Switzerland, Lake of IV. Cantons, Canton Unterwalden; four miles from Saarnen. See routes in table to Lucerne.
Climatic station; whey and milk cure.
The starting point for diligences for Bernese Oberland. It is situated at the foot of Mount Pilatus, whence extends a beautiful panorama of the surrounding mountains. It is 3½ hours ride over a good mule path to top of mountain.
Hotel: The *Hotel Pilatus*, a clean and comfortable house. Telegraph and telephone office. M. Britschgy, proprietor.

Alsö-Sebes.—Austria-Hungary, comitat of Sáros, near Eperies.
Sulpho-magnesian waters, 50° F.
Special indications: Affections of the lymphatic system, scrofula, commencing tuberculosis, catarrh of the respiratory organs, constipation.
Very primitive arrangements.

Alt Haide.—Germany, Silesia, near Glatz, 1,300 feet above sea-level.
Alkaline waters and climatic air station.
Special indications: Scrofula and chronic affections of the uterus.
Doctor: Dr. Otte.
Hotel: The Baths.

Alton-Towers.—England, Staffordshire.
A summer health resort, interesting only for its romantic scenery.

Altsohl.—Austria-Hungary, comitat of Sohl, on the left bank of the Gran.
Earthy alkaline ferruginous waters, 50° F. Indifferent arrangements.

Alt Tura.—Austria-Hungary, comitat of Oberneutra, 3,130 feet above sea level.
Alkaline strongly effervescing waters.

Altwasser.—Germany, Prussia, near Liegnitz, in **Silesia**, 1,300 feet above the sea level.
Highly effervescent ferruginous waters; mud baths.
Special indications : Anæmia, chlorosis, skin diseases, general debility.

Alvenau.—Switzerland, Canton of the Grisons, 11 miles from Coire on the Albula road, 3,150 feet above the level of the sea. For routes see table.
Sulphurous saline waters, 46° F.; saline chalybeate waters, 51° F., at Tiefenkasten; iodurated chalybeate waters, 57° F., at Solis. Owing to its elevated position and wooded surroundings, the town is a mountain air station. The waters are exported.
Season : June—September.
Special indications : Rheumatism, chronic derangements of the digestive organs, chronic catarrh of respiratory and urinary organs, diseases of the skin and bones, female complaints, abdominal plethora, scrofula and rickets, mercurial dyscrasia, syphilis and general debility, chlorosis, anæmia, neuralgic affections, hæmorrhoids, goitre, gout, glandular swellings, &c.
The establishment is thoroughly well appointed in accordance with modern requirements. The situation in the valley of the Albula is very picturesque, the vegetation is rich, and extensive woods afford ample shade. It lies midway between Coire and the Engadine, and a well-known stopping place between for English families on the way to and from the Engadine. Near the Kurhaus is a large garden.
English **Church** *Service.*
Post and Telegraph : In the Kurhaus.
Doctors : Drs. Weber and **Balzer.**
Hotel : The *Kurhaus*, a **first-class well reputed house with 150 beds**, reading, concert, billiard, ladies' and smoking rooms. The house is well managed by Balzer Brothers, proprietors.

Amalfi.—Italy, on the Mediterranean, 12 m. from Salerno.
A winter health resort which deserves to be better known.

Ambleside.—England, Westmoreland, Lake district.
A summer holiday resort.
Doctor : Dr. King.

Ambleteuse.—France, department of Pas-de-Calais.
Sea baths; sandy beach.
A very quiet and retired spot.

Amélie-les-Bains.— France, Eastern **Pyrenees**; 700 feet above the level of the sea. For routes see table.
Sulphurous saline waters, 71° to 172° F., and winter station.
Season : All the year.
Special indications : Herpetic diathesis, and catarrhal affections of the respiratory organs, rheumatism, scrofula, syphilis.
Doctors : Drs. Genieys and Arnal.

B

Hotels: The *Grand Hôtel des Thermes Romains;* **first-class** house, covered promenades, heated in winter; baths in the house; large garden, thoroughly well appointed. Dr. Henry Arnal, directeur.

Amlwch.—Wales, Anglesey, near Bull Bay.
Sea baths; sandy beach.
A fine establishment.
Gaseous spring.

Ammoniseal, Gas Spring or Grotto del Cane.—Italy, near Naples.
Hot vapour baths.
Special indications: Partial paralysis.

Amphion.—France, Savoy, near Evian; 1,310 feet above the sea level.
Cold ferruginous bicarbonate waters. Hydropathic establishment.
Season: June 15th to October 1st.
Special indications: Debility of the digestive organs, gravel, hypochondriasis.

The establishments are situated in an extensive park, well shaded by large trees. Gymnastic and lawn tennis grounds on a large terrace overlooking the lake. The site and environs are very picturesque, and offer a good choice of excursions. Horses and carriages for hire.
Telegraph and Post.
Hotels: The Grand Hôtel des Bains, The Casino, are under the able direction of Mr. George Gougoltz, proprietor of the Hotels *Beau Site* and *De l'Esterel*, at Cannes, and contain reading, billiard, assembly and smoking rooms.

Amroth.—Wales, Pembrokeshire.
Sea baths. Pebbly beach.

Amsterdam.—The former capital of Holland. From London by steamer, *via* Harwich.

On the Amstel, a branch of the Wye, an inlet of the Zuyder Zee. The town is built upon piles driven through from 50 to 60 ft. of peat and sand, and is protected by dykes. The numerous canals, passing between 95 islands, are spanned by nearly 660 bridges. The finest of these is the Hoogsluis, which is 610 ft. long.

PLACES OF INTEREST.—Vondel's Park, New Museum, Crystal Palace, which was intended as an Hôtel de Ville; marble sculptures inside; Van den Hoop gallery of paintings; National Gallery in the Trippenhuis; Zoological Gardens; Fodor museum and Van Six gallery; Antiquarian Society's museum; nearly sixty Churches and Chapels of all denominations.

Docks and Royal Naval Wharf; Naval Academy; Koster's diamond cutting establishment; Wynand Focking's Factory for curaçoa and other liquors. Opera house. Post-office.
Doctors: Drs. Davids and Metzger.
Hotel: Amstel Hotel.

Anacapri.—Italy, Island of Capri.
Sea baths and winter station.
Doctor: Dr. Green.

Anagni.—Italy, near the Pontine marshes.
Bituminous-alkaline-sulphurous waters.
A very unhealthy site. Frequented under the Roman emperors, but now altogether neglected.

Ancona.—Italy, on the Adriatic, a fortified maritime port, with dockyards.
Sea baths, sandy beach. A very fashionable resort.
Season: July—September.
A second-rate establishment has lately been built. During the season, concerts, morning dances and regattas, are among the amusements.

Andabre.—France, Department of Aveyron, near Camarès. For routes, see table.
Effervescing bicarbonate saline, and slightly **ferruginous** waters
Special indications: **Affections of the liver**, gout and gravel.
Doctor: Dr. Martin.

Andeer.—Switzerland, Canton of the Grisons, near Pigneu.
Sulphate of lime waters; whey cure.
Very elegant establishment.

Anguillara.—Italy, Tuscany, near Lago Sabatino.
Bitter waters, 70° F.
Special indications: **Paresis**; hemiplegia resulting from rheumatism; kidney **affections**, pyelitis, blenorrhœa and intestinal debility.

Annecy.—France, Upper Savoy, on the lake of Annecy, 20 miles by rail from Aix-les-Bains; 1,460 feet above the sea level; 12,000 inhabitants.
Climatic station; baths in the lake.
A favourite and pleasant resort. Excursions can be made from Aix. Annecy is a curious old town, beautifully situated at the head of the lake. It is very conveniently situated for excursions to the Gorges du Fier, Semnoz **Alps**, which have been well called the Righi of Savoy, and to various other places on the beautiful lake. Diligences and carriages for **Geneva** and Chamounix. Good boating and fishing.
English church service.

Hotels: The Grand Hôtel d'Angleterre, a first-class house, very comfortable; branch house at the Gorges du Fier and Semnoz Alps, Righi de Savoie. A. Vallin, proprietor.

Antibes.—France, on the Riviera, between Cannes and Nice.
Winter resort, but little frequented; visitors stay more at the Cap d'Antibes; see this.

Antogast.—Germany, Duchy of Baden.
Ferruginous bicarbonate waters, containing arsenic; used in the country only as table waters.

Antrim Spa.—Ireland, County Donegal.
Saline and bicarbonate of lime waters.
Special indications: Scrofula.

Apollinaris.—Germany, Rhenish-Prussia; in the valley of the Ahr, near Remagen.
Effervescing table-waters, used only for exportation.
A Company has appropriated the title "Queen of Table Waters" for these waters. The large amount of carbonic acid gas in Apollinaris water is owing to the fact that the gas is caught in condensers as it escapes from the spring, and then pumped into the bottles. For further information see NEUENAHR.

Appenzell.—Switzerland, capital of the Canton of same name, near Sitton; 2,538 feet above the sea-level.
Carbonate of magnesia waters. Whey-cure station of first importance.

Appledore.—England, West Devon.
Sea baths.
Good surf and bathing appliances.

Aramayona.—Spain, Province of Aláva, district of Vitoria.
Sulphate of lime waters.
Special indications: Chronic skin diseases, bronchial ophthalmia, mercurialism, leucorrhœa.
The establishment and accommodation are modern and very good.
Doctor: Dr. Guedea.

Aranjuez.—Spain, near Toledo.
Acidulous, bitter waters, containing a large quantity of sulphate of soda.
Special indications: Impairment of the bilious and digestive secretions, catarrh of the intestines, chronic liver complaints, pain in the kidneys, gravel and calculus, ophthalmia.

Aranzarre.—Spain, Province of Guipuzcoa.
Sulphate of lime waters, with sedative properties.

Arapataka (Lindorf).—Austria, Transylvania, comitat of Ober Albat, 30 k. from Kronstadt.
Mixed bicarbonate waters.

Arborme.—France, Department of Gironde, Arrondissement of Frontiers.
Saline waters, 53° F.

Arcachon.—France, Department of Gironde, 34 miles from Bordeaux.
For routes see table.
Winter station, with sea baths.
Season: Winter season lasts from **November** till June; summer season from May till October.
Special indications: For invalids requiring a sedative air; for delicate, lymphatic and anæmic persons; nervous complaints; chest and lung affections; scrofula; gout.
The town of Arcachon did not exist before 1857. It has been built especially with a view to the comforts and requirements of foreigners, bathers and winter residents. A promenade, of more than 2½ miles in length, bordered with residences, runs along the sea, while the gardens slope down to the very beach. Scientific men will find a museum and a marine aquarium to which they can be admitted.
Post and Telegraph.
English Church Rev. S. Radcliff; two Services on Sundays; all sittings free.
A college and gymnasium.
House and Estate agent and wine merchant: A. Brannens, who gives all information gratis.
English banker: Fred Audap.
English chemist: G. Soulan, Boulev. de la Plage.
Doctors: Dr. Hameau.
Hotels: The *Hôtel de France*, open all the year round, very comfortable and well situated on the beach; sheltered position. L. Lahore, proprietor.
The *Hôtel Continental* is a first-class house, with English comforts. Arrangements made for families. Open all the year. B. Ferras, proprietor.

Archena.—Spain, Province of Murcia, near Inicia on the Segura.
Sulphurous and saline waters, 124° F. Steam baths.
Special indications: Syphilitic and scrofulous affections.
Very excellent accommodation.
The water is considered to act as a stimulant on the skin and genital organs, and may, it is said, induce actual congestion of the lungs and uterus.
Doctor: Dr. Zabala.

Archevaleta.—Spain, Province of **Guipuscoa**, on the French frontier.
Saline waters, 70° F.
Stimulating and diuretic action.

Special indications: Scrofula, herpetic affections, fluor albus. The establishment and the accommodation are good and moderate in price; the environs are romantic.
Doctor: Dr. Breñosa.

Areidosso.—Italy, Tuscany, near Grosseto.
Bicarbonate ferruginous waters, 45° to 55° F.

Arco.—Austria, **Southern Tyrol**, above the Garda Lake.
Winter station.
Doctors: Drs. **Sehieder**, Schreiber and Althammer.
Hotels: Kurhaus Arco.

Arcs (Les).—France, Department of Var.
Ferruginous waters.

Ardales.—See Carratraca.

Ardrossan.—Scotland, Firth of Clyde.
Sea baths.
The air is bracing, but sometimes very humid.

Aregos.—Portugal, Province of Beira.
Sulphurous waters, 150° F.

Arenosillo.—Spain, **Province of Cordova**, near Montoro.
Chloride and sulphate of sodium waters, 54° F.
Special indications: Scrofula and skin diseases.
The **establishment and** accommodation are inadequate and antiquated.
Doctor: Dr. Reguera.

Arenzano.—Italy, one hour from Genoa on the **Riviera di** Ponente.
Winter station, beginning to attract visitors.
Cogoleto, the birthplace of **Christopher Columbus, is** close by. **Very** agreeable promenades **and drives in** the neighbourhood.

Argentières.— France, Department of Allier, Arrondissement Montluçon.
Bicarbonate of soda water.

Argentona.—Spain, Province of Barcelona, near Mataró.
Acidulous ferruginous waters, 45° F.
Special indications: Gastralgia, dyspepsia, urinary **and** abdominal obstructions.
The establishment is **antiquated;** accommodation cheap, but inferior.
Doctor: Dr. **Barrols.**

Arlanc.—France, Department of Puy-du-Dôme, Arrondissement of Ambert.
Ferruginous bicarbonate waters, containing 0·025 silicate.

Armajolo.—Italy, Tuscany, near Siena.
Sulphate and carbonate of lime waters.
Special indications: Affections of the kidneys and urinary system, gravel.

Arnedillo.—Spain, Province of Logroño; 1,000 feet above the sea-level.
Saline waters, 128° F.; natural steam baths.
Special indications: Rheumatism, syphilis and affections of the urinary system
The establishment and accommodation are good and moderate in price.
Doctor: Dr. Martinez.

Arnside.—England, Lancashire, on Morecambe Bay.
Sea baths, sandy beach.
Doctor: Dr. Chapman.

Arnstadt.—Germany, Province of Saxony, near Erfurt, on the Gera.
Saline waters.
Special indications: Scrofulous diseases.
Doctors: Drs. Niebergall and Oswald.

Arp.—Switzerland, Canton Valais, 2½ hours from Loëche-les-Bains.
Strong sulphurous waters.
Special indications: Skin diseases.

Arpad.—Hungary, Slavonia.
Saline and purgative waters.

Arromanches.—France, Department of Calvados; from Paris via Bayeux.
Sea baths, sandy beach,

Arteijo.—Spain, Galicia, near Coruña.
Sulpho-saline waters, 90° F.
Special indications: Scrofula, caries, gout, rheumatism.
The situation is charming, but the waters are scanty.
The establishment and accommodation are good, and prices not too high.
Doctors: Dr. Mayoral.

Artern.—Germany, Prussia, Province of Saxony, near Merseburg.
Saline waters.
No amusements are offered here, and the place is therefore visited only by invalids.
Hotel: Kurhaus.

Arzilhe.—Switzerland, near Berne. 1,600 feet above the sea level.
Thermal saline sulphurous waters.
Special indications: Skin disease.

Asciano.—Italy, Tuscany, near Pisa.
   Sulphate of lime waters.
   Special indications: Affections of the urinary system, lithiasis, catarrh of stomach and intestines.

Ascoli.—Italy, in the Marches.
   Saline-sulpho-ferruginous waters, 90° F.
   Special indications: Affections of the urinary system.
   Very primitive arrangements, used only by the inhabitants.

Ashby-de-la-Zouch.—England, Leicestershire, 14 m. from Derby.
   Cold saline waters.
   Season: From 1st June till 1st October.
   Special indications: Scrofulous affections.

Asinalunga.—Italy, Tuscany, Val d'Orcia.
   Ferruginous waters.
   Special indications: Leucorrhœa, affections of the urinary system, spleen, and liver.

Asnelles-sur-Mer.—France, Department of Calvados, from Paris *via* Bayeux.
   Sea baths, sandy beach.

Aspio.—Italy, Province of Ancona, one hour from Ancona.
   Ferruginous alkaline waters.
   Special indications: Affections of the liver, scrofulous and glandular diseases.
   Little known and used only by the inhabitants.

Assmannshausen.—Germany, Nassau, in the Rhinegau.
   Saline and alkaline waters.
   Hydropathic establishment.
   Special indications: Rheumatism and affections of the urinary system and nervous centres.
   Doctor: Dr. Mahr.
   Hotels: Kurhaus, Anchor.

Astrakhan.—Russia, on the Caspian Sea.
   Mud baths, containing saline and bituminous substances.
   Special indications: Scrofula, rheumatism and paralysis.

Astroni.—Italy, half-hour from S. Germano.
   Thermal sulphurous waters.
   Owing to its very unhealthy site, this station has been abandoned.

Athimonus.—Ireland, County of Leitrim.
   Strong sulphurous waters.
   Special indications: Rheumatism, paralysis.
   The establishment is comfortable.

Athlone Waters.—Ireland, County of Roscommon.
Cold ferruginous waters.
Only used internally by residents.

Attisholz.—Switzerland, one hour from Solothurn.
Alkaline waters, 60° F.
Special indications : Gout, rheumatism, chronic diarrhœa.

Atya.—Austria-Hungary, comitat of Weissenburg.
Earthy alkaline waters.
The springs form two lakes.

Audinac.—France, Department of Arriège, near St. Girons.
Sulphate of lime and ferruginous waters, 80° F.
Special indications : Diseases of the vesical and digestive organs.

Auerbach.—Germany, Hesse, half an hour from Darmstadt.
Chalybeate waters; more frequented as a summer health resort.
Special indications : Anæmia, atony, and in general all affections requiring ferruginous treatment.

Augnat.—France, Department of Puy-de-Dôme, near Isoire.
Ferruginous bicarbonate waters.
Special indications : Anæmia, enlargement of the liver and spleen, and gravel.

Augustusbad.—Germany, Kingdom of Saxony, near Raveberg.
One saline and six bicarbonate chalybeate springs.
Doctors : Drs. Breunig and Dorner.
Hotels : Palais, Thal.

Aulus.—France, Arriège, near St. Girons. For routes see table.
Alkaline waters, 48° F.
Special indications : Purgative, laxative and diuretic action, according to the dose; in diseases of the liver, in arthritic affections, and in syphilis.
Doctor : Dr Alrig.

Aussee.—Austria, Styria near Graz.
Saline waters and climatic station.
Season : July—September.
A very elegant establishment, with a sanatorium, offering good accommodation.
The situation is most picturesque, and numerous excursions may be made in the neighbourhood.
Doctors : Drs. Schreiber and Pohl.

Auteuil.—France, suburb of Paris.
Cold non-effervescing ferruginous waters, which are difficult of digestion.

Availles or Abrac.—France, Department of Charente, near Conjatens.
Saline **waters.**
Special indications: Intermittent fever and atonic diseases.

Avesne.—France, Herault, Arrondissement of Lodèves, near Montpellier.
Mixed bicarbonate waters, 65° F., sedative and tonic.
Special indications: Cutaneous affections.

Ax.—France, Department of Ariége, Arrondissement of Foix, near Tarascon. For routes see table.
Sulphurous waters.
The sulphurous elements appear here, as at Luchon, in the form of sulph-hydrates. There exists a great similarity between the two waters. There are more than 60 springs, with a temperature varying from 54 to 200° F.
Special indications: Rheumatism especially chronic, scrofula, affections of the skin in **atonic** and inveterate **forms.**
Doctor. Dr. Garrigon.

Axenfels.—Switzerland, Lake of Lucerne, Canton of Schwyz, 2,300 feet above the sea level. For routes see table to Lucerne; thence by rail or boat to Brunnen.
Climatic station; mineral waters of all the best known springs; milk and whey cure.
Axenfels is beautifully situated and well sheltered. **The** views from the hôtel and gardens are very fine, embracing **the** lake and mountains and glaciers. The excursions **are** numerous and well worth making. Carriages can be had **in the** hotel.
A doctor resides in the hotel during season.
Hotels: The *Grand Hôtel Axenfels,* a first-class house with **every** comfort; 200 beds; **conversation, ladies', sm**oking and billiard-**rooms.** Post and telegraph; orchestra. Omnibus **at** station and landing stage. Th. Wirth-Strubin, proprietor, **also** of the Schweizerhof Interlaken.

Axenstein.—Switzerland, Canton of Schwyz, Lake of Lucerne, 2,500 **feet** above the sea level. For routes see table to Lucerne; thence by Gotthard railway, **or by boat to** Brunnen; from Brunnen by omnibus 50 minutes.
Climatic **station;** milk **and** whey cure. Mild, yet bracing Alpine **air; well sheltered from** winds.
Season: May **15th to Octo**ber.
This establishment is situated in **a most picturesque** country, near the small village of Morschach. **An extensive** and beautifully planted park surrounds it, with pine and fir trees, which **contribute very** largely to the health-giving properties of the **air. Queen Victoria** said of this place: "Axenstein is the finest

place I have seen during my Swiss tour." The excursions are very varied and interesting. Two villas situated in the park contain morning, card and billiard rooms, verandahs and bath**s. Telegraph and post-office in the hotel. English church in one of the villas.**

Doctor : There is a resident physician during the season.

Hotels : The *Grand Hôtel Axenstein* is a magnificent **hotel, and** combines **every** comfort **with** moderate prices. Excellent kitchen. Large airy rooms **and** suites. A dairy belonging **to** the proprietors supplies the **establishment with pure** produce. Maps of **the park may be** had of the secretary at the hotel. A. Eberlé Sons, **proprietors.**

Baassen.—Hungary, near Temesvar and Mediah.
Saline bromo-iodurated waters, 50° F.
Season : July, August.
Special indications : Chronic rheumatism, scrofula, disorders of **uterus** and ovary, syphilitic affections **of** the bones and tissues.
Very picturesquely situated; a **quiet but** agreeable place.

Baccanella.—Italy, **Tuscany, near** Alita **and Pondera.**
Sulpho-ferruginous waters, 50° F.
Special indications : Chronic sores and skin affections.

Baczuch.—Hungary, **comitat of Sohl.**
Cold ferruginous **waters.**
Not much comfort; **the town is but little visited.**

Baden.—Lower **Austria,** 1 hour from Vienna. For routes **see table.**
Sulphurous waters, 82° to 95° F. Sitz, douche, shower **and common** baths, open **swimming baths;** climatic station. Two hydropathic establishments.
Season : May 15th to **1st October;** the establishment is open, however, **all** the year **round.**
Number of Visitors *:* **The visitors' list shows** from 14 to 15,000 annually.
Special indications : **Rheumatism, gout, anæmia,** scrofula, more **especially of a subacute character.**
The **town** is very picturesquely **situated on the northern slopes of the** Viennese **mountains. It is surrounded by pine and beech** forests, **and** the **air is remarkably healthy and invigorating.** The surrounding **country is** romantic, **and affords** opportunities for very interesting excursions; the ascent of the mountain "Eisernes Thor" is **one of the best.**. This mountain is 2,500 feet high, and affords a very extensive **view over Vienna** and the Alps.
There is a Casino with club, an Arena, a Museum with very interesting collections ; Theatre ; Protestant church service, &c.

Post and telegraph office; numerous trains and tramways daily to and from Vienna.

The town park, with its shady walks, is scrupulously kept, and affords the chief recreation ground for patients. Here the band plays three times a day.

Doctors: Drs. Barth, von Mühlleitner, J. Schwarz, Heinz, Czuberka, Kosak, and Brandstätter.

Hotels: The *Hotel de la Ville de Vienne, Green Tree, Golden Deer, Golden Lion, Shepherdess.*

Hôtels meublés: *Zeilner, Rechtberger, Riedel,* the *Duke Hôtel, Julienhof.*

Baden.—Switzerland, near Aarau. For routes see table.

Sulphurous waters, 119° F.

Special indications: Neuroses, gout, uterine and vesical catarrh, induration of the thymus gland, scrofula and rheumatism.

Baden-Baden.—Germany, duchy of Baden, 650 feet above sea level, 12,000 inhabitants. For routes see table.

Alkaline chloride of sodium waters, varying in temperature from 110° to 150° F.; climatic station; every description of baths; hydropathic and pneumatic establishments; hospital for eye affections; whey cure.

Season: May—October; but open all the year round.

The number of visitors annually amounts to between 40,000 and 50,000.

Special indications: Uric acid diathesis, gout and kindred affections (due to excess of uric acid); malaria, and certain skin diseases; chronic rheumatism in all its forms, wounds, injuries and fractures, scrofula, constitutional syphilis, chronic catarrh of the mucuous membrane of the respiratory and digestive tracts, certain kidney affections, anæmia and convalescence after serious illness.

The effect of the Baden waters, like that of chloride of sodium springs, generally is to dissipate and dissolve congestions, to stimulate, accelerate and regulate digestion, to increase the secretive power of the skin and kidneys; their most distinctive characteristic is a high temperature and the absence of carbonic acid, which renders them mild and easy of digestion and very suitable to irritable and delicate persons, whose organs are capable of but little reaction. The springs are also useful in cases of weakened and easily disordered digestion.

The amount of mineral ingredients varies for the several springs from 22 to 30 parts in 10,000. The quantity of water supplied in 24 hours by all the springs combined is about a million litres. Of this the main shaft spring supplies one-half. According to the analysis by Dr. Bunsen, of Heidelberg, in 1881, this spring is remarkable for the unusually high percentage of arsenic and lithium which it contains.

Thermal baths can be had at several hotels, but best at the "*Grand Ducal Friedrichs Baths*." This may be looked upon as a model of what a bathing establishment should be. It cannot be surpassed either for the completeness of its arrangements, which have been designed to carry out both hydropathic and thermal treatment, or for its excellent system of attendance. It contains baths of the ordinary kinds, stream or wave baths (wild baths), electric baths, inhalation apparatuses, cold water treatment, Russian or Irish baths, hot air baths, swimming baths at 59°, 80°, 95° and 100° F. Douches of every description, temperature and strength. In the summer of 1884, a new section of medical gymnastics and massage (shampooing) was opened. By these additions the Frederick Baths have become the most complete therapeutic establishment in Europe.

Separate Russian vapour baths; Bos baths. Skilled gymnasts and shampooers attend. It is open all the year round, and is thoroughly warmed in winter.

Medical **Director of the** *Baths*, as **appointed by the Government**, Dr. **Heiligenthal.**

Baden-Baden is a charming town picturesquely situated in a valley, sheltered by the lower ranges of the Black Forest. The winter is very mild, and the summer not excessively hot. It is a resort for idlers, pleasure-seekers and patients from all parts of the world. Its springs were known and used by the ancient Romans.

The centre of Kur-life in the morning is "The Trinkhalle." Visitors and patients come here between 6.30 and 7.30 a.m., to take the waters, promenading in the intervals in the gardens and listening to the band.

The frescoes with which the Trinkhalle is ornamented represent legends of the Black Forest.

The most popular rendezvous, however, is the **"Conversationshaus."** Concerts twice daily are held in the small pagoda in front. The Conversationshaus includes assembly rooms of immense size, reading-rooms, coffee and billiard-rooms, and a restaurant. The theatre and suite of apartments for assemblies and balls are **magnificently** furnished and are close by.

The gardens and promenades are all admirably kept and very beautiful.

The race course at Iffezheim, at a short distance from Baden-Baden, is one of the best, if not the very best in Europe. In the early days of September, the races bring together the *élite* of the German and Austrian aristocracy. They are, perhaps, the best races in Germany or Austria. The Excursions are very varied and interesting. Amongst those at short distances may be mentioned: The old and new Schloss, whence on clear days the spire of the Cathedral of Strasburg is visible; ruins of Eberstein Castle, the Favorite, and the Hermitage. Farther off are the falls of Allerheligenstein Eberstein, the Murg valley, Forbach, Raumünzach, Gernsbach Rothenfels, Kuppenheim, Wildbad, Griesbach, Wolfach, Petersthal, etc.

Amongst the amusements may be mentioned good fishing and hunting, grand concerts, symphonic and quartetto *soirées*, special concerts by the best known artists. Balls, *réunions*, childrens' balls, military concerts, operas and comedies with artists of world-wide celebrity fire works and illuminations.

The educational establishments of Baden are of the highest order

The prices being everywhere fixed by Government, it is scarcely necessary to bargain for anything, the traveller being well protected against extortion.
English church: Rev. S. A. White.
Post Office: Leopoldsplatz.
Telegraph: Station.
Banker and Exchange Office: F. C. Joerger.

Bookseller: At the entrance of the Friedrichs Baths.
Doctors: Drs. Heiligenthal, Berton, Schliep and Shiel
Hôtels: The *Hôtel de Russie*, first class, on the promenade; eighty rooms. A. & G. Moerch, proprietors.

The *Hôtel de France*, excellent position facing the Trinkhalle and Kurhaus, on the promenade. C. Ulrich, proprietor.

The *Hôtel Victoria* can be well recommended as an excellent first-class hotel. V. Grossholz, proprietor.

Badenweiler.—Germany, Duchy of Baden, in the southern parts of the Black Forest, near Freiburg in Baden; 1,300 feet above the sea level. For routes see table.

Warm earthy waters, 80° to 82° F. A climatic station. Surrounded by pine forests.

Special indications: Nervous and catarrhal affections generally, and rheumatism.

The establishment is very elegantly fitted up with marble basins, open swimming baths, douches, and single baths. Cows', goats', and donkeys' milk, whey cure, and all mineral waters may be obtained.

Large, well-kept, and shaded park, with the ruins of an old Roman bath in good preservation; good promenades through the pine-forests; excellent facilities for excursions, ascent of the Hochblauen, 3,550 feet high, with a superb view of the Alps, Lake of Constance, etc. Large reading-room in the Casino.

The amusements are varied; there are daily concerts, theatre, etc. Very comfortable hotels, villas, and private apartments. English church.

Post and telegraph station.

Cabs and carriages by the hour and day.

Doctors: Drs. A. Siegel, Thomas, Mandowski and Kollman.
Hotels: *Hôtel Sommer*, same proprietor as *Hôtel Zaehringerhof*, in Freiburg; the *Römerbad*.

Badstofuhver.—Sweden, district of Suulendinga-Fiordunger, in the valley of Geyss.

Alkaline-saline waters, 189° F.

Special indications: Obesity, disorders of the urinary system, liver complaints.

Bagnaccio di Colombajo.—Italy, Tuscany, district of Nievole, near Valagli.

Sulphurous ferruginous waters, 62° F.

Bagni-a-Baccanella.—Italy, Tuscany, Val d'Agra, on the Arno.

Alkali-saline waters.

Special indications: Disorders of the digestive and urinary system.

Bagnères de Bigorre.—France, Department of Hautes Pyrénées; 1,750 feet above sea level; 12,000 inhabitants. For routes see table.

Saline, sulphurous, ferruginous and arsenical waters, varying in temperature from 72° to 120° F.; upwards of 50 springs; thoroughly good establishments. There is a cold sulphurous spring at Labassère, the waters of which are largely exported.

A winter station.

Season: June to September, especially for the baths. November till May as winter residence, but baths may be taken all the year round.

Number of visitors, 23,000 to 25,000 annually.

Special Indications: Owing to the complex character of the waters and the mildness of the climate, the therapeutic indications are very varied. The waters enjoy a high reputation in cases of tuberculosis and affections of the respiratory organs, in affections of the intestines and urinary system, in plethora abdominalis, anæmia, chlorosis, and various female disorders.

The Labassère spring having lately been acquired by the Company, will add considerably to the therapeutic resources of Bagnères de Bigorre, as it is one of the most powerful cold sulphurous springs of the Pyrenees.

The town is situated in a valley, open to the north and south; east and west it is sheltered by a range of hills. The climate is very mild and humid, but bracing, owing to the high elevation. Mean temperature during the seven winter months, 46·5° F.; mean barometric pressure 752 mm.; mean rainfall, 4·278 inches a month in winter; rainy days, 147; sunny days 123 in the year.

The environs are very attractive, and walks and excursions are abundant, the latter being highly interesting. The town is within five hours of the Pic du Midi, where the highest meteorological observatory in Europe stands. The Casino, a very elegant building, is in an extensive park, and contains a theatre, ball, concert, assembly, billiard and smoking rooms. Concerts daily.

English church: Rev. T. Grundy

Post and Telegraph Office: Place de **Strasbourg.**

Bookseller: Léon Péré, Place de **Strasbourg, No. 7.**

Doctors: Dr. Bagnell (who speaks English) and several French physicians.

Hotels: The *Hôtel de Paris* is a first-class family hotel, centrally situated, facing the **Promenade,** and looking full south, with a mountain view. B. Nogues, proprietor.

Bagnet (Le).—France, Hautes Pyrénées, near Bagnères de Bigorre.

Waters containing sulphate of soda.

Special indications: Debility.

Bagni-a-Acqua.—Italy, **Tuscany,** near Pisa, **Val di Chiana.**

Sulphurous earthy waters, 100° F.

Special indications: Skin diseases, rheumatic gout and paralytic affections.

Bagni (Acqua dei).—Italy, near Naples.
Alkali-saline waters.
Special indications: Jaundice, stomachic, intestinal and vesical catarrhs, diabetes, plethora abdominalis, rheumatism and neuralgic affections.

Bagno-a-Morba.—Italy, Tuscany, between Pomarance and Castelnovo.
Sulphurous alkaline waters, 100° to 130° F.
Special indications: Rheumatism, kidney disorders, calculus, gravel, scrofula, herpes.

Bagni di Crana.—Switzerland, Canton of Ticino, 3,270 feet above the sea level, near Lago Maggiore and Lago di Lugano.
Sulphurous waters, 90° F.
Very efficacious waters, but there is little or no accommodation. Visited by tourists, but seldom by invalids.

Bagno.—Switzerland, Canton Valais.
Sulphurous waters.
Little used.

Bagno Roselle.—Italy, Tuscany, Val di Paglia, near Siena.
Ferruginous effervescent waters, 105° F.

Bagno d'Apollo.—Italy, Tuscany.
Alkaline waters, 90° F.
Special indications: Jaundice and liver complaints.

Bagno-fresco.—Italy, Island of Ischia, in the Val Tamburino.
Alkali-saline waters.
Special indications: Affections of the digestive and urinary system.

Bagnolino dei Rachitici.—Italy, Tuscany, near Levane.
Alkali-ferruginous waters.
No bathing establishment. The peasants in the neighbourhood bathe children affected with rickets in the waters.

Bagno-in-Romagna; also Terme Leopoldine or Sta. Agnese.—Italy, province of Florence, 3,000 feet above the sea level; diligence from Forli, 3 fcs. each person; carriages, 20 fcs. and 30 fcs.; 6 hours' drive.
Carbonic and sulphurous acid waters, 108° to 110° F., used for drinking; sitz and shower baths.
Season: June–September.
Special indications: Rheumatism, sciatica, skin diseases, gout.
There is a well-appointed bathing establishment, affording accommodation in 100 rooms; pension from 4 fcs. to 6 fcs. daily. The situation is pretty and healthy, in the midst of chestnut and fir forests.
A Post and Telegraph Office.
Doctor: Dr. Fanti.

Bagnolles.—France, Department of the Orne. For routes see table.
Saline, sulphurous, lithic, silicate, and arsenical waters. Temperature, 66° F.
Three ferruginous, arsenical, and manganese springs, 26° F.
Special indications: Diseases of the digestive organs, skin, rheumatism and scrofula, chlorosis, congestion of the abdominal viscera, phlebitis, old wounds, and chronic sores.
Bathing establishment very complete
Doctors: Drs. Commandré and Joubert.

Bagnoli.—Italy, near Naples, on the road to Baiæ, between Posilipo and Pozznoli by steam tramway.
Saline and alkaline waters, 112° F. Baths, douches, steam baths.
Special indications: Paralysis, rheumatism, gout, cutaneous affections and scrofula.
Complete bathing establishment.
Springs of Pisciarelli and of the Temple of Serapis.
The principal establishment, where the sea baths are also taken, is called "La Pietra." They are open all the year round.
Doctor: Dr. Pepere.

Bagnols-les-Bains.—France, Department of Lozère. For routes see table.
Sulphate of soda waters, from 50 to 105° F.
Special indications: Cutaneous and pulmonic diseases, rheumatism, scrofula, wounds, and chronic sores.
Two establishments with piscinæ, douches, baths, steam baths, &c. Climate rather severe, and barometric changes rapid.

Bains-les-Bains.—France, Department of Vosges, near Epinal; from Paris, *via* Mulhouse and Epinal.
Sulphate of soda, arsenical waters, from 70° to 115° F.
Special indications: Rheumatism, debility, nervous and uterine affections.
Two thermal establishments. The thermometric changes are considerable during the day.
Doctor: Dr. Bailly.

Bains de la Reine.—Algiers, Province of Oran, near Oran.
Saline springs containing bromine, 130° F.
Special indications: Liver and stomach affections, cutaneous diseases, rheumatism, &c.

Bains, near Arles.—France, Department of Pyrénées Orientales.
Slightly sulphurous alkaline waters, 122° F.
Special indications: Rheumatism, skin diseases, disorders of the digestive organs and urinary system.
The establishments are said to be the original Roman ones.

Bajfalú.—Austria-Hungary, comitat of Száthmar.
Sulphurous waters.
Special indications: Gout, rheumatism, skin diseases, obesity, liver and urinary affections.

Balaruc les Bains.—France, Department Hérault, near Montpellier and Cette.
Strong saline waters; springs contain bromine, 45° to 100° F.
Special indications: Paralysis, softening of the bones, chlorosis, struma, saturnine intoxications, rheumatism, and anchylosis.
Three bathing establishments.
Doctor: Dr. Ducroix.

Baldini (Acqua dei).—Italy, Tuscany, near Rinfresco.
Saline waters. Strongly purgative action.

Baldócz.—Hungary, comitat of Zips, near Kirchdrauf.
Slightly alkaline ferruginous waters.
Special indications: Hypochondria, hysteria, giddiness, sterility, stomachic disorders.

Ballater.—Scotland, Aberdeenshire.
A summer station, also frequented for the waters at Pannanich Wells, which are two miles distant.
Scenery very fine.

Balnea d'Avignone.—Italy, Tuscany.
Chalybeate waters, 120° F.

Ballycotton.—Ireland, county Cork.
Sea baths, sandy beach.

Ballyspellin Spa.—Ireland, Leinster, county Kilkenny.
Saline and ferruginous bicarbonate waters, 60° F.
Special indications: Nervous affections of the digestive organs; also in abnormal conditions of the blood.
The establishments and accommodation are sufficient for a very large number of visitors.

Balzach.—Switzerland, Canton of St. Gall.
Sulpho-alkaline waters.
Special indications: Scrofulous dyscrasia, chronic skin diseases, old sores.

Bangor.—Wales, Carnarvonshire.
Sea baths; pebbly beach.
Doctor: Dr. Hughes.

Banjaluka.—Austria, Bosnia.
Sulphurous waters, from 55° to 105° F.

Bañolas.—Spain, Province of Catalonia.
Sulpho-alkaline waters, 70° F.
Special indications: Neuroses, skin diseases, scrofula, syphilis.
The baths should not be taken for longer than **10 minutes** at a time.
The establishment is good, and the accommodation in the village moderate in price.
Doctor: Dr. Alsina.

Bantry.—Ireland, Cork.
Sea baths; sandy beach.
Doctor: Dr. Swanton.

Banyuls-sur-Mer.—France, Pyrénées Orientales.
Bicarbonate waters and sea baths.

Baraci.—France, Corsica, near Propriano.
Sulphurous saline springs, 85° F.
Scarcely any accommodation, though the waters are efficacious.

Baracza.—Hungary, comitat of Gömör, near Czúko.
Earthy waters, 60° F.

Barambio.—Spain, Province of Alava.
Sulphurous earthy waters.
Special indications: Affections of skin and bones.
The establishment is somewhat deficient.
Doctor: Dr. Martinez.

Barbazan.—France, Department Haute Garonne. For routes see table.
Sulpho-ferruginous waters, 50° F.
Special indications: Cutaneous affections, chlorosis and disorders of the respiratory organs.

Barberie.—France, Loire Inférieure, near Nantes.
Silicated ferruginous waters.
Special indications: Anæmia, chlorosis.

Barbern.—Russia, Finland, district of Mitau, 25 miles from Riga.
Sulphurous waters.

Barbotan.—France, Gers. For routes see table.
Sulphate of soda and ferruginous waters, from 50° to 120° F.
Sulphurous muds.
Special indications: Rheumatism, **paralysis**, cutaneous and uterine affections, rickets and syphilis.
Doctor: Dr. Dupeyron.

Barcelona.—Spain, capital of Province of same name, formerly of Catalonia. By Orleans and South of France Railway, by way of Toulouse and Cette.

Barcelona is the second largest city in Spain, and the first commercially. It contains several well-arranged libraries and collections of natural history and antiquities. Opera house.

OBJECTS OF INTEREST.—Cathedral, Churches of Sta. Maria del Mar and del Pino, Lonja, Deputacion, Casa Gralla, and Dusay.

The Rambla and Paseo de Gracia are the Hyde Park and Rotten Row of Barcelona, from 2 to 5 p.m. daily. Tivoli and Campos Eliscos. Bull-fights, Casino. Gran cafe, and Cafe de las siete puertas.

CABS: one horse up to midnight, 4 reales; two, 6 reales; one horse, midnight to dawn, 6 reales; and two horses, 9 reales. By the hour, the former 8 to 10 reales; the latter, 10 to 15 reales. Every quarter of an hour is extra.

Many excursions in the country on horseback, mules, carriages, &c.

    Post-office: Rambla de la Sta. Mónica.
    Telegraph office: Plaza de Palacio.
    Doctor: Dr. Roberts, Calle Cristina, 8.
    Hotels: De las cuatro Naciones, del Oriente.

Barèges.—France, Hautes Pyrénées; 4,200 feet above the sea level.

    Alkali-saline-sulphurous waters, from 45° to 105° F.
    Special indications: Scrofula, diseases of the bones, herpes and syphilis.
    Season: 15th June to 15th September.
    Doctors: Drs. Armieux, Betons, and Grimmaud.
    Hotels: De France, de l'Europe, des Princes.

Barestrand-Syssel.—Iceland.

    Indifferent waters, 218° F.

Bari.—Austria, Hungary, comitat of Zemplin.

    Sulphurous waters.

Bari (Acqua-di).—Italy, Province of Bari in Calabria; line from Foggia to Brindisi, stop at Bari.

    Sulphurous magnesian waters. Sea baths, sandy beach.
    The springs all lie close to the sea shore. Purgative action exclusively. No establishment.

Barmouth.—Wales, Merionethshire, Cardigan Bay.

    Sea baths. Smooth sandy beach.
    Doctor: Dr. Edwards.

Bartfeld.—Hungary, near Eperies.

    Cold ferruginous bicarbonate waters.
    A well-arranged bathing establishment, much frequented in summer.

Barzun.—France, Department of Basses-Pyrénées; 4,000 feet above sea level.

Sulphurous alkaline waters, 85° F.

Special indications: as at Barèges.

Basel.—Switzerland, on the Rhine, from Paris by East of France Railway, *viâ* Mulhouse, or by North of France Railway to Cologne, and *viâ* Coblentz, Mayence, and Strasburg.

Native town of Holbein the painter. Cathedral founded A.D. 1019. Byzantine church of St. Elizabeth. Museum; Town Hall, Fountain, and Fish-market. Training institution for Protestant missionaries. Circulating library. Very fine excursions in the environs.

Doctors: Drs. De Wette and Bruckner.

Hotels: The *Hôtel Hofer*, near the Swiss railway station, is a small and very comfortable clean house; prices moderate; no omnibus required. C. Hofer, proprietor.

Bastennes.—France, Landes, near St. Sever.

Cold saline and sulphurous waters.

Bath.—England, Somersetshire, 12 miles from Bristol.

Saline-sulphate of lime and carbonate of iron waters, from 104° to 120° F.

Special indications: Gout and rheumatism, chronic skin diseases, anchylosis.

The climate of Bath is comparatively mild. The season is chiefly in spring, but the baths are also frequented in winter and autumn.

Bath affords ample and varied recreation and amusements.
In addition to balls, concerts, theatres, &c., visitors can also join in very interesting meetings held at the various literary and philosophic institutions. Victoria Park is the fashionable rendezvous in the afternoon.
The number of visitors to Bath has been fewer of late years, though the reason for this is not very evident, unless, indeed, it be due to a change in fashion. Living is cheap, and the hotels are very comfortable.

Doctor: Dr. Kerr.

Hotels: York House, Grand Hotel.

Bath.—Jamaica.

Thermal sulphurous waters, which have been employed with success in those cases of intestinal derangements which are frequent in tropical countries.

Bath.—United States of America, Virginia.

Thermal waters visited for amusement rather than medical treatment.

Batignolles.—France, Paris.

Cold sulphurous waters.

Battaglia.—Italy, Venetia, near Padua.
Sulphurous saline waters, 190° F.; mud baths; natural steam baths.
Special indications: Rheumatism, scrofula, skin diseases, inflammation of the uterus and ovary, syphilitic bone and skin affections.
The establishment, a veritable palace, is situated on the hill of St. Elena.
Doctors: **Drs. Rosanelli** and **Pezzolo.**

Bauche (La).—France, Savoy, *viâ* Chambéry and Les Echelles; 1,500 feet above the sea level.
Highly effervescent ferruginous waters.
Special indications: Chlorosis, anæmia, dyspepsia and muscular debility.
Doctor: Dr. Guilland.

Baza or Zujar.—Spain, Province of Granada.
Sulphurous waters, 100° F.
Contains large quantities of sulphuretted hydrogen, and is therefore disagreeable to the taste and smell.
Special indications: Hysteria, anæmia, amenorrhœa and chronic skin diseases.

Beaucens.—France, Hautes Pyrénées, near Cauterets and Barèges.
Sulphurous saline waters, 48° to 50° F.; little known.

Beaulieu.—France, Department of Puy de Dôme, near Issoire.
Cold chalybeate waters, said to be intermittent.

Beaulieu.—France, Alpes Maritimes, near Nice.
Winter station. A small but charmingly situated village in a very sheltered position. It is one of the best stations on the Riviera for the winter months

Beaumaris.—Wales, Anglesea.
Sea baths, fine sandy beach.
Situate at the entrance of the Menai Straits. The views over sea and mountain are very romantic.
Doctor: Dr. Owen.
Hotel: Liverpool Arms.

Beaupreau.—France, Maine and Loire.
Cold ferruginous bicarbonate waters.

Bebra.—Germany, Saxony, near Eckhardsberge.
Chloride of magnesia waters, 35° F.
Special indications: Neuroses, chlorosis, and menstrual derangements.
Doctor: Dr. Rühlmann.
Hotel: German Emperor.

Beggenried.—Switzerland, **Canton** Unterwalden, on the south shore of Lake Lucerne.
Climatic station.
Owing to the low temperature and pure mountain air, this place is much visited in July and August.

Bejar.—Spain, near Valladolid, near Salamanca.
Thermal silicate (0.059°) waters, from 50° to 100° F.
A thermal establishment, much frequented in Spain.

Bekees.—Austria, Hungary, comitat Zemplén, near Szerencs.
Indifferent waters, 75° F.
Special indications: Gout and rheumatism. Establishment small.

Bela.—Austria-Hungary, comitat of Zips, on the Poprád.
Sulphate of lime ferruginous waters, containing much sulphur.
Special indications: Chronic skin diseases, gout and rheumatism.

Belalp.—Switzerland, **Canton Valais,** 3½ hours by an excellent mule path from Brigue.
Mountain air and whey cure.
Season: From June 1st to October 31st.
English Church.
A starting point for many most picturesque excursions
The *Hotel Belalp*, overlooking the Aletsch Glacier, 30 minutes' walk. Resident physician; English church service; post and telegraph office in the hotel. G. Klingele, proprietor.
The *Hotel Jungfrau* at Æggischhorn, 7,200′, at the foot of Æggischhorn, 2½ hours above Fesch, a mountain air station; magnificent panoramic view of the Eggishorn, one hour above the hotel; excursion to the Aletsch Glacier and the Lake of Mergelen. The Cabane of Concordia is 4½ hours distant. Grand and interesting excursions in the **Cernoise** Alps. Em. Cathrein, proprietor.

Belascoin.—Spain, Province of Navarra, near Pamplona.
Alkaline waters.
Special indications: Lithiasis, calculus, gastric and hepathic colics.
The establishment and accommodation are inferior.
Doctor: Dr. Ortega.

Bellagio.—Italy, on the Lake of Como.
Climatic station.
Season: Spring and autumn.
Special indications: Nervous and lung complaints.
Its pure air, picturesque site, and perennially clear sky, have gained for Bellaggio the title of "Garden of Lombardy."
Hotels: The *Hôtel de la Grande Bretagne* is a large and magnificent hotel, exceedingly well appointed. Meyer & Arrigoni, are proprietors also of the Grand Hotel at Sestri Ponente.

Bellas Agoas.—Portugal, near Lisbon.
Sulphurous waters.

Bellême.—France, Orne, near Montagne.
Ferruginous bicarbonate waters.

Bellerive.—Switzerland, Canton of Berne on the Birs, 7 hours from Basle.
Sulphurous alkaline waters.
Special indications: Intestinal obstructions, congestions, scrofula, hypochondriasis, hysteria, jaundice.

Belleville.—France, in the *enceinte* of Paris.
Cold sulphato of lime waters.

Belloc.—France, Gironde, near Bazas.
Cold ferruginous bicarbonate waters.

Bellus.—Spain, south of Valencia.
Sulphate of magnesia and carbonate of lime springs, $65°$ to $70°$ F.
Little frequented. Establishment and accommodation old fashioned.
Roman and Arab ruins in the environs.
Doctor: Dr. Cerezo.

Bellus.—Austria-Hungary, comitat of Trenczin, on the Waag, and 1½ hours from Trenczin.
Cold sulphurous waters.

Benavente.—Spain, Province of Leon, 23 m. from Zamora.
Ferruginous bicarbonate waters, $40°$ to $45°$ F.; rich in carbonic acid gas.

Benedekfalva.—Hungary, comitat of Liptau, ½ hour from Liptó-Szent-Miklos.
Alkaline waters.

Benetutti.—Italy, Piedmont, near Incontrado-Logoduro.
Numerous sulphurous alkaline springs, $85°$ to $95°$ F. No bathing establishment.
The arrangements are so inadquate that the simple tastes of the Italians only could tolerate them.

Ben Haroun.—Africa, Province of Algiers, near Dra el Nizan.
Ferruginous bicarbonate waters, highly gaseous; used only for exportation.

Benimarfull.—Spain, Province of Alicante, near Alcoy
Sulphuretted lime waters, 40 to $45°$ F.
Special indications: Cutaneous affections, vesical catarrhs, muscular contractions.
The establishment is modern and complete. The prices are reasonable and the accommodation good. The situation is romantic.
Doctors: Dr. Genoves y Tio.

Ben Rhydding.—England, Yorkshire, **Wharfedale, near Leeds 480 feet; above the sea.**
Hydropathic establishment.

Bentheim.—Germany, Hanover, near **Osnaburg.**
Sulphate of lime waters.
Doctors: Drs. Stroth and Stollenkamp.
Hotels: Kurhaus.

Benyus.—Hungary, comitat of Sohl.
Alkaline waters.

Berencze.—Hungary, comitat of Neograd.
Cold effervescent ferruginous waters.
There being no bathing nor drinking arrangements, the waters are used only by the inhabitants.

Berg.—Germany, Würtemberg, **near Stuttgart.**
Saline springs and ferruginous waters, of 45° to 50° F.; similar to the Cannstadt waters, with which their therapeutic properties are also identical.
A very excellent bathing establishment.

Bergallo.—Italy, Tuscany, near Castelnuovo-Berrardengn.
Bicarbonate of lime springs, 33° F.

Beringerbad.—Germany, Saxony, near Alexisbad.
Bromurated saline cold waters, containing 12 grammes calcic chloride, and 9 grammes sodic chloride per litre.
Special indications: Scrofulous and chronic skin affections, diseases of the bones.

Berka.—Germany, Duchy of Saxe-Weimar, near Weimar.
Sulphate of lime and chalybeate waters.
Special indications: Rheumatism and paralysis, chronic mercurial intoxication.
Doctors: Drs. Ebert and Freygang.
Hotels: Kurhaus, German Emperor.

Berlin.—Capital of Germany and Prussia, from **Paris by** North of France Railway, *viâ* **Lehrte, Hanover, or** Cologne **Minden.**
One of the largest and handsomest cities in Europe. Garrisoned by 20,000 soldiers of the guard. The river Spree is crossed by over fifty bridges. Principal street, Unter den Linden. Some very fine palaces.

OBJECTS OF INTEREST.—Brandenburg gate. Equestrian statue of Frederick the Great by Rauch. Palaces of the Emperor, Crown Prince, and Queen of Holland. Academy of Fine Arts, Opera-house, Arsenal, Academy of Engineers and Artillery, Schauspielhaus, Victoria, Wallner, and Woltersdorf Theatre, Kroll's establishments, Orpheum, Colosseum. Royal museum of pictures, New Museum, National Gallery and Historical

Museum, Royal Palace and Royal Library, Raszynsky's Picture Gallery, Exposition of Artists; Zoological and Botanical Garden; Exchange and Town Hall, Parliament Buildings. Guards and military bands play opposite the Court-house; Sculpture Gallery, the Schlossbrücke, Royal Foundry, Borsig's Engine Works, Industrial Museum.

EXCURSIONS: To Charlottenburg, Tegel; Potsdam and Sanssouci Babelsberg; Pfingstberg and Ghenicke.

CABS: One or two persons, 1½ mile, 1 mark or 1 shilling. Three or four persons, 1½ marks.

Post-office: Leipzigerstrasse.

Telegraph-office: Französischestrasse.

American Agency The American Exchange in Europe, reading rooms, &c, Unter den Linden, 45.

Dentist: Dr. Suersen.

Doctors: Drs. Wendel, Baer, Gustorf and Hahn.

Hotels: *Kaiserhof*, *de Rome*, *Central*, *Royal*.

Berne.—Capital of Switzerland and Canton of Berne, from Paris by East of France Railway, *viâ* Basle.

OBJECTS OF INTEREST.—Church of the Holy Ghost, Hospital, Morat gate, Federal Palace with Picture Gallery, Museum, Cathedral, Bronze statue of Eulach, Platform or Terrace, Town Hall, Bear Pit.

Fine excursions in the neighbourhood.

Post and telegraph office at the railway station.

Bookseller: J. Dalp, Place de la Gare.

Doctors: Drs. Bourgeois, Nichaus and Schoeder

Hotels: *Bernerhof*, *du Faucon*, *Belle Vue*, and *du Maure*.

Bernières-les-Bains.—France, Department of Calvados, from Paris, *viâ* Caen.

Sea baths, sandy beach.

Bernos.—France, Department of Gironde.

Bicarbonate ferruginous cold waters.

Bernstein.—Hungary, comitat of Eisenburg.

Sulphurous, ferruginous and cupric waters; little used and scarcely known.

Berthemont.—France, Department of Alpes Maritimes, near Nice.

Highly effervescent, indifferent waters.

Special indications: Lithiasis and vesical catarrh.

Bertrich.—Germany, Rhenish Prussia, circuit of Trèves on the Moselle, stop at Bullay Moselle Railway.

Alkaline, saline and sulphate of soda waters, of 75° F.

Special indications: Rheumatic affections, liver and spleen disorders and affections of the urinary system.

Excellent bathing establishments.

Doctor: Dr. Cüppers.

Hotels: *Wering*, *Adler*.

Bertua.—Spain, Guipuzcoa, near Corunna, and the Convent of St. Miguel.
Sulphurous thermal waters.
Special indications: Goitre, sciatica, rheumatic paralysis.

Berwick (North).—Scotland, Haddington, on the Firth of Forth.
Sea baths, firm sandy beach.
Doctor: Dr. Crombie.

Beschtau or Maschukabaths.—Russia, Caucasus, near Pjätigorsk.
Alkaline-sulphurous waters, varying in temperature from 110° to 170° F.
Under this denomination are classed all the springs which rise in the district of Eisenburg and Pjätigorsk. Protected by Peter the Great, they owe their development mainly to the Emperor Nicolas. Residence here is very expensive, which may be one reason why the baths are not more frequented. The establishments are first-rate, and both comfort and amusement are amply provided for. The scenery is of the most romantic character. Abundant shooting and fishing.

Besenyöfalva.—Hungary, comitat of Liptau, on the Waag.
Cold effervescent ferruginous waters.
Special indications: Disorders of the urinary system, intermittent affections and affections of the abdominal viscera.

Besse.—France, Puy de Dôme, near Issoire.
Cold ferruginous waters.

Betaille.—France, Department of Corrèze, near Tulle.
Cold ferruginous waters.
Special indications: Catarrhs, anæmia, chlorosis, atony of the stomach, fluor albus and female complaints.

Beuren.—Germany, principality of Hohenzollern-Sigmaringen, in the Black Forest; 1,580 feet above sea level.
Climatic, air and whey-cure station. Climate very mild and steady.
Special indications: Tuberculosis, chronic catarrh of respiratory organs, bronchial and laryngeal catarrhs.

Betelú.—Spain, Province of Pamplona, near Tolosa.
Sulphurous saline waters, 50° F.
Special indications: Chronic skin diseases, scrofula, syphilis, mercurialism, laryngeal and bronchial catarrhs.
The establishment is modern and complete; the situation charming, and the vegetation luxurious. The accommodation is very good, and prices moderate, ranging *en pension* from 6 francs to 10 francs a day. Betelú is therefore much frequented.
Doctor: Dr. Gurrucherri.

Beuzeval.—France, Department of Calvados, from Paris, *via* Caen.
Sea baths, sandy beach.

Bex.—Switzerland, Canton Vaud, near Lausanne and Vevey; 1,380 feet feet above sea level. For routes see table.

Saline waters strongly bromo-iodurated, brine baths inhalation, hydropathic establishment, whey and grape-cure climatic air station, electric, shower, vapour and steam baths, fumigations.

Season: 15th May till October, but open all the year.

Special indications: Scrofula, rhachitis, chronic rheumatism, uterine affections, plethora abdominalis, chronic affections of larynx and pharynx, tendinous and articular retractions, paralysis, neuralgia.

The situation is very fine, being on the road to the Canton Valais. The climate is mild, but stimulating. The establishment is fitted up with all appliances requisite for modern methods of treatment. All the internal arrangements are luxurious and comfortable.

Doctor: Dr. Decker, who speaks English.

Hotel: The *Grand Hôtel des Bains*, situated in a large shady park, is a first-class house, well managed, and highly recommended. It is comfortable and pleasantly situated, and has well shaded lawn tennis ground. Chas Hieb, proprietor.

Bexhill-on-Sea.—England, Sussex, between Hastings and Eastbourne.
Sea baths, sandy beach.

Biarritz.—France, Department of Basses Pyrénées, close to the Spanish frontier, on the northern shore of the Bay of Biscay; 5,500 inhabitants. For routes see table.

Sea bathing and winter station; climate mild and humid; fine sandy beach.

Season for baths: The late summer and autumn months.

Winter Season: From November till May.

Above 10,000 visitors annually.

Special indications: Chlorosis, anæmia, chest and lung complaints, laryngitis, pharyngitis.

Biarritz is a small but picturesque town, which was brought much into notice by the Empress Eugenie, who used to reside there every year. It is distinguished for its very select society, which assembles for sea bathing and also during the winter season. The scenery is beautiful, although the vegetation in the immediate neighbourhood of the town is somewhat scanty. The town is built like an amphitheatro on the rising cliffs of the Pyrenees, and has some fine modern buildings, of which the Villa Eugenie is one of the most striking.

There is a casino with ball, concert, card and reading room.

English Church.
Post and Telegraph Office.
Banker and Exchange Office: E. H. Bellairs.
Bookseller: Victor Benquet.
English Chemist: J. Moureu, 5, Place de la Mairie.
Dentist: Edwards.
Doctors: Drs. Adhéma, Girdlestone, Augey, Malpas and Wellby.
Hotels: The *Hôtel de France* is a very comfortable house, in a quiet part of the town. J. Fourneau, proprietor.

Bikszad.—Hungary, comitat of Szathmár.
Saline and bicarbonate waters, 4 gr. mineralisation.
Special indications: Feeble digestion, constipation, chronic disorders of the respiratory organs, incipient phthisis, menstrual irregularities and other affections of the genito-urinary organs.
The water is exported. In its chemical composition it is analogous to Gleichenberg, but stronger.
Notwithstanding very moderate accommodation and little comfort offered, Bikszad is much frequented; a good proof of the efficacy of its waters.

Bilazais.—France, Department of Deux Sèvres, near Poitiers.
Sulphate of calcium waters, 56° F.

Bilén.—Austria-Hungary, comitat of Marmarós, district of Rahó.
Alkaline waters.
Very primitive arrangements.
Special indications: feeble digestion, constipation, liver and spleen affections.

Bilin.—Austria, Bohemia, near Teplitz.
Alkali saline waters; their composition is similar to those of Vichy.
Bilin is well known under the name of the "cold Vichy."
Special indications: Catarrh, hæmorrhoidal affections, morbus Brightii, disorders of the urinary system, obesity.
Doctor: Dr. Reuss.
Hotels: Lion and *High House.*

Birchington.—England, near Dover.
Sea baths, sandy beach. Very picturesque site.
Doctor: Dr. Neame.

Birkenfeld.—Germany, Grand Duchy of Oldenburg, near Trèves and Saarlouis.
Ferruginous, magnesian, sodic and calcic bicarbonate waters—tonic and resolvent.
Special indications: Affections of the glandular system.
Doctor: Dr. Ricken.

Birlenbach.—Germany, Duchy of Nassau.
Ferruginous bicarbonate table water.

Birmansdorf.—Switzerland, Canton of Aargau, near Baden.
Acidulous or bitter waters, 31 gr. of mineralisation.
These waters are only used for exportation and as purgative agents.
Doctor: Dr. Pfeuffer.

Birresborn.—Germany, Rhenish Prussia, in the Eifel, near Gerolstein.
Ferruginous bicarbonate waters, exported and brought into the market by the Gerolstein Mineral Water Co., Limited, London.
A station little frequented.

Bishop's Teignton.—England, Devonshire, 2½ miles from Teignmouth.
Hydropathic establishment and health resort.
A charmingly situated residence at the foot of the "Holdon," overlooking the Teign and hills beyond. The baths are perfect, and cures numerous. From the top of Holdon the views are among the finest in the country. Chas. Carpenter, proprietor.

Bistritz.—Austria, Moravia, near Ollmütz and Hullein.
A celebrated whey cure station; The Appenzell of Austria. Sheep's milk only is used.
Season: June—September.

Bisztra.—Austria-Hungary, comitat of Marmarós, near Majdanka.
Alkali-ferruginous waters.
Mineral constituents very scanty.

Bivuto di Termini.—Italy, Sicily, between Palermo and Sciacca.
Thermal sulphurous waters.
Very suitable bathing establishments, and the accommodation in the Convent of St. Kalogerus is comfortable.

Blackpool.—England, Lancashire.
A much frequented sea-bathing town.

Blanchemont.—Belgium, near Liege.
Cold ferruginous bicarbonate waters.

Blankenberghe.—Belgium, near Bruges.
Much frequented sea baths; more retired and quiet than Ostend.
Doctors: Drs. Cosyn, Notebaert and Van Mullem.
Hotels: Cursaal, Godderis de la Paix.

Blankenburg.—Germany, Principality of Rudolstadt, near Weimar and Halberstadt.
Pine-cone sap and resinous baths.

Special indications: Herpes, scabies, excessive perspiration, muscular pains, paresis.

The accommodation is comfortable, and living **exceedingly cheap.**

Bleichbad.—Switzerland, Canton of St. Gall, near Altstätten; 1,500 feet above the sea level
Ferruginous saline waters, with strong odour of sulphur.
Special indications: Gout, rheumatism, paralysis, neuralgias.

Bleville.—France, Seine Inférieure, near Le Havre.
Cold ferruginous bicarbonate waters.

Blue Sulphur Springs.—U.S., North America, Virginia.
Sulphurous waters.
Special indications: **Rheumatism, scrofula, syphilis.**

Blumenstein.—Switzerland, near Berne.
Cold ferruginous alkali-saline waters.
Special indications: **Constitutional debility.**
A fairly good bathing establishment.

Blumenthal.—Switzerland, near Solothurn.
Saline-alkaline waters, only used by the inhabitants.

Bobbio.—Italy, Lombardy, Province Pavin; 60 k. from Genoa.
Thermal, sulphurous and saline springs.

Bobotsch.—Roumania.
Sulphurous waters.
Owing to the lack of energy on the part of the inhabitants there is little or no accommodation; the waters are very efficacious.

Bocheggiano.—Italy, near Siena.
Five springs, some ferruginous and some sulphate of lime, 35° F.
Special indications: Anæmia, chlorosis, constipation, liver, spleen, and chronic skin diseases.

Bocklet.—Germany, Bavaria, near Kissingen.
Cold, sulphurous and chalybeate waters, rich in carbonic acid gas; mud baths.
Frequented by patients at Kissingen.
Special indications: Scrofula, plethora, diabetes, hyperæmia of the liver, anchylosis, chronic muscular affections, torpid sores.
Doctor: Dr. Scherpf.

Bocskó Rahó.—Hungary, comitat of Marmaros, 7 miles from Szigeth.
Iodurated saline waters.
Special indications: Affections **of spleen and liver,** scrofula and **rhachitis.**

Bodajk.—Hungary, near Weissenburg.
Sulphate of lime **waters**, rich in silicic acid.
Special indications: Hæmorrhoids, scrofula, gout, rheumatism.

Bodendorf.—Germany, Rhenish Prussia, near Neuenahr.
Climatic **station**; grape cure.
Especially frequented by **invalids** suffering from chest and nervous complaints.

Bodenfelde.—Germany, Hanover.
Saline waters.

Bodok.—Transylvania.
Cold alkaline waters, very similar to those of Bilin.

Boesing.—Austria-Hungary, **comitat** of Pressburg.
Ferruginous alkaline **waters**, effervescing.
Special indications: Chlorosis, blenorrhœa, **hysteria**.

Bognor.—England, Sussex.
Sea baths, firm sandy beach.
Doctor: Dr. Todd.

Boisse (La).—France, **Savoy**, near Chambery.
Cold ferruginous **waters**.

Boll.—Germany, Würtemberg, near Göppingen; 1,300 feet above the level of the sea.
Cold alkali-saline-sulphurous waters; milk **cure**; mild climate.
Special indications: Affections of the larnyx, skin diseases, contractions, stiffness of the joints.

Boltenhagen.—Germany, Mecklenburg-Schwerin, equally distant from Lübeck **and Wismar.**
Much frequented for its sea-baths; fine sandy beach, with pine forests behind.
Doctor: Dr. Kelling.
Hotels: Kurhaus, Luckmanns.

Bonar.—Spain, Galicia, near Leon.
Ferruginous saline waters, 75° F.
Special indications: **Constipation, amenorrhœa,** chlorosis, **hypochondriasis.**
These waters were used by the Romans. Grass withers under their influence, the meadows in the neighbourhood being all more or less blighted.

Bondonneau.—France, Department of Drôme, near Montélimar, 500 feet above sea level.
Alkaline iodurated waters; effervescent.
Special indications: Anæmia, chlorosis, dyspepsia, diarrhœa.

Bonn.—Switzerland, Canton of Fribourg, two hours from that town.
Cold sulphurous waters.
Special indications: Skin diseases.

Bonnefontaine.—France, Department Meurthe, near Metz.
Cold bicarbonate of iron waters.

Bonneval.—France, Savoy, near Bourg St. Maurice.
Thermal sulphurous waters, analogous to those of St. Gervais, 85° F.

Bonnington.—Scotland, near Leith and Edinburgh.
Strong ferruginous and cold sulphurous waters.
Special indications: Anæmia, chlorosis, scrofula, rhachitis.
Every comfort in the bathing establishments, which are very good.

Boräs.—Sweden, near Elfsborg.
Acidulous or bitter waters; exported.

Borbye.—Germany, Province of Schleswig, near Eckernförde.
Sea-bathing place on the Baltic coast; little frequented.

Bordeaux.—France, on the Gironde, the capital of that Department, from Paris by Orleans Railway.
The third port of France, and a chief centre for wines and brandies; the Roman *Burdigala*.
OBJECTS OF INTEREST.—Remains of an amphitheatre. The town belonged to England from 1152 to 1453. Bell tower, Porte du Palais, and other old gates.
Cathedrals of St. André, St. Croix, St. Severin, and St. Bruno.
The Mairie in the old palace has a library of 150,000 volumes, a collection of antiquities and a picture gallery. Large Prefecture, Town Hall, Law Courts, Exchange and Custom House, Commercial Court, Bank, Mint, Theatre. Place des Quinconces, Quai des Chartrons, and botanical gardens.
CABS: By the drive 1 fr., for the hour 2 fr.
Doctor: Dr. Breen.
Hôtels: Des Princes and de la Paix. Very good house. Queuille and Darie, proprietors.

Bordighera.—Italy, on the French frontier and on the Riviera.
Winter station, well frequented, but very quiet.
Doctor: Dr. Goodchild.
Hotels: D'Angleterre, Grand, Windsor.

D

Borgo Maro.—Italy, Piedmont, near Oneglia.
Cold sulphurous waters.
Special indications: Scrofula and skin diseases.

Borines.—Spain, Province of Oviedo, near this town.
Saline sulphurous waters, used only internally on account of the limited supply from the spring.
Special indications: Gastralgia, dyspepsia, skin diseases and affections of the mucous membranes.
The establishment is modern, and is visited chiefly by the surrounding inhabitants. The site of Borines is romantic.
Doctor: Dr. Ayegui.

Borkum.—Germany, Oldenburg. An island in the North Sea.
Much visited for sea-bathing, especially by Hanoverians.
Doctor: Dr. Schmidt.
Hotel: Kurhaus.

Borkút.—Hungary, comitat of Màrmaros, District of Rahó.
Saline and strongly iodurated waters.
Special indications: Scrofula, lymphatism, goitre, adenitis.
A small bathing establishment.

Borla.—Italy, also called Acqua del Mortagone.
Saline waters, 85° F.; highly effervescent.
Special indications: Constipation, lymphatism, adenitis.

Bormio.—Italy, Lombardy, Upper Valteline, on the Adda; 4,016 feet above the sea level.
Inert thermal waters, from 90° to 120° F.; a climatic station.
Doctor: Dr. Battersby.

Borra.—Italy, Tuscany near Arrezzo, in the valley of the Arno.
Ferruginous alkaline waters, effervescing.

Borrone.—Italy, Tuscany, near Montalto.
Ferruginous waters, 65° F.
Special indications: Disorders of the digestive and urinary organs.

Borsa.—Hungary, comitat of Màrmaros.
Cold alkaline ferruginous waters.
Special indications: Jaundice, and disorders of liver and spleen.

Borsaros.—Transylvania.
Ferruginous bicarbonate waters, 45° F.
Special indications: Rheumatism and herpes.

Borschom.—Russia, Russ. Georgia, on the Kura; 2,630 feet above the sea level, near Pjätigorsk.
Alkali-saline waters, 75° to 95° F.
Special indications: Gravel, affections of the mucous membrane of the respiratory, digestive and urinary organs; hyperæmia of liver and spleen; hæmorrhoids and plethora abdominalis, as also disorders of uterus and ovaries.
Borschom is very quiet, and the surrounding country wild. The accommodation is good, and residence agreeable.
Doctor: Dr. Remmert.

Borszek.—Austria, Transylvania, on the Moravian frontier in a valley of the Carpathians; 2,400 feet above the level of the sea.
Alkali-saline waters, highly charged with gas; mud baths.
Special indications: **Affections of the uterus**, atony, paralysis, liver and spleen **disorders**, gravel and vesical complaints.
A good bathing establishment.
The climate is very trying and variable. Borszek is the most important watering place in Transylvania. The water is largely exported. Accommodation abundant, and in accordance with modern requirements.

Bottaccio.—Italy, Tuscany, near Castelnuovo-Berrardenga.
Sulpho-ferruginous waters.
Special indications: Affections of the **uterus, amenorrhœa, chlorosis.**

Botzen.—Austria, Tyrol.
Air cure station; grape cure.
Hotel: Victoria.

Boudes.—France, Puy de Dôme, near Issoire.
Bicarbonate alkaline waters, 35° F., mineralisation, 4 to 9 grs.
May possibly in the future become an important spa.

Boulogne-sur-Mer.—France, Pas de Calais.
Sea baths, and one ferruginous saline spring.
The Casino faces the bathing establishment. There is a Theatre. Fine view from belfry. There is an English college and many English schools and churches; in fact, the town may be said to be half English.
Bookseller: Merridew, 60, Rue de l'Ecu.
Doctors: Drs. Harvey and Walker.
Hotel: The *Hôtel Dervaux*, in the Gde. Rue des Vieillards, will be found a good family hotel. L'heureux, Proprietor.

Boulou (Le).—France, Pyrénées Orientales, near the Spanish frontier.
Alkaline waters similar to those of Vichy, and ferruginous to those of Spa.
Special indications: Anæmia, obesity, chlorosis, gravel.
Doctor: Dr. Massot.

Bouquéron.—France, Isère, near Grenoble.
Alkaline waters similar to those of Evian.
One of the best appointed hydrotherapeutic establishments.
Terebinthine baths.
Special indications: Affections of the urinary organs, bladder, liver and kidneys.

Bourasol.—France, Haute Garonne, near Toulouse.
Ferruginous bicarbonate waters, 40° F.

Bourbon l'Archambault.—France, Department of Allier. For route see table.
Bromo-iodurated saline waters, 125° F.; bicarbonate ferruginous magnesian waters, 120° to 160° F., highly gaseous.
Special indications: Scrofula, rheumatism, paralysis, nervous affections.
The bathing establishment is the property of the Government. There is a civil and military hospital
Doctor: Dr. Regnault.

Bourbon-Lancy.—France, Saone and Loire. For routes see table.
Six thermal saline ferruginous springs, from 66° to 135° F.
Special indications: Scrofula, chlorosis, paralysis, syphilis.
These waters stimulate mucous secretion; when taken in large doses they are purgative; also tonic and alterative.
Doctor: Dr. Valentini.

Bourbonne-les-Bains.—France, Haute Marne; 850 feet above the sea level. For routes see table.
Bromo-iodurated saline waters of 119°, 138° and 180° F.
Special indications: Scrofula, paralysis, rheumatism, catarrh of the stomach and constipation.
The new establishment is replete with all appliances.
There are some fine promenades, and the excursions are very interesting.
Doctor: Dr. Bougard.
Hotels: *Hotel des Bains.* Lacordière, proprietor.

Bourboule (La).—France, Auvergne, Department of Puy de Dôme; 2,850 feet above sea level; 1,400 inhabitants. For routes see table.
Effervescent saline arsenical waters (28 milligr arseniate of soda per litre), 140° F.
Season: 25th May to 30th September.
Number of Visitors: Above 6,000 annually.
Special indications: Anæmia, lymphatism, general debility, affections of the skin and respiratory organs, diabetes, rheumatism and intermittent fevers.
There are three bathing establishments thoroughly well fitted up and ably managed. They offer all the appliances of

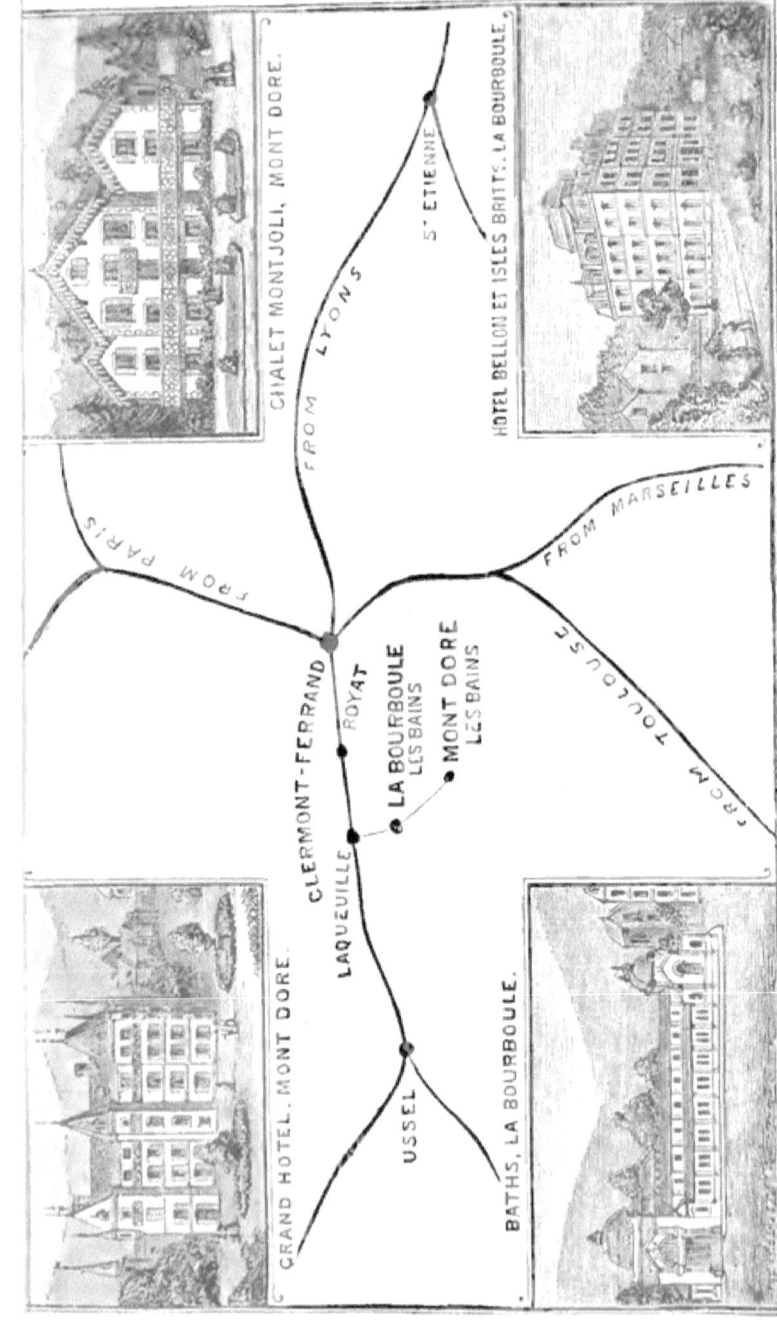

modern times: single sitz baths, general baths, steam baths, inhalation and spray rooms, douches of every description, shampooing, &c.

The waters keep well in bottle, and excellent results have followed their use at home. It is advisable to begin taking them before coming to La Bourboule for the cure; the usual dose is from ½ to 3 glasses a day, taken before breakfast, or with wine during that meal.

The country around La Bourboule is very mountainous, as may be supposed from its elevated position. It is situated in one of the most picturesque parts of Auvergne, and the excursions which may be made from it are very varied. There is a beautiful park, well suited for promenades.

The Casino is in course of construction; concerts are given daily; all sorts of amusements, including lawn tennis.

English Church service.

Doctors: Drs. Danjoy, Morin and Verité.

Hotels: The *Grand Hôtel Bellon and des Iles Britanniques* is a large first-class family hotel; in an excellent position, close to the establishment, table d'hôte and restaurant. J. Donneaud, proprietor. *See* plan.

The *Grand Hotel*, a first-class establishment, near the baths. A. Ferreyrolles, proprietor.

The *Hôtel de Paris*, a very comfortable, well-kept family hotel. A. Lequime, proprietress.

The *Grand Hôtel de l'Etablissement*, first-class house, near the establishment and facing Casino; châlet for families; gardens and private saloons, Table d'hôte and restaurant. Vimal-Chousse, proprietor.

The *Splendide Hôtel et d'Angleterre réunis*. A private villa in the garden of the above hotel, containing 25 rooms, is let for the season; apply to the proprietor, A. Lemerle.

Bourg d'Ault.—France, Department of Somme.
Sea baths, near Abbeville.

Bourg d'Oisans.—France, Isère.
Cold neutral waters.

Bournemouth.—England, Hants.
Cold ferruginous waters and sea baths.
Hydropathic establishment. A much frequented watering-place.
Sandy beach five miles long.
Especially recommended for chest and lung complaints. Several establishments for the treatment of consumption. The climate is dry and mild. Average rainfall 30 inches, and mean winter temperature 42° F.

Doctors: Drs. Hirson and White.

Bourtscheidt.—Germany, adjoining Aix-la-Chapelle, and forming one town with it.
Chloride of sodium and sulphurous waters, from 107° to 190° F.
Special indications: Rheumatism, scrofula, syphilis and neuroses.
Doctors: Drs. Lüth, Laaff and van Erkelenz.
Hotels: De la Rose, Charles, Newbath.

Braga.—Portugal, Province of Minho.
Sulphurous and ferruginous waters.

Bramstadt.—Germany, Holstein.
Three springs; one saline, one sulphate of soda, and one chalybeate.

Braque.—Spain.
Bitter waters, 55° F.; mineral muds.
Special indications: Chronic skin diseases, scrofula, leucorrhœa, amenorrhœa, anchylosis and chronic effusions.

Bräsa.—Roumania.
Saline sulphurous waters, 55° F.
Little visited.

Braunfels.—Germany, Rhenish Prussia.
Resinous steam baths.
Special indications: Rheumatism, gout, chronic skin diseases, mucous discharges, especially in females; menstrual difficulties.
The bathing arrangements are of a superior kind. Accommodation very good. The surrounding country is charming.

Bray.—Ireland, Counties Dublin and Wicklow.
Sea baths, sandy and gravelly beach.
There is also a mineral spring, which, however, is little used, its exact composition not being known.
Doctor: Dr. Ward.

Bréb.—Hungary, comitat of Marmarós, district of Sugatagh.
Saline sulphurous waters, 60° F.
Special indications: Muscular contractions and gout.

Bregenz.—Austria, Tyrol, on the eastern extremity of the Lake of Constance, 1,903 feet above the sea level. 5,500 inhabitants.
Climatic station in summer, with mild and equable air. Chiefly a centre for excursions.
The old town is very interesting from its ancient buildings. It is surrounded by pine and beech forests, which fill the air with their odour.

Bathing establishment on the lake, near the "Reichstrasse." The walks in the immediate vicinity of the **town** are well kept and longer excursions of endless variety may be made. There is a museum in the Riedgasse, containing many interesting specimens of the fauna and flora of the districts, with **many** antiquities and prehistoric **relics**.

Post Office: Near the station and the Port.
Telegraph Office: Römerstrasse.
Bookseller: Tentsch Sohn.
Doctors: **Drs. Bär, Von Houstetter, Müller** and Heusler (Homœpath).
Hotel: The *Hotel d'Autriche* faces the landing-place of the steamers, and is near the station; a very comfortable hotel, and **very central** for excursions **into** the surrounding country. **A. Thönen**, proprietor.

Bresigala.—Italy, near Faenza.
Sulphurous waters.

Brides-les-Bains.—France, **Savoy, near Moutiers**; 1,750 feet above **the** sea level.
Alkali-saline springs, 85° F., with purgative and tonic **action**.
Special indications: Dyspepsia and stomach diseases.
Doctors: Drs. Desprez and Philbert.

Bridge of Allan.—Scotland, Stirlingshire.
Climatic **station in summer**; also saline waters at Airthrice Wells, **45°** F.
Special indications: Stomach and liver complaints, **skin** diseases and scrofula.
Very many interesting **excursions can** be made in the environs. Picturesque site; **climate mild and equable**; air dry.
Doctor: Dr. Paterson.

Bridge of Earn.—Scotland, Perthshire, **near** Perth.
Highly effervescent saline waters, 45° F., at Pitkeathly.
Special indications: **Plethora, calculus**, syphilitic and liver affections.
Doctor: Dr. Laing.

Bridlington-Quay.—England, Yorkshire.
Sea baths and chalybeate waters.
Sandy and shingly beach, **air pure** and bracing.
Doctor: Dr. Hutchinson.

Bridport.—England, Dorsetshire.
Sea baths, firm, sandy beach, bracing climate; the bathing **is**, however, dangerous for those unable to swim.
Doctor: Dr. Evans.

Brieg.—French Brigues—Switzerland, Canton of Valais, on the right bank of the Rhône, at the foot of the Simplon and Furca Pass. For routes see table.

Sulphate of lime springs, 105° F. The waters are essentially similar to those of Leukerbad, but somewhat more potent.

Special indications: Blenorrhœa, constipation, chronic rheumatism and skin diseases.

Notwithstanding the proximity of glaciers, the climate is very mild, and in the valley even tropical fruits ripen. Visited so long ago as the fourteenth century, it has been neglected of late, owing to the proximity of Leuk.

Hotel: The *Hotel Crown* and *Post*, facing telegraph and post-office. English Church service in the hotel during summer; carriages to be had in the hotel for excursions, and for Italy. Jos. Escher, proprietor.

Brighton.—England, Sussex.

The best sea-bathing place in the United Kingdom. In the neighbourhood is a sulphate of lime and iron spring.

Special indications: Anæmia, chlorosis, affections of the respiratory organs, skin diseases, scrofula, general debility, and nervous affections.

This bathing-place has risen steadily in public favour since George IV.'s time, and has competed successfully with Bath in attracting fashionable and aristocratic society. The town offers every luxury and comfort. The climate is mild and very favourable to patients. But little visited by foreigners. For persons subject to lung and throat affections the air is too bracing, although the temperature of the sea-water is from 60° to 65° F.

Warm baths, especially the large swimming baths, are used more than the open sea bathing. Indian shampooing baths have been erected by a native of India, and are replete with every luxury and comfort. They are, as regards the shampooing, very similar to Russian steam baths, but differ in that aromatic vapours are used instead of steam. Brighton is a well-known station for yachting and, in fact, for amusements of every kind.

Doctors: Drs. Addison, Barrett and Barker.

Hotel: *Grand*.

Haxells. Magnificent sea view facing the Aquarium; excellent house, with moderate charges.

Brindisi.—Italy, Province of Terra d'Otranto.

Sea baths; sandy beach.

The arrangements are very defective.

Brixen.—Austria, Tyrol, 30 m. from Brenner Pass, 1,934 feet above the sea level.

Considered an air-cure station.

Hotel: Elephant.

Brixham.—England, Devonshire, near Torquay.
Climatic station. Season, October—April.
Doctor : Dr. Colston.

Broadstairs.—England, Kent, between Ramsgate and Margate.
Sea baths; sheltered and sandy beach.
Climate healthy and bracing.
Season : June to October.
Doctor : Dr. Walter.

Bronia.—Italy, Tuscany, district of Montale.
Indifferent waters, containing chiefly sodium.
Little visited; no bathing establishment.

Broughty Ferry.—Scotland, County Forfar.
Sea baths.
The good establishments, comfortable accommodation, &c., attract large numbers of visitors.

Brownstown Spa.—Ireland, Meath.
Ferruginous waters.
The bathing establishments are said to be the best in Ireland. The town of Killarney offers every comfort to visitors.

Bruca (La).—Italy, Sicily, near Catania.
Cold sulphurous waters.
Little used and no establishment.

Brückenau.—Germany, Bavaria, near Kissingen.
Chalybeate springs.
Special indications : Anæmia, uterine diseases, affections of the mucous membranes.
Doctor : Dr. Wehner.
Hotels : Kurhäuser; all are under Government management.

Brughéas.—France, Allier, near Vichy.
Bicarbonate of soda waters.

Bruneval.—France, Department of Seine Inférieure, from Paris *viâ* Les Ifs and Le Havre.
Sea baths; sandy beach.
A quiet and aristocratic place.

Brusah.—Turkey, Asia-Minor, one of the most flourishing towns in the Ottoman Empire.
Alkali, saline and sulphurous waters, varying in temperature from $55°$ to $170°$ F.
Season : All the year round.
Special indications : Skin diseases, gout, rheumatism, syphilis, anchylosis, chronic sores.

Brusah is the most interesting and the cleanest city in the East. It is situated in the ancient Kingdom of Bythynia, in a very beautiful country at the foot of Olympus, which is 4,500 feet high. The most interesting object in the town is the baths; as many as 3,000 springs are said to exist in and round the city.

Amongst the bathing establishments, "Ieni-Kaplidscha" is the most interesting and best appointed, both in its interior and its architecture. It is generally called the "New Bath." The cisterns, basins, walls and seats are all of white marble; light enters from cupolas. There are seven public baths, all fitted with such splendour and elegance, that this establishment may fairly be said to have no rival either in Europe or the East. The arrangements are, however, limited to the Turkish, *i.e.*, perspiration baths.

The waters are exclusively used in bathing and are not taken internally. Their high temperature and the steam in which the bather is enveloped during the whole process make these baths pre-eminently "Turkish" or "Steam baths." The greatest luxury is every where met with.

There are two or three other establishments with very good arrangements, and a large number of inferior ones.

Brussels.—Capital of Belgium, from Paris by North of France Railway, *via* Maubenge; from London *via* Ostende in 8½ hours.

Brussels is a very old town, and offers many attractions to visitors. The higher town is inhabited by the wealthy classes, and the lower by the artisan. The former is exclusively French and modern, while the latter retains its mediæval character much as Bruges, Ghent, and Antwerp do. Brussels is often spoken of as "le petit Paris."

OBJECTS OF INTEREST.—Cathedral St. Gudule, King's Palace with picture galleries, Palace of the Comte de Flandres (King's brother). Palace of the Nation, Palais Ducal of the Prince of Orange. Avenue Louise and Bois de la Cambre; the latter is considered finer than the Bois de Boulogne in Paris. Church of Notre Dame de la Chapelle, Exchange and Town Hall, Palais de Justice (Law Courts), Palais des Beaux-Arts. National Bank, museum, Palaces of the Dukes of Arenberg and De Ligne; Wiertz gallery of pictures, Mint, Gallery St. Hubert, Théâtre de la Monnaie. The Squares and Boulevards, Zoological and Botanical Gardens, Guild house. Many statues and fountains, amongst the latter the Mannekin Pis. Conservatory of Music, St. John's Hospital.

EXCURSIONS: To Laeken and Waterloo.
Post and Telegraph Office: Temple des Augustins, Boulevard Anspach.
CABS: 1 fr. for first half-hour, and every ¼ hour 40 c. more, for one-horse carriage; two-horse carriages 2 frs. first ½ hour, every ¼ hour 50 c. more.

Doctors: Drs. Collignon, Thompson and Kelly.

Hotels: Mengelle, de Saxe, de Suède.

Bruszno.—Hungary, District of Sohler.
Sulphate of soda waters, 45° to 50° F.
A much frequented watering place.

Brüttalen.—Switzerland, Canton Berne; 1,360 feet above the sea level.
Alkali saline ferruginous waters.
Special indications: Constipation, congestion, hypochondria, hysteria, jaundice.

Bubendorf.—Switzerland, near Basle; 1,200 feet above the sea level.
Earthy waters.
A very good establishment, with sitz, shower and vapour baths.

Buca dei Fori.—Italy, Tuscany.
Sulphurous saline waters, 97° F.
No establishment. No accommodation.

Buclesore.—India, Bengal.
Springs forming a small sulphurous brook, 50° F.

Bude.—England, Cornwall, near Barnstaple.
Sea baths, sandy beach.
No bathing machines, caves in the rocks being used instead.

Budimir.—Austria, Bosnia.
Saline waters, of high therapeutic value.
Accommodation is only afforded in the Khan, a building in the style of an Eastern Karavanserai. Provisions must be brought by visitors.

Budis.—Hungary, comitat of Arva-Turócz.
Very strong alkaline chalybeate waters. Only used as table waters. Exported.

Budleigh-Salterton.—England, Devonshire, near Exmouth.
Sea baths, gravelly beach.
Air very pure. Climate mild but bracing.
Doctor: Dr. Mercer.

Budos.—Austria-Hungary, Transylvania.
Thermal sulphurous springs and sulphurous exhalations.

Büdöstö.—Hungary, Transylvania.
Sulphur waters, 40° F.
A popular bathing resort.

Builth.—Wales, Breconshire, near Llandrindod.
Saline, sulphurous and chalybeate waters.
Doctor: Dr. Vaughan.

Bujak.—Austria-Hungary, comitat of Nesgrad.
Sulphurous waters, containing much free sulphur.
No establishment and no accommodation.

Bulgnéville.—France, Vosges, near Contrexéville.
Mixed bicarbonate waters.

Bullyeome (Le).—France.
Sulphate of lime waters, 145° F. This spring emits large quantities of carbonic and hydrosulphuric acid gases.

Buneome.—U.S.A., North Carolina.
Thermal sulphur springs, 95° F.

Bundoran.—Ireland, Donegal, near Enniskillen.
Sulphurous waters at Kinlough.
Sea baths, sandy beach, with excellent bathing.
Doctor: Dr. Flood.

Buntschibad or Weissenburg, Switzerland.—(*See* Weissenburg.)
Alkali-saline waters.

Burgbernheim.—Germany, Bavaria, 16 m. from Anspach.
Sulphate of magnesia waters.

Burnham.—England, Somersetshire, on the Bristol Channel.
Saline ferruginous waters; sea baths and fine sandy beach.
Doctor: Dr. Matthews.

Burrone.—Italy, Tuscany, near Arceno.
Ferruginous bicarbonate waters, 40° to 45° F.

Buschbad.—Germany, Saxony.
Cold sulphurous springs.

Bushey.—England, Hertfordshire, near Watford, one hour's ride from Euston Station, London; coach daily from the "Green Man," Oxford Street, at 11 a.m.
Hydropathic establishment of world-wide repute; a climatic station in summer and spring; pure and bracing air; inhalation.
Season: All the year.
Special indications: Morbus Brighti, rheumatism, affections of the respiratory organs, constitutional diseases, anchylosis, muscular contractions and articular swellings, debility, nervous circulatory disorders and lumbago.

This establishment is a handsome and substantial building, erected at considerable cost. It occupies a charming situation in the midst of a healthy and beautiful country. The pleasure grounds surrounding it are in keeping with the mansion, and extend over more than 30 acres. They comprise lawn, tennis grounds, terraces, gardens and shrubberies, with shady avenues and pine wood walks. The Hall stands in a park, well wooded, of 240 acres, on one side skirted by the River Colne, which affords fishing and boating.

As regards its Baths, Bushey is one of the best supplied establishments in Europe. There are Turkish, Russian, vapour, electro-chemical, electro-magnetic, salt water, sulphur and other medicated baths; a large swimming bath, wave, shower, spray, cataract, spinal and other water applications. Pine baths and massage by experienced shampooers.

The surrounding country has been termed "The Garden of England." It affords many pretty drives and interesting excursions. Many ancestral parks and historic mansions of the aristocracy are within easy distance of Bushey.

The Hall has drawing, dining, library, billiard and Smoking, and upwards of 120 bedrooms; corridors and assembly rooms are all of noble proportions, artistically and luxuriantly decorated. The buildings are thoroughly heated by hot water apparatus. Post and telegraph in the house. Inclusive terms, 3½ to 8 guineas per week, or from 12s. to 24s. per day, including Turkish, swimming and ordinary baths.

Doctor: Dr. Robert J. Banning.

Hotel: The Hall, a very well conducted and thoroughly comfortable house; ventilation and drains irreproachable; the water supply is the purest in England; it is under the direct supervision of C. Hawkins, late manager of the Continental Hotel in Paris.

Busignargues.—France, Department of Hérault.
Ferruginous bicarbonate waters, 40° F.

Busk.—Russia, Poland, near Cracow.
Saline, sulphurous, acidulous waters, 45° F. Three springs.
Special indications: Disorders of liver and spleen, mercurial intoxication, chronic rheumatism.

Bussang.—France, Vosges. For routes see table.
Ferruginous bicarbonate waters, for table and bath use.

Butterby.—England, 2 miles from Durham, on the river Wear.
Saline and sulphurous waters, 50° F.; there is also a bicarbonate of lime spring.
Special indications: For the sulphurous waters, skin diseases, scrofula, rachitis; for the bicarbonate spring, kidney and bladder affections.
A picturesque site, with mild climate.

Buyères de Nava.—Spain, Province of Oviedo, at the foot of the Peña Mayor, on the left bank of the Pla; 700 feet above sea level.
Sulphurous and carbonate of lime waters, 62° F.
Season: July—October.
Special indications: All kinds of skin diseases, scrofula, syphilis, mercurialism, vaginal and vesical catarrhs, bronchial and laryngeal affections.

The establishment is one of the most sumptuous of Spain, and fitted with all modern appliances and luxury. The gardens are magnificent. Accommodation is abundant, luxurious and moderate in price.

Doctor: Dr. Gavilanes.

Buxton.—England, Derbyshire. From Whaleybridge by carriages to Buxton, 30 minutes.
Carbonate of lime waters, 66° F.
Special indications: Anæmia, chlorosis, affections of kidney and bladder, blennorrhœa, eczema, chronic diarrhœa, chronic sores, &c.
The environs are very picturesque. Many old castles are still existing in the neighbourhood, such as Chatsworth, Haddon Hall, Old Hall, &c.
Doctors: Drs. Bennet and Turner.

Buzias.—Hungary, District of Temesvar.
Five springs, all cold, containing chloride of sodium, iron, carbonic acid, &c.
Special indications: Scrofula and failing vitality.
Buzias is one of the best frequented stations in Hungary.

Buzot.—Spain, Province of Alicante, 2 hours from Alicante.
Sulphate of magnesia springs, 100° F.
Special indications: Rheumatism, herpes, sciatica, paralysis and leucorrhœa.
There are natural sweating baths at Buzot.
Season: May 1—November 1.
The site is very picturesque, and the slopes of the hills are covered with lovely gardens and villas. The scenery is remarkably fine.
The establishment, which belongs to Count de Casas Rojas, is magnificent, and is fitted with every modern appliance, and must be ranked among the best in Spain. The accommodation is in keeping, and notwithstanding the degree of luxury, prices are moderate. A residence very agreeable.
Doctor: Dr. Lopez.

Bykowicz.—Russia, Ukraine, Government and Province of Charkow.
Alkali-saline waters. Very powerful, and used chiefly for exportation.

Cabeco-de-vide.—Prov. Alemtejo, Portugal.
Bicarbonate of soda waters, 68° F.
Special indications: Diabetes.

Cabourg.—France, Department of Calvados.
Sea baths; good beach.
A fine casino, and well frequented station.

Caccio Cotto.—Italy, Tuscany, district of Volterra, on the northern slopes of Monte Cerboli.
Sulphurous-alkaline waters, 80° to 90° F., rich in sulphur.
A large and elegant bathing establishment, with hotel adjoining.

Cadéac.—France, Department of Hautes Pyrénées.
Cold sulphurous waters.
Special indications: Rheumatism, anæmia.
Doctor: Dr. Fontau.

Cadenabbia.—Italy, Lake of Como.
A climatic health resort.
Very pure and healthy air.

Cadiz.—Spain, on the Straits of Gibraltar.
Sea baths; indifferent beach.
The establishments are fairly well arranged.

Caille (La).—France, Haute Savoie, near Annecy.
Sulphate of lime waters, 72° F.
Special indications: Struma, herpes and rheumatism.

Cairo.—Capital of Egypt.
A winter station; much resorted to by sufferers from diseases of the chest owing to the dryness of its air. A very interesting town.
Doctor: Dr. Williams.
Hotel: The *New Hotel*.

Calais.—France, Department of Pas de Calais, opposite Dover.
Sea-baths.
Calais is a neat little town, built in the shape of an oblong parallelogram. It has wide streets, with houses built mostly of brick.
Post and Telegraph Office: Rue des Thermes.
Doctor: Dr. Daringhem.
Hotels: The *Hôtel Dessein* is a good old first-class hotel for families, well situated and thoroughly well managed.
The *Hôtel de Londres*, a good second-class house in close proximity to the railway station. Sainsard, proprietor.

Caldanella di Campiglia.—Italy, Tuscany.
Saline waters.

Caldaniccia.—France, Corsica, near Ajaccio.
Sulphate of soda waters, 90° F.

Caldas da Nossa Senhora do Pronto.—Portugal, Province of Beira.
Sulphurous waters, from 80° to 90° F.

Caldas (Las) de Besaya.—Spain, Province of Santander.
Saline springs, containing chlorides, 90° to 95° F.
Special indications: Rheumatism, scrofula, gastralgia, blenorrhœa, leucorrhœa, ophthalmia.
The establishment is moderately good. Accommodation abundant and rather dear in price. There is a kitchen in which visitors can prepare their own meals.
Doctor: Dr. Villafranca.

Caldas (Las) de Bohi.—Spain, Province of Lérida.
Thermal springs, 100° to 110° F.
Composition various, chiefly sulphate of lime.
Doctor: Dr. Pano.

Caldas (Las) de Cuntis.—Spain, Province of Pontevedra. For routes see table.
Twenty sulphate of soda springs, from 95° to 145° F.
Special indications: Herpes and rheumatism.
Doctor: Dr. Ortega.

Caldas (Las) de Estrac.—Spain, Province of Barcelona, near Mataró.
Saline springs from 95° to 105° F.
Special indications: Rheumatism and gravel.
The establishment is comfortable and elegant, and well fitted with modern appliances.
Doctor: Dr. Calvo.

Caldas (Las) de Gerez.—Portugal, Province of Entre Douro and Minho.
Sulphurous waters, 110° to 115° F.

Caldas (Las) de Malavella.—Spain, Province of Jiron.
Alkaline waters, 145° F.
Special indications: Rheumatism and paralysis, chronic catarrhs, asthma, nervous affections of the stomach.
The establishment is adequate and accommodation abundant.
Doctor: Dr. Rabel.

Caldas (Las) de Mombuy.—Spain, Province of Barcelona, near Granollers.
Saline waters, from 145° to 160° F.
Special indications: **Paralysis**, rheumatism, gout, syphilis and leucorrhœa.
There are eight different establishments, some of **which are very elegant and comfortable.** Accommodation very **abundant, and reasonable in price.** The site is **charming, and the place is much frequented by** the inhabitants of Barcelona.
Doctor: Dr. Lloget y Cailá

Caldas (Las) de Oviedo.—Spain, **Province of Oviedo.** For routes see table.
Carbonate of lime waters, containing nitrogen, 105° F.
Special indications: Pulmonary catarrh, consumption, rheumatism.
The establishment is elegant and modern. The environs are charming, and the climate is delicious. The accommodation is limited, but good, and prices are moderate. The site is very picturesque.
Doctor: Dr. Bouilla.

Caldas (Las) de Rainha.—Portugal, Province of Estremadura.
Saline waters, 80° F. One of the most largely frequented of **Portuguese watering places.**

Caldas (Las) de Reyes.—Spain, Province of Pontevedra. For routes see table.
    Sulphurous saline waters, 80° F.
    Special indications: Rheumatism, paralysis, and herpes.
    Doctor: Dr. Mosquera.

Caldas (Las) de Tuy.—Spain, Province of Pontevedra, Galicia, on the Minho.
    Sulphurous-saline waters, 112° F.
    Special indications: Rheumatism, paralysis, gout.
    The establishment is backward, but the site charming.
    Doctor: Dr. Acosta.

Caldas (Las) Novas.—South America, Brazil, near Santa Cruz.
    Thermal waters.

Caldiero.—Italy, between Vicenza and Verona.
    Alkali-saline waters, 75° to 80° F.
    Special indications: Constipation, **obesity**, liver and spleen disorders.
    A very picturesque site, with many excursions. The springs are **surrounded by walls, and form** a small lake. No accommodation, and but **few visitors.**

Caldillas de San Miguel.—Spain, Province of Salamanca.
    Saline waters, 65° F.

Caldini.—Italy, Tuscany.
    Effervescent alkaline ferruginous waters, 90° to 95° F.
    No establishment.

Calliano.—Italy, Piedmont.
    Sulphate of lime waters, 40° F.

Calliarhoë.—Palestine, North of the Dead Sea.
    Thermal sulphurous waters.
    Special indications: Elephantiasis.

Calvello, or Acqua dei Poggetti.—Italy, Tuscany, in the valley of the Ombrone.
    Effervescent alkaline waters, 90° F.
    The country around is barren and uncultivated. The springs are little used, and that only by the surrounding inhabitants, as a protective against malaria.

Calw.—Germany, Würtemberg.
    Sulphurous ferruginous waters.
    A small drinking establishment.
    Very little known.

Camarès.—France, Department of Aveyron, near Sylvanès.
    Alkali-saline-ferruginous waters. Exported.
    There is a good establishment and comfortable accommodation.

Cambo.—France, department of Basses Pyrénées three hours from Bayonne per diligence; 2,000 inhabitants; 180 feet above sea-level. For routes see table.

One sulphurous alkaline spring, with a temperature of **80° F.** is rich in carbonic acid, sulphate of lime, magnesia, soda, arsenic **and lithium**. Ferruginous waters, at 60° F., effervescent, containing arsenic **and phosphates**, are more **acid**. Climatic and winter station.

Season : April to June, and September and **October**; but open all the year.

Special indications: **Chronic** affections of the **respiratory organs, bronchitis, emphysema**, latent phthisis, **catarrh of of** stomach and urinary organs, lithiasis and liver affections, **herpes**, scrofula, chlorosis, anæmia, cachexia, and during **convalescence**.

The situation of Cambo is charming, and the excursions and promenades numerous. The establishment is complete, containing a large swimming bath, douches, spray and inhalation rooms. It **is** open all **the** year round.

Post and Telegraph.

Doctor : Dr. Lescamel.

Hotels : **Grand Hotel d'***Angleterre*, open all the year round, and recommended for its **comfort**. A. Saint-Martin, proprietor.

The *Hotel de Paris* **and** *de Londres*, first-class houses, facing **the** establishments. **Constant** Colbert, proprietor

Camoins.—France, **Bouches du Rhone, near** Marseilles.

Sulphate of lime springs, **35° to 45°** F.

Campagne.—France, Department of Aude, near Limoux.

Two ferruginous springs, **from** 60° to **70°** F.

Special indications : Dyspepsia, intermittent fever.

Campbeltown.—Scotland, Argyllshire.

Sea baths.

Much frequented on account **of its** mild climate and well-arranged **establishments**.

Campfer.—Switzerland, Engadine, half-hour's drive from St. Moritz village, and quarter **of** an hour from St. Moritz baths. **5,792** feet above sea level.

Climatic **station** in **summer, noted through its** tonic and stimulating **air**.

Special indications: Anæmia, **chlorosis, nervous complaints, failing powers of nutrition, phthisis**.

The place is well sheltered against **all northerly** winds. It is frequented chiefly by patients taking the waters at St. Moritz. **It is** especially suitable to persons of nervous temperament, being quiet and retired.

Doctors: Drs. Drummond and Duncan.
Hotel: The *Hotel d'Angleterre* is a very comfortable first-class house; lawn tennis grounds. All rooms are fitted with fire-places. Well suited to patients requiring quietude. A. Slanina, proprietor, as also of the *Hotel Square Brougham*, at Cannes.

Canal-grosso.—Italy, Tuscany, one hour from Calice.
Saline sulphurous waters, 50° F.
A small bathing establishment.

Canaveilles, or Grans d'Olette.—France, Department of Pyrénées Orientales, near Prades.
Sulphate of soda waters, 130° F.

Candé.—France, Department of Vienne, near Loudun.
Waters contain sulphate of iron and silica (0·2988 siliceous acid). Their action is laxative, resolvent, tonic and diuretic.

Candin.—Spain, Province of Leon.
Bicarbonate of iron waters: action purgative.

Canena.—Spain, Province of Jaen.
Bicarbonate of iron waters.

Canillejas.—Spain, Province of Madrid.
Saline waters.

Cannes.—France, Department of Alpes Maritimes, on the Bay de la Napoule, 19,000 inhabitants. For routes see table.
A Winter station of first importance. Climate tonic and stimulating near the sea; sedative in the direction of Le Cannet and at other points removed from the sea. Sea baths in spring.
Season: October till May.
Number of Visitors: Above 16,000 annually—6,000 being the average number staying at any given time.
Special indications: Nervous debility, anæmia, chlorosis, phthisis, laryngitis, pharyngitis, rheumatism, paralysis, gout and diabetes.

The climate of Cannes is more stimulating and invigorating than that of either Mentone or San-Remo. The air is very dry and fogs are rare. The mean winter temperature is 50° F.; but from 10 a.m. to 3.30 p.m., on even the coldest day, the thermometer never falls below 55° F. Humidity of air is 65%; mean rainfall, 88; mean barometric pressure, 760 mm. Rainy days during season (seven months), 44; sunny days, 108. Snow is exceptionally rare, and remains only a few hours even in the coldest winters. Cannes is one of the least windy stations on the Riviera.

The reputation of Cannes as a winter resort has been made since 1830—at least from an English point of view—when

Lord Brougham fixed his residence there. A few English families soon followed, and made the place known. To-day the original town is surrounded for a radius of three miles by villas, with beautiful gardens. One of the most fashionable quarters is still the West-end, with Lord Brougham's villa. The East-end, **with** some very **exten**sive gardens, is newer, but has also **some** magnificent residences, amon**gst** which the **Chateau Scott** may be especially mentioned. Both quarters are **in partial** proximity **to the** sea. Towards Le Cannet is the inland **quarter, and that is** the more sheltered of the two as regards atmospheric disturbances. The west and the extreme east are on porphyry soil, which retains its heat far into the winter, and **consequently** produces earlier vegetation in spring.

Cannes, in a word, may be said to be the *"pied à terre,"* **during** winter, of the English and French aristocracy. They have given the town a peculiar position among Mediterranean stations. No fewer than 80 English families hold property in **the** locality, and every year witnesses an increase in their number.

The drinking water has become excellent **since the construction of the Siagno canal.** The drainage, a **sore** point of late years, **has been the object** of considerable attention, both from the **municipality and the** medical faculty. These bodies, fully aware **of the importance of the matter, have** combined, with the **assistance and** advice **of** Captain Douglas Galton, in subjecting it to very material improvements, so that before long the drainage of Cannes will be as perfect as that of any English **town**.

The environs of Cannes are very interesting, and excursions to the Esterel mountains, the Lerins, &c., are well worth making. **Amuse**ments are not **in general of** a public character, but are, **to a great extent,** confined to family parties and *soirées*, and, **in these respects,** Cannes has no rival on the Riviera.

For all information address Mons F. Moutou, president of the Society of Local Interest and of the **race** course committee at his agency, Place des Isles

The artistic pottery manufactory **of** Clement Massier is at Golfe-Juan.

English churches: Christ Church, at the West-end; Holy Trinity, at the East-end; and St. Paul's, on **the** Boulevard du Cannet.

Post and Telegraph office: Rue Bossu.

Estate Agents and Bankers: John Taylor & Riddett, 43, Rue Fréjus, wine merchants. Established 1864.

Dentist: A. Preterre.

English **bookseller:** Fotheringham, 59, Rue d'Antibes.

Chemist: Isaac B. Ginner, 40, Rue d'Antibes.

Bankers: Crédit Lyonnais, Boulevard de la Croisette.

English **Grocer** *and* *Provision Dealer:* Buchillon & Co., Rue Central.

Doctors: Drs. Frank, Williams, Bright, de Valcourt, Battersby, Stephens, Whitely and Guimbert.

Hotels: The *Beausite* and *Esterel* hotels, 300 rooms, beautifully situated in an extensive garden; first-rate lawn tennis ground. Geo. Gougoltz, proprietor.

The *Hôtel Continental* is a large and first-class house, with a winter garden, beautifully situated, overlooking the bay. A sumptuous establishment. Chas. Bodemer, manager.

The *Hôtel de la Californie*; first-class house; extensive view from its large gardens. A. Chabassière, proprietor.

Cannstadt.—Germany, Würtemburg, near Stuttgart; 750 feet above the level of the sea. For route see table.

Forty different springs of saline and chalybeate waters, 50° F. These waters are stimulating, laxative, tonic, and diuretic.

Dr. Fischer's sanatorium for nervous complaints.

The climate of the valley of Cannstadt is mild and equable. Accommodation good.

Special indications: Anæmia and general weakness of the digestive organs.

Doctors: Drs. Blezinger, Pantlen, Rühle and Veiel.

Hotel: Grand Hôtel Herman.

Cap d'Antibes.—France, Department of Alpes Maritimes; 30 minutes from Cannes and 1½ hours from Nice; ½ hour's drive from Antibes Station.

Winter station and sea baths, with fine beach.

Season: November—June, especially autumn and spring months for the sea bathing.

Special indications: Anæmia, chlorosis, emphysema, tuberculosis.

The situation is superb, and the gardens are very fine. A perfect and quiet winter resort.

Doctors: As at Cannes.

Hotel: The *Grand Hôtel du Cap*, first-class; 200 rooms and salons. Omnibus at all trains; private carriages. Ph. Jaceard, director.

Caprenne.—Italy, Tuscany, district of Castiglione Ubertini.

Ferruginous alkaline waters, 50° F.

No bathing arrangements, and the waters are used only in a primitive fashion by the surrounding peasantry.

Capri.—Italy, island in the Bay of Naples. For routes see table.

Winter station.

Air very pure and healthy.

Special indications: Bronchitis, phthisis, chronic pneumonia.

Doctor: Dr. Cerio.

Hotels: Quisisana, de France.

Caprifico di Valaspra.—Italy, Tuscany, between Casale and Fercole.
Alkali-saline-ferruginous waters, 50° F.
A very rough and barren country. No establishments.

Capvern.—France, Hautes Pyrénées. From the station to the establishment ½ hour by omnibus. For routes see table.
Sulphate of lime waters, from 55° to 60° F.
Special indications: Gravel and gout.
Doctor: Dr. Delfau.
Hotel: The *Grand*.

Carbagnal.—Spain, ¾ hour from Valencia.
Sea baths, sandy beach.
The most fashionable of Spanish sea bathing places. What Biarritz was under the Second Empire, Carbagnal is to-day in Spain. The baths form but a minor feature, fashionable people and elegant toilettes being the main attractions.

Carballino.—Spain, Province of Orense.
Thermal waters, containing sulphate of soda, 80° to 85° F.
The establishment is very primitive.
Doctor: Dr. Barrio.

Carballo.—Spain, Province of Coruña, near Santiago.
Sulphurous saline waters, 90° F.
Special indications: Skin diseases, rheumatism, scrofula, chronic gastralgias and vesical catarrhs.
The establishment is moderate. Accommodation abundant, ranging from 10 fs. to 20 fs. a day, *en pension*.
Doctor: Dr. Castells.

Carcanières.—France, Department of Ariége, near Foix.
Thirteen sulphate of soda springs. The management of these springs is by no means perfect.

Carlsbrunn.—Austria, Silesia, near Freudenthal and Troppau; 2,380 feet above sea-level.
Highly effervescing, ferruginous manganese waters, 45° F. climatic station.
Season: June, September.
Special indications: Debility of the generative organs in both sexes, sterility, impotence, affections of the brain due to overwork.
The situation is romantic and charming. Vegetation very luxuriant. The accommodation and bathing establishments are well up to the level of modern requirements.
Doctor: Dr. Steinschneider.

Carlsruhe.—Germany, capital of the Grand Duchy of Baden, a few miles east of the Rhine. From Paris by East of France Railway, *via* Strasburg and Kehl; or by North of France Railway, *via* Erquelines, Namur, Cologne, Mayence and Mannheim.

One of the most attractive capitals so far as situation and external appearances go. The Ducal palace is the centre, from whence the streets radiate in all directions.

OBJECTS OF INTEREST.—The Palace, with a Winter Garden 420 feet long, Conservatories and Botanical Gardens. The Kunsthalle, Polytechnic school, Friedrich's Platz, with united collections of paintings, Museum of natural history, &c. Catholic church.

Doctors : Drs. Battlehner, Gutsche, Maier (oculist), and Tenner.

Hotels : Erbprinz and *Englischer Hof.*

Carnarvon.—Wales, Carnarvonshire.
Sea baths, sandy beach.
Doctor : Dr. Davies.

Carratraca or Ardales.—Spain, Province of Málaga, near Campillos.
Sulphurous waters, 64° F., contains arsenic.
Special indications : Skin diseases and affections of the respiratory organs.
A delicious climate, fine situation but imperfect arrangements.
Doctor : Dr. Salgado y Guillermo.

Carteret.—France, Department of La Manche, from Paris *via* Valognes.
Sea-baths ; sandy beach.

Casale.—Italy, Tuscany, valley of Cecina, near Bibbona.
Acidulous bitter waters. Exported.
Special indications : Affections of the urinary system, congestions of the head, giddiness, noises in the ears, deafness, affections of the eyes and chest, palpitation, dyspnœa, liver and spleen complaints, hypochondriasis and hæmorrhoids.

Casamicciola.—Italy, Island of Ischia.
Indifferent waters of Gurgitello, 100° to 160° F.
Climatic station.
Season : May—September.
Special indications : Diabetes, nervous complaints, scrofula.
Doctor : Dr. Reitemeyer.
The bathing establishment and entire town were completely destroyed by the earthquake of July, 1883.

Casa Nuova.—Italy, Tuscany, near Triana.
Sulphurous waters, 50° F., with traces of iron.
No arrangements whatever for visitors.

Casares.—Spain, Province of Malaga, three hours from Gibraltar.
Effervescent sulphurous saline waters, 55° F.
There is an hospital, and also some bathing establishments, but they are wholly devoid of comfort. Another serious drawback to the growth of the town is the bad roads and inferior means of communication.

Casa-Stronchino.—Italy, Tuscany, valley of Modigliano.
Saline bromo-iodurated waters, 49° F.
Very strongly mineralised, and only used externally. Poor accommodation.

Caseiano (San).—Italy, Tuscany, Val d'Era, district of Montajone.
Sulphurous alkaline waters, 55° F.
Only used internally; they cannot be exported, as they soon become turbid.

Casino-delle Currigliane.—Italy, Val d'Arno, near Pontedera.
Saline waters, containing chloride of calcium in such large quantities that they can be used only externally.

Casola.—Italy, Tuscany, near the sources of the Magra.
Sulphurous-saline waters, 45° F.
The bathing establishment is in good condition, but visitors must not expect too much in the way of accommodation.

Casnefouls.—France, Department Aveyron.
Strong bicarbonate of iron waters.

Cassel.—Germany, former capital of the Electorate, and now the capital of the Prussian Province of Hesse-Nassau. From Paris by North of France Railway, *via* Cologne and Giessen.
Founded by the Landgrave Frederick II. The town is beautifully situated on both banks of the Fulda.
OBJECTS OF INTEREST.—The Ducal palace, Museum with its library and picture gallery, several churches, and many charitable and educational institutions.
EXCURSION: To Wilhelmshöhe, where Napoleon III. was detained in 1870-71. It is the Versailles of Germany, and is now owned by the Prussian Crown. This spot is well worth visiting. The waters play every Wednesday and Sunday. The highest fountain on the Continent is here, one column 12 in. in diameter being projected 200 ft. high. One of the most magnificent residences in Europe. The palace at the bottom of the hill. Colossal statue of Hercules and cascade of Karlsburg.
Dentist: Chas. Zimmer.
Doctors: Drs. Schmidt, Schotten and Lange.
Hotels: Hotel *König von Preussen* and *Schirmer*.

Castel-Connell.—Ireland, Limerick and Tipperary.
Strong chalybeate waters.
Special indications: Jaundice, bilious disorders, ulcers.
Doctor: Dr. Ryan.

Castel-Doria.—Island of Sardinia, near Busachi.
Sulphate of lime waters, 160° F.

Castel-jaloux.—France, Department Lot and Garonne, near Neyrac.
Bicarbonate of iron waters.

Castellamare di Stabia.—Italy, Province of Naples, on the southern coast of the Bay of Naples, looking northwards. 45 minutes by rail from Naples; 28,000 inhabitants. For routes see table.

Autumn and winter station. Sea baths. Cold chloride of sodium, bitter and sulphurous chalybeate waters.

Seasons: Sea and mineral bathing, principally from May till October. Winter season, October—April; the bathing establishments are open all the year.

Number of visitors: More than 10,000 visitors are reported annually.

Special indications: Obstructions of the liver and spleen, affections of the mesenteric glands, biliary and vesical calculi, jaundice, dropsy, hæmorrhoids, chronic ophthalmia, herpes, catarrh of the digestive organs, hypochondriasis, urinary calculi, vesical catarrh, scrofula, lymphatism, congestion of the uterus, leucorrhœa, blenorrhœa, &c.

Castellamare, built upon the ruins of the ancient Stabia, is beautifully situated at the foot of Monte St. Angelo, overlooking the Bay of Naples, Ischia and Capri. Pliny died here, 79 A.D. It has been recognised from very early times as a place of residence for sufferers from affections of the respiratory organs. Summer, spring and autumn are very dry, but the winters rather damp. As an illustration of the health-giving power of its climate, may be mentioned the fact that Ferdinand I., of Bourbon, built a castle here, which he named "*Qui si sana.*" The Bosco di Quisisana, or park surrounding this villa, has delightful walks and is beautifully laid out. It is open to visitors. The view from the terrace is charming.

Owing to its being open to the north wind, the hottest days in summer are as cool as October days are in Naples. The winter is mild. The sea and well-wooded slopes of Monte St. Angelo render the atmosphere of Castellamare tonic and stimulating. The town was called Stabiae as late as 1226, when Charles I. of Anjou built two castles close to the sea for purposes of defence, since which it has received its present name. The town consists principally of one main street, running for almost four miles along the sea, with a second one parallel to and behind it.

Numerous excursions may be made in the neighbourhood. The ascent of Monte St. Angelo, 5,000 feet high, and clothed almost to the summit with chestnut trees, is one of the most interesting. From the top a view of the whole bay and surrounding country is obtained (4 hours); 1½ hours to Sorrento; 20 minutes to Pompeii; 45 minutes to Naples; 1½

hours to Salerno; 3½ **hours to** Pæstum, and 3 **hours** to Amalfi. Any special information which **may be** required will **readily** be given by the Mayor of the town.
Amusements: Theatro Principe **Umberto;** three clubs (Union, Nazionale, dell' Independenza); public gardens, with daily orchestra; "Stabia Hall," the **rendezvous** of society, with Theatre, Ball, Concert, Reading, Card and **Refreshment rooms;** a verandah looking **seawards;** boating and fishing.

Post and **Telegraph** *Office:* Town Hall. see plan.

Bankers: **National** Bank, Corso **Vittoria** Emanuele 18.

English Church in Winter: Rev. **D.** Clark.

House and Estate Agent: Giornale di Stabia, Corso Vittoria Emanuele.

Bookseller: Canzanella, largo Municipio, 5.

Doctors: **Drs.** Olivieri, Gentile, Somma, Scherillo **and Fusco.**

Hotels: Hotel **Quisisana;** *Hotel* Stabia.

Castelleto-Mascagni.—Italy, Tuscany.
Bicarbonate, sulphurous and earthy waters, **55° F.**
Only used externally; accommodation poor.

Castelletto **d'Orba.**—Italy, Piedmont, Province of Acqui.
Sulphurous alkaline waters.

Castelnuovo.—Italy, Piedmont, near Asti.
Sulphurous and iodiue waters; the richest in iodine of any of the Piedmont springs.

Castel San Pietro dell' Emilia.—Italy, Province of **Bologna,** one hour from Bologna, carriages and omnibuses at the station.
Three springs of ferruginous, sulphurous and saline waters, 75° to 85° F. Natural steam baths.
Season: July—August.
Special indications: Chlorosis, anæmia, rheumatism, **stomachic and** intestinal disorders.
The establishment is deficient **in comfort,** but efforts have lately been made to put **it on a** better footing. There are some well-appointed hotels **in the town.** The country around **is romantic,** and abounds in **well-kept walks.**
Post and telegraph office, **theatrical** performances and concerts.
Doctors Drs. Nella and Gatti.

Castenar dei Bor.—Spain, Province **of Caceres.**
Sulphate of iron waters. Mixed **waters, 45° F.**

Castera-Verduzan.—France, Department of Gers, near Auch.
Sulphate of lime springs, 60° F.; one is a cold chalybeate spring.
Doctor: **Dr. Dupeyron.**

Castiglione.—Italy, island **of Ischia, near** Casamicciola.
Saline waters, 175° F.
Only used internally. Very fine bathing establishments existed here in ancient times, as shown by their ruins.

Castro-Caro.—Italy, Province of **Florence.** Rail to Forli, thence one hour's drive. Omnibus, 2 fcs. each **person.** Carriages, one horse, **5 fcs**; two horses, 7 fcs.
Saline-iodo-bromide waters, very rich in **iodine.** In this **respect,** indeed, they **are first among European iodurated** waters. Douches, **sitz, shower and steam** baths, **and** inhalations.
Season : June—September ; **but open** all the year round.
Special indications : Goitre, **scrofula,** rhachitis, obesity, diseases of the female organs.
Two establishments, with all modern appliances, afford **com**fortable accommodation at a pension of 7 fcs. per diem.
The village has 2,000 inhabitants, and is charmingly situated. Mild and healthy climate. The variety of amusement which forms the great attraction of many other bathing-places **is not to be** found here, but instead of it quietude and repose **which are** so valuable in restoring health when impaired by **city life.** The number of visitors is very considerable.

Catania.—Italy, Sicily, on the east coast. For routes see table.
A winter station of first importance.
Ferruginous waters at Paterna.
Season : Middle of November till end of March.
Mean temperature in winter, 55° to 60° F. ; mean dampness, 72.5 m. ; amount of rainfall, 42.5 ; rainy days during season, 42 ; cloudless days, 103 ; air dry. The thermometer never falls **below freezing** point.
The **drainage of** the town is emptied **into the sea.** The smells are **intolerable,** and the **sanitary condition of the place very** defective. Visitors generally reside at the back of the town.
Doctor : Dr. Joris.

Catena.—Italy, between Pisa and **Florence, on the** Evola.
Alkali-saline waters, **55° to 60° F.** Only used for baths.

Cati.—Spain, **near** Castellon de la Plana.
Carbonate of lime waters, 45° **F.**

Cattenaja.—Italy, **district** of Subiaco.
Highly effervescent ferruginous waters, 55° F.
No establishments and no accommodation.

Cauquedes.—Chili, near Santiago, by rail viâ Rancágua, 2½ hours by diligence.

Sulphurous and chalybeate waters, cold, tepid and hot; also climatic mountain air cure station.

The spot is very picturesque and rocky, being surrounded by the majestic ranges of the Andes, which are here most imposing. The establishment consists of two square buildings, are well constructed, and surrounded by a large garden with shady walks. They are well managed and largely frequented.

Cauterets.—France, Department of Hautes Pyrénées; 2,109 inhabitants; 3,050 feet above sea level. For routes see table.

Sulphurous saline waters, 55° to 145° F. Climatic mountain station; milk and whey cure; the waters are rich in silicates.

Season: 15th May to 15 October.

Number of Visitors: Above 20,000 annually

Special indications: Affections of the mucous membranes, catarrh of the respiratory organs, phthisis in first and second stage, plethora abdominalis, laryngitis and glandular disorders, affections of the digestive organs, skin diseases, hepatic and splenic, as also uterine derangements, rheumatism, mercurial intoxication, struma.

The balneal appliances of the three principal establishments, "Thermes des Œufs," "Grand Etablissement," and "La Raillère," are of the most modern and complete kind on the Continent. The chemical constituents of the waters being, moreover, very complex, the list of ailments treated in Cauterets is a long one.

The small town itself is composed of about 300 houses, most picturesquely situated in a valley surrounded by lofty mountains. The climate is changeable and somewhat humid, but the temperature is steady and the nights are delightfully cool and bracing. The sojourn is more agreeable here than at Luchon, where the heat in the day is almost unbearable. The waters and balneal appliances of Cauterets are also more efficacious and more complete.

There are two Casinos, with theatre, reading, ladies, concert, smoking and billiard-rooms, the one in the "Etablissement des Œufs" being the most popular.

There is a very large field for excursions of interest. The promenades are well kept and abundant.

English Church service.

Bookseller: A. Cazaux.
Chemist: Broca, Place St. Martin.
Doctors: Drs. Duhourcau, Bordenave and Bouyer.
Hotels: Hôtel Continental, de France.

Cauvalat-le-Vigan.—France, Department of **Gard**; 700 feet above sea level. For routes see table.
  Sulphate of lime springs.
  Doctor: Dr. Champtier.

Cave.—Italy, Tuscany, district of Vico Pisano, near Oliveto.
  Earthy saline waters, 86° F., highly effervescent.
  Special indications: Disorders of the urinary organs, gravel, calculus, chronic rheumatism, gout, skin diseases.

Caxamarca.—South America, Peru.
  Thermal waters, not yet analysed.
  It was at these baths that the Inca Atahualpa had fixed his residence when Pizarro first appeared in Peru.

Cayeux.—France, Department of Somme, near **Valéry sur Somme**.
  Sea baths, sandy beach; very quiet place.

Cayla (La).—France, Department of Aveyron, near Andabre.
  Cold bicarbonate of iron waters.
  Doctor: Dr. Martin.

Caz-di-Bagno.—Italy, **Velteline**, valley of Masino.
  Sulphurous alkaline waters, 95° F.
  The bathing establishment is old-fashioned, and the place but little visited.

Cecinella.—Italy, Tuscany.
  Sulphurous ferruginous waters, 55° F.
  No bathing establishments, and waters used only by the surrounding peasantry.

Cedres (Sources des).—France, Algiers, Department of Oran.
  Sulphate of iron waters.

Cefalu.—Italy, Sicily, 12 miles from Sclafani.
  Acidulous bitter waters, 135° F.
  Very efficacious, though but little used.

Celles.—France, **Department of** Ardèche, near **Privas**. For routes see table.
  Four bicarbonate springs, from 40° to 60° F. One sulphate of iron spring. Carbonic acid gas is here employed in douches and inhalations.

Ceresole.—Italy, Piedmont.
  Bicarbonate of iron waters, very rich in carbonic acid.

Cervera del Rio Alhamo.—Spain, Province of Logroño, near Castejon.
Sulphurous earthy waters.
Special indications: Skin diseases, scrofula, syphilis, affections of the mucous membranes of the respiratory organs, vesical catarrhs, blenorrhœa.
The establishment is modern and the accommodation comfortable and moderate in price. The situation is fine.
Doctor: Dr. Escudero.

Cesalpine.—Italy, Tuscany, near Arrezzo Val di Chiana.
Ferruginous waters, 55° F.

Cestona-guezalaga.—Spain, Province of Guipuzcoa.
Saline springs, rich in nitrogen. Tonic and laxative.
Special indications: Dyspepsia and gastric catarrh, sciatica, hypochondriasis.
The establishment is very good, with all modern appliances. The accommodation and food are equally good. Large park.
Doctor: Dr. Calderon.

Cetona.—Italy, Tuscany, near Chiusi.
Ferruginous waters, 55° F.

Ceylon.—Island of.
Several thermal springs, from 90° to 100° F., composition unknown.

Chabetout.—France, Puy de Dôme, near Issoire.
Bicarbonate of iron waters.
Mineral constituents, 3 grammes.

Chaldette—France, Lozére, near Chaudes-aigues.
Saline waters, 95° F.
A very good establishment, but the accommodation is indifferent.

Challes.—France, Savoy, near Chambéry, ¾ hour from Aix-les-Bains.
Cold sulphurous iodurated waters.
These waters are frequently prescribed by medical men at Aix-les-Bains, as adjuvants to those of Aix. They have proved specially useful in scrofula and syphilis, as also in locomotor ataxy.
Doctor: Dr. Royer.

Chambon.—France, Department of Puy de Dôme, near Ambert.
Cold mixed bicarbonate waters.

Chamounix or Chamonix.—France, Upper Savoy, at the foot of Mont-Blanc; 3,450 feet above sea level; 4,500 inhabitants. For routes see table.

Climatic mountain air station and holiday resort; some earthy waters and the whey cure.

Season: July 1st—September 15th.

Number of visitors about 20,000 annually.

Chamounix, since the fire of 1855, is quite a new place, with Museum, bathing establishment, and a handsome church. The scenery is among the grandest in Switzerland. The town is the starting point for almost all ascents of Mont Blanc. Excursions to the Montanvert, Jardin and Mer de Glace, the Flégère, Brevier, and Glacier des Bossons, Tête Noir, Col du Géant and Martigny.

The new Casino, opened recently, contains ball, concert, theatre, card, reading and smoking rooms.

English church.

Post and Telegraph office.

Photographs, Bookseller and curiosity shop: Jos. Paccard, Magasin des Souvenirs, route de Genève.

Doctor: Dr. Duchosal.

Hotels: The *Hôtels de Londres et d'Angleterre*, first-class, well appointed for families. T. Crepaux fils ainé, proprietor.

The *Alpine Club Hotel et Pension Union*, an excellent house, open all the year. Edw. Jutz, director.

The *Hôtel et Pension Couttet* is a very comfortable hotel; the proprietor, Francois Couttet, known as "Baguette," is an old guide, and has made the ascent of Mont Blanc 27 times.

The *Hôtel de la Couronne* at Argentières, one hour from Chamounix is a modest house, well situated for excursions, and faces the Glacier of Argentières. J. Tissay, proprietor.

Champoleon.—France, Department of Hautes Alpes, near Embrun.

Sulphate of lime water.

Chapelle-Godefroy (La).—France, Department of Aube, near Nogenton, left bank of the Seine.

Cold bicarbonate of iron water, 6 grs. of mineral constituents.

Charbonnieres.—France, Rhone, near Lyons.

Cold bicarbonate of iron, sulphurous and iodurated waters.

Charlottenbrunn.—Germany, Silesia, district of Breslau; 1,300 feet above sea level.

Cold bicarbonate of iron waters.

Doctors: Drs. Bujakowsky and Neisser.

Hotel: Kurhaus.

Charlottenburg.—Germany, Prussia, Province of Brandenburg, half-an-hour from Berlin.
Earthy, saline and ferruginous waters Hydropathic establishment. Open all the year round.

Chateau-Gontier.—France, Mayenne.
Effervescent ferruginous waters.

Chateauneuf.—France, Department of Puy de Dôme, near Riome; 1,200 feet above sea level.
Cold, tepid and warm ferruginous waters; 14 springs, varying from 40° to 110° F. Their composition is similar to that of the springs of St. Alban.
Special indications: Dyspepsia, scrofula, rheumatism.
Doctor: Dr. Rainvillers.

Chateldon.—France, Department of Puy de Dôme, near Thiers and Vichy.
Cold bicarbonate of iron waters; five springs, containing 3, 4 and 5 grs. of mineral constituents.
Special indications: Dyspepsia.

Chatel-Guyon.—France, Puy de Dôme, near Riom.
Saline and ferruginous waters.
Seven springs, from 55 to 110° F.
These waters are slightly purgative.
Special indications: Liver complaints, splenic and glandular derangements.
Doctor: Dr. Baraduc.

Chatenois.—Germany, Alsatia, near Schlettstadt.
Cold saline waters, 7 gr. of mineral matter, very similar to the waters of Niederbronn; traces of arsenic and fluoride have been found in them.
Doctor: Dr. Levy.

Chaudefontaine.—Belgium, near Liége.
Thermal springs, assimilating to those of Spa, 80° to 90° F.
A charmingly situated quiet little place, with a great variety of excursions. The country around is very picturesque.

Chaudes-Aigues.—France, Department of Cantal, near St. Flour.
Alkaline waters, containing iron, sulphur and arsenic. Several hyperthermal springs, from 135° to 195° F.
Special indications: Articular rheumatism, and generally same as those of Bourbon Laney.

Chaumont.—France, Department of Marne and Loire.
Cold bicarbonate and other waters.

Chaves.—Portugal, Province of Traz-os Montes.
Thermal sulphurous spring, 135° F.

Chazam.—Algiers, Province of Oran, near Mostanagem.
Saline waters.

Cheltenham.—England, Gloucestershire.
Several springs, saline and sulphurous, sulphate of magnesia, mixed saline, carbonate and chalybeate waters. Climate mild and agreeable

Season: November to April.

Special indications: Anæmia, dyspepsia, abdominal obstructions and hepatic enlargement.

The bathing establishments and assembly-rooms are justly renowned for their elegance and comfort. They are among the best in England.

The town is laid out in boulevards, crescents and squares, planted with fine trees. It more resembles a collection of country seats than a town in the general sense of the word.

Dentist: Mr. Cull.
Doctors: Drs. Abercrombie, Bennett and Cottle.
Hotels: Royal, Queen, Belle Vue.

Chemille. France, Department of Marne and Loire.
Ferruginous waters little known.

Cherbourg.—France, Department of La Manche.
Sea baths; sandy beach.

Chianciano.—Italy, Tuscany, District of Montepulciano, in the valley of the Chiana Railway from Siena to Rome, ½-hour's carriage from Asciano; 1,850 feet above sea level; 3,500 inhabitants. For routes see table.

Sulphurous waters, 100° F.; ferruginous effervescent waters, 45° F.; alkaline, acidulous waters, 85° F.; hydropathic establishment and cool and delightful climatic station in summer.

Season: June to September.

Special indications: Disorders of liver, spleen, stomach and intestines, for the alkaline waters; anæmia, atonic sores, chronic catarrh of the mucous membranes, difficult menstruation, leucorrhœa and convalescence, for the ferruginous; catarrhal affections of the respiratory organs, bronchial catarrh, scrofula, rheumatic paralysis and skin diseases, fractures, anchylosis and paraplegia for the sulphurous waters.

Chianciano is situated in a delightful valley, about half-an-hour from Chiusi. The environs offer a great many attractions with numerous walks and drives.

Carriages attend the trains during the season. Post and Telegraph Office. All applications should be directed to Luigi Innocenti, lessee of the baths, who gives information gratis. The waters are to be had at all the best hotels and mineral water merchants throughout Italy.

Doctor: Dr. Burresi.
Hotels: The prices for accommodation and comfortable living may be had by applying to Luigi Innocenti as above.

Chiatamone.—Italy, Naples.
Ferruginous waters, 68° F.
Special indications: Anæmia, general debility, chronic affections of the **nervous** system, neuralgia, chlorosis.
The establishment is fitted with all modern **requirements;** one of its best features being the Hammam or Turkish baths. Swimming bath.
A café is attached to the baths. Concerts every evening.

Chiavari.—Italy, Genoa, two hours from Genoa.
A climatic station. Especially frequented in the later summer months and in autumn.

Chichimequilo.—Mexico.
Hyperthermal waters, 230° F., not analysed.

Chiclana.—Spain, near Cadiz.
Bitter waters, 45° F. Very highly mineralised.
Special indications: Herpes, mercurialism, skin diseases.
Good establishment, and moderately frequented.
Doctor: Dr. Costina.

Chinciano.—Italy, Tuscany, **Val d'Elsa,** near Siena.
Highly effervescent alkaline and ferruginous waters.
Special indications Amenorrhœa, anæmia, chlorosis, gravel, vesical calculi, and generally affections of the digestive and urinary **system.**
Establishments indifferent.

Chios.—An Island in the Grecian Archipelago.
Sulphate of lime waters, purgative. Thermal waters of Tsesmé, from 110° to 130° F., contain chlorine, bromine and iodine.
No bathing establishment.

Chitignano.—Italy, Arezzo.
Effervescent alkali-saline waters, containing some iron, 55° F.

Chiusa dei Monaci.—Italy, near Arrezzo.
Alkali-saline and ferruginous waters, 55° F.
The establishment is poor and the accommodation bad.

Choranche.—France, Department of Isère.
Cold sulphurous lime-springs.

Christenhofsbad.—Germany, Würtemberg, District of the Jaxt, near Mögglingen.
Sulphurous waters.
Internally, diuretic and laxative; externally, stimulating to the skin.

Chulilla.—Spain, Province of Valencia, near Villar del Arzobispo.
  Sulphurous earthy waters, 95° F.
  Special indications: Rheumatism, skin diseases, gout, paralysis.
  The establishment is simple but thorough. Accommodation good. The country is mountainous.
  Doctor: Dr. Merregner.

Ciechocinek.—Russia, Poland, near the Prussian frontier, one hour from Thorn.
  Very strong saline waters and brine baths.
  Season: May 20th—September.
  Visited by upwards of 2,000 visitors.
  Special indications: Scrofula, rheumatism, catarrh of the larynx and bronchi, neuroses and hysteria.
  The bathing establishments are completely fitted with all modern appliances.
  The casinos and assembly rooms are elegant and comfortable. The accommodation is first class.

Cipollo.—Italy, Tuscany, Val di Nievole.
  Saline waters, 70° F.
  Only applied externally.

Citára.—Italy, Island of Ischia, near the Capo Imperatore.
  Sulphurous alkaline waters.
  Special indications: Sterility.
  Known to and much visited by the ancient Romans, who erected a temple here to Venus Genitrix.
  The earthquake of July, 1883, made great havoc here.

Civilliano.—Italy, Venetia, between Vicenza and Verona.
  Cold ferruginous and sulphurous waters.
  Special indications: Affections of the mucous membranes, scrofula and chronic diarrhœa.
  One of the strongest sulphate of iron waters at present known.

Civita Vecchia.—Italy, the seaport of Rome.
  Thermal sulphurous waters, natural sweating baths.

Clacton-on-Sea.—England, Essex, near Walton-on-the-Naze.
  Season: July—October.
  Sea baths; shingly and sandy beach. Climate mild and bracing.
  Doctor: Dr. Maine.

Claremont Park.—England, Lancashire, near Blackpool.
  Hydropathic establishment.

Cleethorpes.—England, Lincolnshire, on the Humber.
  Sea baths; sandy beach. Climate healthy.
  Season: June—September.

Clermont.—France, Department of Puy de Dôme.
Numerous ferruginous bicarbonate springs, from 35° to 55° F. Mineralisation from 4 to 5 grammes. The spring " Puits de la Poix " is said to be bituminous.

Cleves.—Germany, Rhenish Prussia, on the Rhine, near the Dutch frontier, a very old town.
Chalybeate waters, fine Kurhaus, and well-appointed bathing establishment.
Special indications: Nervous complaints.
Doctor: Dr. Arntz, who speaks English.
Hotels: Bath Hotel, Styrum.

Clevedon.—England, Somersetshire, on the Severn.
Sea baths; pebbly and muddy beach; bathing not very good.
Season · June—September
Doctor: Dr. Pizey.

Clifton.—England, Gloucestershire, a beautiful and fashionable suburb of Bristol; 27,000 inhabitants.
Autumn and winter residence; alkaline waters 75° F.; much frequented by Indian and colonial families coming home to recruit their health, and for educational purposes.
Season: September till May, but frequented all the year.
Climate warm and dry: most part well sheltered and open to the south; variations of climate may be obtained according to locality chosen, the Downs being considered as bracing as the seaside, being the most elevated part, and only a few miles from the sea.
Special indications: Diabetes, liver obstructions, splenic and urinary disorders, for the waters; consumption, pulmonary complaints, anæmia, chlorosis, debility, bronchial, laryngial, and pharyngial and kindred ailments.
Clifton is specially noted for the longevity of its inhabitants and lowness of death rate, which ranges from 3·6 to 11 per thousand, being the more remarkable, owing to the district being the resort of a very large number of invalids. It is, therefore, looked upon as one of the most healthy parts of England.
The grand plateau of Clifton and Durham Downs are no less than 442 acres in extent; Clifton Downs 230, and Durham Downs 212 acres, extending for upwards of two miles along the banks of the picturesque river from the suspension-bridge to the sea-walls. The prevailing westerly breezes are health-giving direct from the Channel.
The late Mr. Stoddart, County Analyst, found ozone in the air of the Downs. The waters of the Channel can be seen

CLIFTON.

from the sea-walls, where the surrounding scenery is most extensive and beautiful, extending far down the Welsh coast and hills, and in the other direction beyond the Mendip Hill.

These magnificent Downs have been secured to the city by Act of Parliament in 1861. They are the favourite resort for driving, riding, cricketing, and other amusements.

Along the edge of St. Vincent's Cliff, the old Camp on Observatory Hill, and the delightful plateau beyond, appropriately named "Fairyland," are many points whence the grand scenery of the gorge of the Avon is enjoyed. But it is the suspension-bridge, upon which visitors can stand, as it were, in mid air, between the stupendous cliffs on the one side, and shelving rocks, mantled with soft foliage on the other side of the chasm, and take in the surroundings of a scene which never fails to inspire the beholder with a wondering sense of the Titanic forces that must one day have been exerted to form this mighty cleft in the limestone rocks. Suspended amid such giant cliffs, at the giddy height of 287 feet above the river bed, the aërial pathway linking rock to rock at a span of 703 feet, suspended by gossamer-like rods, and looking at a distance as fragile as the strings of a lady's harp, it seems more like the magic work of a fairy's dream than a reality. Numerous excursions can be made in the neighbourhood to most picturesque and charming spots; the long drives being particularly noted for their beauty and extensive scenery.

The amusements consist of theatres, balls, concerts; the Monday Popular Concert being noted for its almost unrivalled band. There are the Zoological Gardens, with a large selection of animals and beautiful grounds. Few places can boast of greater advantages as to educational establishments. The Clifton College, in the heart of Clifton, a noble building, with large grounds. It has won a high rank among similar institutions in the country for high-class training. The Grammar School, a large building with extensive grounds, is also noted for its classical and commercial training. The High School for Girls is a new institution, and already distinguished. A large number of private schools for boys and girls.

Turkish Baths and hydropathic establishment, 14, **Royal Promenade**. Chas. Carpenter, proprietor; see also **Bishops Teignton**, Devon.

Chemists : Ferris & **Co., Chemists** by appointment to the Queen, 4 & 5, Union Street, **Bristol**.

Bookseller and Stationer : Thomas Thatcher, 44, College Green, **and Suspension Bridge**.

Estate and House Agents : De Ridder & Co., 54, White Ladies Road, from whom all information may be obtained.

Doctors : Dr. Eubulus Williams, Victoria Square.

Hotels : The *Clifton Down Hotel*, over-looking the Downs and suspension-bridge. All produce for this establishment is supplied from the farm belonging to the hotel. Address, manager.

St. Vincent's Rocks Hotel, Sion Hill, Clifton Downs.

Boarding Houses : Lyndhurst—22, Pembroke Road, Mrs. Hancock ; Arlington House, Pembroke Road, Miss Smith.

Clynnog Vawr.—Wales, Carnarvonshire.
  Sea baths; beach and bathing good.
  Season : Summer.
  Doctor : Dr. Roberts

Coblentz.—Germany, capital of the Province of Rhenish-Prussia on the confluence of the Moselle and Lahn with the Rhine. From Paris by North of France Railway, *viâ* Cologne, 12½ hours.
  A fortified **town** of great reputation Ehrenbreitstein **(the** Gibraltar of the Rhine) is one of the strongest fortresses in **the** world. Very charmingly situated.
  OBJECTS OF INTEREST.—The Archbishop's Castle, or Burg, **now** a factory, Church of St. Castor, Metternich House and Teutonic House, now a magazine. The Royal Palace, the favourite resort **of the** Queens of Prussia.
  Casino and promenades on the banks of the Rhine.
  One half-hour's walk along the banks **of the Rhine is the celebrated establishment of** Lanbbach (see **page 175).**
  EXCURSIONS: To Stolzenfels, Ehrenbreitstein, to Sayn, **to Moselweiss, and** up the Moselle, **to Ems,** *viâ* the Asterstein and Oelberg, to **Neuwied.**
  CABS: One-horse, for one or two persons, 50 pfg.; for two-horse, 75 pfg; two-horse carriage, three **or** four persons, 1 mark 25 pfg; one-horse carriage, 1 mark.
  Post and Telegraph Office : Brückenstrasse, near the Schlossplatz.
  Bookseller : W. Fischer, railway station.
  Dentist : F O. Saal.
  Doctors : Drs. Kirchgaesser, Timmé, **and** Hermann.
  Hotels : Belle Vue, Anchor.

Cocomiso.—South America, New Granada.
  Sulphurous and bicarbonate of soda waters, **175°** F.
  This spring emits such quantities of carbonic and sulphurous acids, that a long stay in its vicinity becomes dangerous.

Coire, or Chur.—Switzerland, Canton of Grisons; **7,550** inhabitants; **1,936 feet above** sea level. For routes see table.
  Climatic station, but more as **a** transitory station from and to the Engadine.
  The town **is a very picturesque old** place, on the Splügen road, and **two miles from the Rhine.** There are some very fine old buildings, amongst which the most noteworthy is the Cathedral of St. Lucins, **with an** adjoining palace of the 8th century; town hall, goods house, museum, hospital, &c. It is the terminus of the railway from Zürich to the Engadine. Some very interesting excursions may be made in the environs.
  The diligences start daily for Davos, Samaden, St. Moritz, **Maloja,** Dissentis, Andermath, &c.
  English Church service.
  Doctors : **Drs.** Deuz and Henni.

Hotel: The *Hotel Steinbock*, a really first-class house in all respects, and deservedly popular. Hauser, Keim & Co., proprietors.

Coise.—France, Savoy, near Chambéry.
Ammoniacal, bromo-iodurated waters.
Special indications: Goitre and kindred affections
Doctor: Dr. Dubouloz.

Colberg.—Germany, Pomerania, near the Baltic.
Saline waters and sea baths.
The establishment is very well fitted up. Accommodation first rate. The walks and drives are well kept and shady.
Doctors: Drs. Bodenstein, von Bühnau and Hirschfeld.
Hotels: *Strandschloss, Society House.*

Colico.—Italy, upper end of Lake Como.
A climatic air station for those returning from Italy. Very fine position and grand scenery.

Collioure.—France, Department of Pyrénées Orientales.
Sulphate of lime waters, 45° F.

Cologne.—Germany, Rhenish-Prussia, on the left bank of the Rhine; from Paris by North of France Railway, 10 hours, 60 frs. 1st class.
Although Coblentz is the official capital of Rhenish Prussia, Cologne is the most important town in the province. It is the third most important in Prussia, and is strongly fortified. The old middle age fortifications are now demolished, and new ones built. Their circuit is nearly 30 miles. The circuit of the outer forts is about 50 miles. Cologne is the Roman "Colonia Agrippæ."
OBJECTS OF INTEREST.—The Cathedral of Cologne, the highest edifice in the world, was finished in 1880. The iron Railway Bridge, Museum and Library, Town Hall, Churches of St. Maria Capitol, St. Peter, Apostles, St. Ursula, St. Gereon; Casino Gurzenich, Chamber of commerce and Exchange, Theatre, Aquarium, Zoological and Botanical Gardens.
Post and Telegraph Office: Glockengasse.
CABS: **One** or two persons, ½ mark; four persons, 1 mark. By the hour, one or two persons, 2 marks; three **or** four persons, 3 marks; parcels outside, 25 pfgs. each.
Dentist: G. A. Buechner.
Bookseller: Aug. Lesimple, High-street.
Doctors: Drs. Bardenheuer, Schmitz (specialist for respiratory organs, eye and ear), Birnbaum, and Robt. Hall.
Hotels: The *Hôtel du Nord.* A first-class house near the Cathedral, 300 rooms, post, telegraph, railway tickets, and registration of luggage in the hotel; English church and central garden. J. Friedrich, proprietor.

Colwyn Bay.—Wales, Carnarvonshire.
Sea baths; fine sandy beach. Very mild climate.
Doctor: Dr. Brooks.

Combe-Girard.—Switzerland, Canton of Neuchâtel, near La Chaux de Fonds.
Effervescent ferruginous waters. The spring affords a very scanty supply, so that the water is available only for drinking.

Como.—Italy, on the southern bank of Lake Como.
An intermittent spring, chiefly visited as a transitory station in spring and autumn, for patients coming and going to Italy.
Hotel: Volta.

Concepcion de Peralta (La).—Spain, Province of Madrid, near Alcalá de Henares.
Alcali-saline waters, 45° F. contains arsenic.
Special indications: Skin diseases of a chronic character, vesical and urinary catarrhs, calculus and chronic rheumatism.
The establishment and accommodation are good.
Doctor: Dr. Osuna.

Condillac.—France, Department of Drôme.
Bicarbonate of lime waters, which have been called by some French medical authorities "the Queen of table waters;" exported.
Doctor: Dr. Coche.

Coney Island.—New York, United States of North America.
Sea baths, much frequented by the inhabitants of New York.

Contrexéville.—France, Department of Vosges, from London direct in 19 hours; from Paris in 9 hours; 800 inhabitants; 1,090 feet above sea level. For routes see table.
Slightly alkaline waters, with lime, lithium and iron, 55° F. The export of the waters is considerable.
Thoroughly well-arranged establishment.
Season: May 20th to September 15th.
Number of Visitors: Above 3,000 during the season.
Special indications: Gravel, biliary and vesical calculi, and catarrh, diabetes, gout and gouty rheumatism, disorders of the urinary system, affections of the uterus, hepatic complaints generally.

The name "Contrexéville" almost connotes the cure of gravel. In regard to the treatment of this disease its waters are without a rival; its Pavilion-spring reigns supreme. The highest French medical authorities consider Contrexéville water as *the* remedy in all such cases.

The village itself is situated in a charming valley, which

widens out into a large plain. Visitors will do well to bring some warm woollen clothing, as the thermometric changes are at times sudden.

The surrounding country is hilly and well wooded, and affords an ample field for excursions.

The principal spring, "the Pavilion," is situated in a large and beautifully laid-out park. Surrounding the spring is a covered gallery, which affords patients taking the waters the opportunity of out-door exercise in wet weather. The gallery is connected with the hydropathic and other bathing establishments and with the Casino. There is a very comfortable hotel in the park close by, kept by A. Morel.

The Casino is an elegant structure, and contains assembly, reading, ball, concert and billiard rooms. In the theatre, in the same building, performances of opera-comique and comedies are given during five days of the week.

There are also two daily concerts by a well-trained band.

English Church Service.

Postal Telegraph Office, with four deliveries a day.

Ample accommodation may be had in furnished apartments and private villas.

Carriages and Horses: Ed. Grandvallet.

Doctors: Drs. Debout d'Estree (inspector of the baths, who speaks English), Aymé, Broichox, Brongniart, Graux, Pierre and Thiery.

Hotels: The *Grand Hotel de Paris*, near the railway station, a large house, clean, with moderate prices. Schuhkraft, proprietor.

The *Hotel de la Providence*, a family house, celebrated for its good attendance and cooking. F. Chevet, proprietor.

Corcoles.—Spain, Province of Guadalajara.

Saline waters, 50° F.

Special indications: Syphilis and rheumatism.

Cordeac.—France, Department of Isère, near Grenoble.

Sulphurous sodaic waters.

Cordova.—Spain, capital of the province of same name. From Paris by Orléans and South of France lines, *viâ* Bordeaux, Irun, and Madrid.

A very interesting old town, founded by the Romans. There are many antiquities, most of them Moorish. Under the Moors it was one of the most flourishing cities in Spain, but it is now much reduced.

OBJECTS OF INTEREST.—The Cathedral or "La Mezquita," with its Court of Oranges and Belfry, the Archiepiscopal Palace, Alcazar, Collegio de la Ascension, La Correda, Hospital, Octagon tower, the Bridge, El Triunfo.

Excursions to the ruins of the Palace of Azzihra and Arrizafa, Hermitages of Valparniso, and to the wine cellars at Montilla.

Hotels: *Fonda Suiza* and *Fonda Rizzi*.

Corenc.—France, Department of Isère, near Grenoble.
Sulphurous sodaic waters, 35° F.

Corfu.—Greece, one of the Ionian Islands in the Ionian Sea; capital of 24,000 inhabitants; from Trieste 53 hours, and Brindisi 12 hours by steamer; ceded by England to Greece in 1864.
Winter station of first importance. Sea baths.
Season: November—May.
Special indications: Scrofula, chlorosis, anæmia, chronic neuroses, paralysis, nervous debility, and convalescence, phthisis and chronic rheumatism, affections of the respiratory organs, bronchial, laryngeal and pharyngeal catarrh.

The mean temperature of the seven season months is 57° F., mean barometric pressure 762 m.m., mean relative dampness 76½ per cent., rainy days 79 during the seven months, rainfall 1,068 m.m., death rate 32·5 per thousand.

The town is well sheltered and the bay open to the full south and south-east winds, and consequently well protected against due northerly winds. The Bora is not felt here. North-west and north-east winds are tolerably frequent, though the latter are objectional only as being cold. The atmosphere is never stagnant, owing to a sea breeze up to three o'clock, after which the land breeze sets in. Rapid atmospheric disturbances, such as are met with on the Riviera, occur but seldom, and there is almost complete immunity from dust and high winds.

The climate is warm but bracing. At the end of January and in February, hyacinths, roses and almond trees begin to blossom. By March, spring has set in, and in April cherries are ripening. The temperature rarely falls to freezing point, and still more rarely below it. Fogs occasionally occur and last some time.

The town itself offers many attractions. It is a link between the East and the West. Its streets are very lively. Close to the town on a small hill is a villa belonging to the King of Greece. The esplanade and the Strada Nuova are the centre of Corfu life. The environs are mostly covered with olive groves. The interior of the island offers a fine field for excursions and sports. Naples has been called the Eldorado of Hypochondriacs, and Corfu may claim the same honour. The drinking water and the means of communication are excellent.

Post: Near the gate at the port.
Telegraph: Near the Royal Palace.
There is an *English Church* at the *House of Parl*.
Banker and Exchange Office: Mr. Fells.
Chemist: Collas.
Doctors: Dr. Politi, Dr. Neranzi.
Hotel: The *Hotel St. George*, comfortable and moderate in price.

Cormus.—Austria, Illyria.
　Saline waters, 35° F.

Corneille de la Rivière.—France, Department of Pyrénées Orientales, near Perpignan.
　Ferruginous bicarbonate waters, 40° F.

Cornigliano-Ligure.—Italy, three-quarters of an hour from Genoa.
　A spring and autumn station.
　The village is exposed to the north-east winds, and has a considerably lower temperature than Montone or San-Remo.

Cortejada.—Spain, Province of Orense, near Celanova.
　Thermal sulphurous springs, 70° to 90° F.
　Ferruginous spring, 60° to 70° F.
　Special indications: Gastro-intestinal affections; rheumatism.
　Doctor: Dr. Ortega.

Corticella.—Italy, Province of Bologna, one hour from this town.
　Ferruginous waters, not yet thoroughly analysed.
　The country around is very picturesque, but there is no establishment.

Cos, Islands of, in the Ægean Sea.
　Saline springs.

Cotone.—Italy Tuscany, near Empoli, between Florence and Pisa.
　Sulphuro usalkaline waters, 55° F.　No accommodation.

Cotto.—Italy, island of Ischia, near Monti.
　Alkali-saline waters, 105° F.
　Special indications: Obesity, constipation, disorders of the urinary system.
　The establishments and accommodation are good, and the place is fairly well visited.
　Suffered by the earthquake of July, 1883.

Couchon.—France, Department of Pyrénées Orientales.
　Ferruginous bicarbonate waters, 35° F.

Coudes.—France, Department of Puy de Dôme.
　Cold bicarbonate of soda waters.

Courmayeur.—Italy, Piedmont, near Aosta; 4,098 feet above the sea level.
　Ferruginous, sulphurous saline waters, from 35° to 85° F.
　Special indications: Herpes and scrofula.

Courpière.—France, Department of Puy de Dôme.
　Cold ferruginous bicarbonate waters.
　Special indications: Same as Vichy.

Cours.—France, Department of Gironde.
Cold ferruginous bicarbonate waters.

Courseuilles.—France, Department of Calvados; from Paris, viâ Caen.
Sea baths, shingly beach, oyster beds.

Courtmacsherry.—Ireland, county Cork, near Lisley.
Sea baths; good beach.

Courtomer.—France, Department of Orne, near **Bagnolles.**
Cold ferruginous water.

Coutainville.—France, Department of La Manche, from Paris, viâ Coutances.
Sea baths; a quiet place.

Cowes.—England, Isle of Wight.
Sea baths: extensive and sandy beach.
Season: May—August. Yachting season till November.
Doctor: Dr. Hoffmeister.

Crailsheim.—Germany, Würtemberg; **1,150 feet above** sea level.
Ferruginous and **sulphurous waters.**
Season: June—September.
The number of visitors has increased since the building of an establishment, with a Kurhaus and hotel.

Cransac.—France, Department of Aveyron, near Rodez; **930 feet above** sea level. For routes see table.
Cold sulphate of lime water. Tonic and restorative.
Good establishment fairly well frequented.

Craveggia.—Italy, Piedmont.
Sulphate of soda waters, 65° F.
Special indications: Goitrous and arthritic affections.

Crêche.—France, Department of Saone and Loire, near Mâcon.
Cold ferruginous bicarbonate waters.

Credo.—France, Department of Gironde.
Cold ferruginous bicarbonate waters.

Criccieth.—Wales, Carnarvonshire.
Sea baths, shingly and sandy beach.
Doctor: Dr. Roberts.

Crieff.—Scotland, Perthshire.
Saline waters, hydropathic establishment.
Season: July—September.
Air pure and bracing; climate comparatively mild.
Doctor: Dr. Gairdner.

Criel.—France, Department of Seine Inférieure; from Paris viâ Dieppe.
 Sea baths, sandy beach.

Cristo (Aequa di).—Italy, Province of Trani, line from Foggia to Brindisi, stop at Trani.
 Highly effervescent earthy waters.
 The spring is close to the seashore on the east of the town.
 Season: June–September.
 Special indications: Gravel, cystitis, obstructions of liver and spleen.

Croisic (Le).—France, Department of Loire Inférieure, near St. Nazaire and Nantes.
 Sea baths.
 A well-appointed bathing establishment.
 Doctor: Dr. Macario.

Crol (Le).—France, Department of Aveyron.
 Cold sulphate of iron and magnesia waters.

Cromer.—England, Norfolk.
 Sea baths, sandy beach. Bracing climate.
 Season: June–October.
 Very fine scenery; esplanade and jetty.
 Doctor: Dr. Flueder.

Crosshaven.—Ireland, county Cork.
 Sea baths, sandy and shingly beach.
 Season: June–September.

Crotoy (Le).—France, Department of Somme.
 Sea baths.

Cudowa.—Germany, Silesia; 1,200 feet above the sea level.
 Ferruginous bicarbonate waters.
 Doctors: Drs. Jabob and Scholz.
 Hotels: *Kurhaus* and *New World*.

Cuervo.—Spain, Province of Malaga, near Medina-Sidonia.
 Sulphurous ferruginous waters.
 There are a number of springs in the neighbourhood, which, owing to their efficacy, have been named after saints by the peasantry. No accommodation, and no establishment.

Cullercoats.—England, Northumberlandshire, parish of Tynemouth.
 Sea baths; sandy beach. Climate bracing and fresh.
 Season: June–September.
 Doctor: Dr. Rennie.

Cusset.—France, Department of Allier, near Vichy; 900 feet above the sea level. For routes see table.
Bicarbonate of soda water, 40° F.
Complementary to Vichy.
Establishment and accommodation good. The environs, however, are said to be malarious, and travellers are recommended to be careful. The waters are exported.

Cuxhaven.—Germany, Hanover, near Hamburg.
Sea baths.
Contain 30 per cent. less salt than the Heligoland, and 30 per cent. more than the Baltic sea baths.

Czaeko.—Austria-Hungary, comitat of Hónt, near Rimaszets.
Effervescent ferruginous waters.
Used only internally by the surrounding peasantry.

Czarskow.—Germany, Silesia.
Ferruginous bicarbonate waters.
Special indications: Gout, anæmia.

Czigelka.—Austria-Hungary, near Bartfeld.
Cold bromo-iodurated alkali-saline waters.

Daetlingen.—Switzerland, Bernese Oberland, four hours from Thune.
Saline waters.
Very poor establishment, and but little visited.

Dagh-Hamman.—Turkey, Asia Minor, about 2 miles from Yalova, on the Sea of Marmora, opposite Constantinople, whence it can be reached in 2 hours.
Sulphurous saline waters, 190° F., containing much iodine; winter station.
Season: 1st October to 1st May; but open all the year round.
Special indications: Skin diseases of all kinds, rheumatism, syphilis, mercurialism, and kindred affections, anchylosis and muscular contractions.
The situation is superb, and the vegetation luxurious. Good hunting and shooting can also be had.

Dale.—Wales, Pembrokeshire, near Milford Haven.
Sea baths; sandy beach.

Dalkey.—Ireland, county of Dublin.
Frequented sea baths.

Daneverd.—Sweden, near Upsala.
Cold ferruginous bicarbonate waters.

Dangast.—Germany, Oldenburg, near Varel.
Sea baths; sandy beach.
Very good establishment for cold and warm sea baths. Accommodation abundant and comfortable.

Dardagni.—Switzerland, Canton of Geneva; 4,500 feet above sea level.
Bituminous waters. Very little used, although they are some of the best waters in Switzerland.

Dartmouth.—England, Devonshire.
Sea baths; indifferent beach and bathing. Climate mild.
The scenery is very fine.
Doctor: Dr. Allnut.

Daruvar.—Austria-Hungary, Slavonia, north of Szigeth.
Mixed bicarbonate waters, from 100° to 115° F.
Slightly mineralised; mud baths.
The situation is very romantic, and the vegetation luxurious.

Davos-Platz.—Switzerland, Engadine; 5,200 feet above sea level.
Climatic station. The air is very dry, and therefore suitable for the earlier stages of consumption. Invalids must be careful, however, as the climate is treacherous. As the influx of visitors increases, the drainage becomes less perfect.
Doctors: Drs. Peters, Beeli, Boner, Schimpf and Spengler.
Hotels: Kurhaus, Rhœtia, Strela.

Dawlish.—England, Devonshire, on the Channel coast.
Much frequented sea baths.

Dax.—France, Department of Landes, near Bayonne. For routes see table.
Sulphurous waters, 120° to 145° F. Mud baths.
Special indications: Articular or rheumatic affections.

Deal and Walmer.—England, Kent.
Sea baths; indifferent beach.
Season: June—September. The climate is less mild than that of St. Leonards or Hastings.
Doctor: Dr. Hulke.

Deauville.—France, Department of Calvados.
Sea baths; fine, firm sandy beach. Climate bracing.
Season: June—September.
Many excursions in the neighbourhood.

Derrindaff.—Ireland, Ulster.
Sulphate of lime waters, with a considerable quantity of sulphuretted hydrogen.
Special indications: Chronic catarrh of the respiratory organs, rheumatism, gout, skin diseases, sores, paralysis and syphilis.

Desaigues.—France, Department of Ardèche, near Tournon.
Bicarbonate of soda water, similar to the Vals water; mineralisation, 5·24 grms.

Desvres.—France, Department of Pas de Calais.
Cold ferruginous waters. Only used by the surrounding inhabitants.

Deutsch-Kreutz.—Austria-Hungary, comitat of Oedenburg.
Cold and pure alkaline waters.
They are largely exported, and are very efficacious in all complaints, where the stronger and more exciting action of effervescent alkaline waters is contra-indicated. In some cases they are more effective than the Gieshübl.
Special indications: Dyspepsia, affections of the mucous membranes, disorders of urinary and digestive organs, vesical calculi, gravel.
The establishments are indifferent, and the accommodation limited.

Deux-Lots.—France, Department of Landes, near **Dax**.
Cold sulphurous waters, 60° F.

Devonport.—England, near Plymouth.
Sea baths.
Establishment good.

Diedenow.—Germany, Pomerania.
Sea baths on the Baltic.

Diemeringen.—Germany, Upper Alsace.
Saline waters; 3 grms. of mineral constituents.

Dieppe.—France, Department of Seine Inférieure. London and Paris, *via* Newhaven, Dieppe, and Rouen, shortest and cheapest route. Through Fares—single tickets, 1st class, 33s.; 2nd, 24s.; 3rd, 17s. Return tickets, 1st class, 55s.; 2nd, 39s.; 3rd, 30s., available one month. These tickets are available for seven days from the date of issue, thus giving passengers who do not desire to proceed direct to Paris, the opportunity of staying for a short time at Dieppe or Rouen. They are available by any of the ordinary trains, as well as by the special and express trains running in connection with the steamers.
Much frequented sea baths.
Passengers will find the buffet open at the arrival of every train from Paris, or boat from London.
Dentist: A. Preterre.
Bookseller: Rainville, 52, Grande rue.
Doctors: Drs. Williams and De la Rue, fils.
Hotels: The *Royal Hotel*, first class, facing the sea; the casino and bathing establ'shment. Open all the year. Larsonneux, proprietor.

The *Grand Hôtel des Bains*, splendid view of the sea, 150 rooms; very comfortable. Table d'hôte breakfast, 4 francs; dinners, 5 francs; pension from 10 francs a day. Morgan, proprietor.

Dieulefit.—France, Department of Drôme, near Montelimar.
Cold bicarbonate of lime waters; little used.

Dievenow.—Germany, Prussia on the Baltic in Pomerania.
Sea baths; sandy beach.
Season: June—September.
Good bathing and fishing. Very interesting excursions.
Assembly rooms; hotel and other accommodation is first class. The place is much frequented by Poles and Germans. Amusements are abundant, and in great variety.

Diezgo.—Spain, Province of Cindad-Real.
Bicarbonate of soda waters, 30° F.

Digne.—France, Department of Basses Alpes.
Saline and sulphate of lime waters, 80 to 110° F. The composition of these waters is not thoroughly understood.
Season: May—August.
Used by the ancient Romans.

Dinan.—France, Department of Côtes du Nord.
Cold ferruginous waters.
Doctor: Dr. Brabant.

Dinard St. Enogat.—France, Department of Ille and Vilaine, from Paris *via* St. Malo.
Sea baths; good beach.

Dingolfing.—Germany, Bavaria, district of the lower Danube.
Sulphurous waters.
There is a bathing establishment which serves also as an hotel.

Dinkhold.—Germany, Nassau.
Cold bicarbonate of lime waters.
Special indications: Abdominal plethora.

Dinsdale, Low. England, Durham.
Cold sulphate of lime waters.
Special indications: Dyspepsia.

Dios-Györ.—Austria-Hungary, comitat of Borsod, near Misklocz.
Sulphurous alkaline ferruginous waters, 70° F.
The therapeutic properties of these waters are as yet unknown. There is a small establishment, with a few rooms for visitors. The situation is fine, and the place could be greatly improved in proper hands.

Dios-Jenö.—Austria-Hungary, comitat of Neograd.
Saline ferruginous waters.
A wooden bathing establishment, and no accommodation.

Dirce.—Greece, near Thebes.
Sulphurous saline waters, 170° F.
No establishments, and the surrounding country is unhealthy owing to extensive marshes.

Dirsdorf.—Germany, Prussia, near Nimpsch.
Cold ferruginous waters.
A recently built bathing establishment and other buildings afford accommodation for about 200 visitors.

Disznopatak.—Austria-Hungary, comitat of Marmarós.
Ferruginous alkaline, sulphurous waters, 45° F.
A small bathing establishment.
Waters are used only externally.

Dives.—France, Department of Calvados.
Sea baths and ferruginous waters.

Divonne-les-Bains.—France, Department of Ain, 1½ hours from Geneva. For routes see table.
Cold water springs; the water is exquisitely pure, and less charged with minerals than any other; 44° F.
Special indications: Chronic rheumatic arthritis, lumbago, pleuro-dynia, gout, sciatica, neuralgia, hypochondria, neuroses, chlorosis, gastralgia, bronchial catarrh, dyspepsia, liver complaints, affections of the bladder, haemorrhoids, paralysis, chronic affections of the spinal chord, scrofula and female disorders.
A very healthy station for those who have resided long in hot climates.
Hydro-therapeutic establishment of great importance. The waters have a very wide application. Every variety of douche and bath, inhalation, electric baths, steam and swimming baths.
Post and telegraph office in the establishment.
Beautiful walks and excursions.
The accommodation in the establishment is very good and prices moderate.
Doctor: Dr. Vidart.
Hotels: The *Establishment* is very comfortable, and the charges moderate.

Dizenbach.—Germany, Württemberg.
Carbonate of lime waters.

Doberan.—Germany, Mecklenburg-Schwerin, on the Baltic shores.
Much frequented sea baths, in a very pleasantly situated valley. Hydromineral station. Several cold mineral springs: one saline, one saline and magnesian, and one ferruginous bicarbonate.
Special indications: Anæmia and general debility.
Doctors: Drs. Kartum and Lange.
Hotels: *Du Nord, Lindenhof, Logenhaus.*

Doccio.—Italy, Tuscany, near Siena.
This spring, which has not been analysed, gives off carbonic and hydrosulphuric acid gases.

Dofana.—Italy, Tuscany.
Saline waters, 80° F.

Doktorka.—Austria, Bohemia, district of Prachin, near Prachatitz.
Alkaline waters.
Very little known. Only used for drinking by persons in the immediate neighbourhood.

Dolha.—Austria-Hungary, comitat of Marmarós, district of Huszth.
Ferruginous alkaline waters. Four springs.
This place is renowned for its mud baths. It has an establishment intended to encourage the use of these baths. The muds are also exported.

Dombhát.—Austria-Hungary, comitat of Bistritz-Naszod; railway to Dees; 20 minutes from Rodna.
Strongly effervescent alkaline, magnesian waters.
The bathing establishment is fitted with all modern comforts and requirements. Hunting, fishing and racing supply outdoor amusement.

Domène.—France, Department Isère, near Grenoble.
Sulphurous and saline springs, 110° F.; mineralisation, 4·76 grms.

Domeray.—France, Department Maine and Loire, near Bougé.
Cold ferruginous bicarbonate waters.

Domèvre sur Vezouse.—France, Department of Meurthe, near Lunéville.
Purgative waters.

Dorfbad.—Switzerland, near Appenzell; 2,200 feet above the sea level.
Alkaline earthy waters.
The bathing establishments are mediæval.

Dorfgeismar.—Germany, Hesse, near Cassel.
   Cold ferruginous waters.
   There is a bathing establishment with a Trinkhalle. The climatic conditions are favourable.

Dorna-Kandrény.—Austria, Gallicia, on the Dorna.
   Many cold, ferruginous, bicarbonate springs.
   Season: June—September.
   Special indications: Nervous affections, anæmia.
   The bathing establishments are efficient, which is however, more than can be said for the accommodation.

Dorres.—France, Department of Pyrénées Orientales, near Prades.
   Sulphurous sodaic springs, 95° to 100° F.

Dotit.—Hungary, comitat of Gömör.
   Thermal sulphur spring.

Douai.—France, Department du Nord.
   Alkaline ferruginous waters. They are not used beyond the district of Douai.

Douarnenez.—France, Department of Finistère, near Quimper.
   Sea baths.

Douglas.—England, Isle of Man.
   Sea baths; beach good. The caves are utilised in the place of bathing machines.
   Climate and air bracing but mild.
   Season: June—September.
   Doctor: Drs. Adair and Woods.

Douville.—France, Department of La Manche, from Paris, viâ Granville.
   Sea baths; sandy beach.

Dovadola.—Italy, Tuscany, near Castrocaro.
   Saline waters; little used.

Dovedale.—England, Derbyshire.
   Summer health and holiday resort. Visited on account of the surrounding scenery.

Dover.—England, Kent, on the Channel Coast.
   Sea baths; shingly beach.
   octor: Dr. Barton.

Dragomérfalva.—Austria-Hungary, comitat of Marmarés, district of Izavolgy.
   Very strong sulphurous springs, containing naphtha, which is exported.

Drahowa.—Austria-Hungary, comitat of Neutra, near Lutrabje.
   Cold ferruginous waters
   Strong diuretic action.

Draitschbrunnen.—Germany, Rhenish Prussia, **close to** Godesberg.
Earthy alkaline ferruginous waters. The waters are used in the establishments at Godesberg.

Drennon Springs.—North America, Kentucky.
Thermal sulphurous saline water.
Much frequented; beautiful site.

Dresden.—Germany, capital of the kingdom of Saxony; **from Paris by** North of France Railway, viâ Cologne, Frankfurt and **Leipzig, or** Cologne, Berlin **and** Leipzig, 28¼ hours. 137½ frs. 1st class.
A very fashionable place of residence for English and American families, who winter here. It is styled the "German Florence." Cheap living, cheap music, theatres, **and** works of art, select and polished society, are some of the advantages Dresden offers.

OBJECTS OF INTEREST.—Royal palace, Green vaults, Picture gallery, Museum of natural history, Körner museum, Brühl terrace, Court church, the Court theatre, the Zwinger; Military museum or Historical Hall, Grand Opera house, Church of Our Lady, Grosser garten, Rietschel Museum, Zoological Gardens.

EXCURSIONS: To Saxon Switzerland, Königstein, Bastei, Pillnitz, Ottowalder, Grund, Kuhstall and Winterberg.
Post and Telegraph Office: Waisenhausstrasse.
CABS: One to four persons, 40 to 90 pfgs.; by hour, 1 m. 60 pfgs. to 2¼ mk.
English club: Bürgerwiese.
Dentist: Otto L. Schoch.
Doctors: Drs F. Elb, Pierson, Vaust and Seltzner.
Hotels: Great Union, Webers, Victoria, Belle Vue, Saxe, *de Rome, de Russie, de Londres*

Dreykirchen.—Austria, Tyrol, **valley of the Eisack, near Kollmann.**
Alkali saline waters.
Very pure and healthy air. Picturesquely situated, with numerous excursions.
The establishment is well appointed.

Driburg.—Germany, Westphalia, near Paderborn.
Earthy, **saline** chalybeate waters. Sulphurous muds.
Notwithstanding its numerous therapeutic resources, Driburg is little frequented. The establishment and accommodation are very good.
Doctors: **Drs. Riefenstahl, Brück and Huller.**
Hotels: *Warnecke, Brockmann.*

Drise.—**Switzerland,** Canton **of Geneva, on the road to** Annecy.
Cold ferruginous **water.**

Drogheda.—Ireland, County Louth.
Sea bathing.

Drohobicz.—Austria-Hungary, Gallicia, district of Sambor, near Truscnwicze.
Saline waters, 25 per cent. of salt. Only used externally.
The bathing establishments are good.

Droitwich.—England, Worcestershire. 3½ hours from London.
Saline waters and brine baths.
Season: May till October, but open all the year.
Special indications: Rheumatic gout, rheumatoid arthritis, muscular rheumatism, sciatica, neuralgia, nervous debility, pelvic cellulitis, uterine derangements. Bright's disease, bronchitis, rheumatic fever, cachexia, skin diseases, scrofula, congestions of liver and spleen.

The potency of the hot salt baths of Droitwich was recognised in 1831, during a visitation of the cholera. The baths and the hotel in connection therewith are now the property of a former medical student, who availed himself of the occasion of the British Medical Association meeting at Worcester in 1883, to invite the whole of the profession, 800 of whom accepted the invitation; a full account of their inspection was given in the British Medical Journal of 19th August, 1883.

The large swimming bath, 100 by 50 feet, containing brine at a temperature of 80° F., has been entirely renovated, and is quite unique, being the only inland salt water swimming bath. The excursions are varied, interesting and very picturesque.

Hotel: The *Royal Brine Baths Hotel* in connection with the above is now replete with all modern improvements, and being under an experienced and very efficient manageress, the comforts of home are secured to invalids and others. The grounds in which the baths are erected are extensive and strictly private. It has coffee, drawing, billiard, private sitting and smoking-rooms. Board with excellent cuisine 2½ guineas per week. Bedrooms, per tariff. For all particulars apply to A. Roe, Secretary.

Drumgoon.—Ireland, Ulster, County Cavan.
Saline sulphurous waters, 50° F.

Drumrastel.—Ireland, County Cork, near Mallow Spa.
Ferruginous waters; used only internally.

Drumsna.—Ireland, County Leitrim.
Sulphurous waters, 50° to 55° F. The waters have an intense odour of sulphuretted hydrogen.

Dubograedsk.—Russia, Government of the same name.
Two bitter water and two alkali-saline springs
Only used internally. Arrangements are of a primitive kind, and although the waters are effective, they are scarcely known beyond the immediate neighbourhood.

Dubowa.—Austria-Hungary, comitat of Arva-Turócz.
   Alkaline waters.
   Primitive arrangements; the waters are used only in the environs, or exported in stone bottles.

Dubrawa.—Austria-Hungary, comitat of Zips.
   Cold alkaline waters.
   An establishment, with a Trinkhalle.

Duivon.—France, Department of Loire, near Molières.
   Bicarbonate waters analogous to those of St. Alban.

Dunbar.—Scotland, Haddingtonshire.
   Sea baths, much frequented.
   Doctor: Dr. Turnbull.

Dunblane.—Scotland, Perthshire, near Stirling.
   Saline waters, purgative and diuretic.
   Special indications: Scrofula and herpes.

Dundrum.—Ireland, Down, near Lisburn.
   Sea baths; good beach.
   Doctor: Dr. Nolan.

Dunkirk.—France, Department of Nord.
   Sea baths, much frequented; sandy beach.
   Season: June—September.
   Dunkirk is duly celebrated for its extensive rabbit-shooting facilities. During season, races of importance are held. Theatre, casino, boating and fishing.
   Hotels: The *Hôtel du Chapeau Rouge*, 5, Rue St. Sebastien, first-class and comfortable house, open all the year; omnibus meets all trains. M. Roux, proprietor.

Dunmore-East.—Ireland, County of Waterford.
   Sea baths.

Dunoon and **Mellan.**—Scotland, Argyleshire.
   Sea baths; sandy beach.
   Doctor: Dr. Reid.

Dürenhof.—Russia, Livonia, district of Wolmar.
   Sulphurous waters, containing a quantity of sulphuretted hydrogen. No establishment, and but little used.

Dürkheim an der Hardt.—Germany, Palatinate.
   Saline bromo-iodurated springs, similar to the Kreuznach waters, but not so rich. Grape cure.
   Doctors: Drs. Löchner and Kauffmann.
   Hotels: Four Seasons.

Durness.—Scotland, Sutherland.
   Sea baths; stony beach.
   Very little frequented and accommodation poor.

Durrheim.—Germany, Baden, near **Donaueschingen.**
Saline waters.

Dürrwangen.—Germany, Würtemberg, district of Bahlingen.
Sulphurous waters, containing a considerable quantity of sulphuretted hydrogen. Little frequented.

Durtal.—France, Department of Maine and Loire, near Baugé.
Ferruginous bicarbonic waters.

Dusternbrook.—Germany, Holstein.
Sea baths on the bay of Kiel.

Eastbourne.—England, Sussex, between Brighton and Hastings.
Sea baths; sandy beach; air very pure and healthy. Mortality 16 per 1,000. Much frequented.
Doctor: Dr. Gould.
Hotel: The *Grand*.

Eaux-Bonnes.—France, Department of Basses Pyrénées; 2,500 feet above the sea level; half-hour from Laruns by carriage. For routes see table.
Sulphurous saline and alkaline waters, 90° F. Inhalation saloon. Hydropathic establishment.
Season: June to September.
Frequented by more than 3,000 visitors.
Special indications: Angina pectoris and laryngitis, bronchitis and chronic catarrhs, asthma, chronic pleuritis, pulmonary phthisis, anæmia, lymphatism and scrofula.

"The thermal sulphurous and sulphurous saline waters of the Pyrénées, and especially the Eaux-Bonnes waters," says Dr. Cazaux, "are specific remedies in all chronic affections of the chest."

The late celebrated Dr. Pidoux wrote:—"In cases of pulmonary tuberculosis, the waters of Eaux-Bonnes possess a force with which no medicine of the chemist can compare. Thermal Europe has no remedies, we venture to assert, worthy to be placed on the same level as the Eaux-Bonnes waters. Even phthisis in the third stage can be cured here, unless indeed the whole system be already undermined."

The town of Eaux-Bonnes is sheltered from winds by the surrounding forest-clad mountains, and can be recommended as an agreeable place of residence. It is also favourably known as an air cure station, owing to its elevated position.

Post and telegraph. Protestant and Catholic churches.

Casino, club and municipal band. Many points for excursions. Villas and furnished apartments.

Doctors: Drs. M. Cazaux and Cazenave de la Roche.
Hotels: The *Hôtel des Princes*, a first-class house in every respect; comfortable and elegant; situated in the park. E. Murret-Labarthe, proprietor.

Eaux-Chaudes.—France, Department of Basses Pyrénées, near Eaux-Bonnes ; 2,020 feet above sea level. For routes see table.
  Sulphate of soda waters, from 22° to 90° F. The Baudot l'Espuriette and Bey springs are taken internally.
  Special indications: Rheumatism, catarrhs, scrofula.
  Doctor: Dr. Anglada.
  Hotels: De France, Baudot.

Ebeaupin.—France, Department Loire Inférieure, near Nantes.
  Ferruginous bicarbonate waters, 30° F.

Ebed.—Austria-Hungary, comitat of Gran.
  Sulphurous waters, 70° F. The volume of the spring is very large.
  Very little known, and no establishment.
  The waters are used to drive a mill, a good proof of an abundant supply.

Eberbach.—Germany, Würtemberg.
  Saline waters.
  A small bathing establishment.

Ebingen.—Germany, Würtemberg, district of Bahlingen.
  Sulphurous waters.
  A small and fairly appointed bathing establishment.

Ebriach.—Austria, Carinthia, district of Klagenfurt, near Kappel.
  Alkali-saline and ferruginous waters.
  The water is exported. The municipal authorities are earnestly endeavouring to provide the accommodation and amusements expected by visitors in the present day.

Echaillon (L').—France, Department of Isère, near Grenoble.
  Sulphate of lime waters, 45° F.

Echaillon (L').—France, Savoy, near St. Jean de Maurienne.
  Saline and sulphate of soda springs, 100° to 110° F.; highly purgative and diuretic.

Echzell.—Germany, Hesse, near Neuschwalheim.
  Sulphurous waters. Used only by the peasantry.

Eckartsbrunnen.—Germany, Nassau.
  Ferruginous waters.

Eckerberg.—Germany, Pomerania.
  Air cure station and hydropathic establishment.
  The place is situated in the midst of a pine forest.
  Doctor: Dr. Vicek.

Eckernförde.—Germany, Schleswig, 5 hours from Kiel.
   Sea baths; sandy beach.
   The bathing establishments are very efficient, and well appointed.

Ecquevilly.—France, Seine at Oise.
   Cold saline spring.

Ecuillé.—France, Department of Maine and Loire.
   Ferruginous bicarbonate water.

Edenkoben.—Germany, Bavaria.
   Sulphurous waters.
   Only used in the immediate neighbourhood.

Egartbad.—Austria, Tyrol, near Meran.
   Cold sulphurous waters.
   The bathing establishment, one of the best in the Tyrol, is much frequented during summer.
   Air sedative and mild; Egartbad is well suited to nervous temperaments as a climatic air station.

Egbell.—Austria-Hungary, comitat of Neutra.
   Sulphurous waters, little used.

Egelhof.—Austria, district of Traun, near Windisch-Garsten.
   Cold, sulphurous waters.
   A small bathing establishment, with a limited number of visitors annually.

Egerdach.—Austria, Tyrol, near Innspruck; 1,000 feet above sea level.
   Alkaline waters.
   The establishment affords everything necessary, and is much frequented.

Eggishorn.—Switzerland, Canton Valais; 7,200 feet above sea level; 6 hours from Brieg, viâ Fesch, by good mule paths and carriage road.
   Mountain air and whey cure station.
   Season: June to October.
   A great centre for excursionists to the Eggishorn, Glacier of Aletsch, lake of Mergelen, 4½ hours to the Cabane of Concordia and the most interesting peaks of the Cernoise Alps.
   Hotels: The *Hotel Jungfrau* commands a magnificent view of the Eggishorn, 2½ hours above Fesch. E. Cathrein, proprietor.

Eghegh.—Austria, comitat of Gran.
   Strong, cold ferruginous waters.

Ehrenbreitstein.—Germany, Rhenish Prussia, opposite Coblentz, at the foot of the fortress.
   Earthy waters, used in the environs as table waters, and also exported.

Eichwald.—Austria, Bohemia, district of Leitmeritz.
Hydropathic establishment, much visited.

Eilsen.—Germany, principality of Schaumburg-Lippe.
Sulphate of lime springs, 35° F. Mud baths.
Special indications: Inhalation in chronic catarrhs and phthisis of the larynx.
Doctors: Dr. Müller, Dr. Wegner.
Hotels: Kurhaus.

Eimbeck.—Germany, Hanover.
Bicarbonate of lime water.

Einöd.—Austria, Carinthia, district of Judenburg.
Ferruginous waters.
A very bare and monotonous country, as the name indicates.

Eisenbach.—Austria, Hungary, near Schemnitz.
Thermal, ferruginous and sulphurous waters, 105°.
Very good establishment and accommodation.

Eisenberg or Zelesna-Wodsk.—Russia, Caucasia, near Pjätigorsk.
Ferruginous waters, 50° to 110° F.
The establishments and general accommodation are very poor, considering the large number of visitors which the efficacy of the waters attracts.

Elba.—Island of, Italy, in the Mediterranean.
Ferruginous sulphur and saline waters; not used therapeutically.

Elche.—Spain, Province of Alicante, and one hour rail from that town.
Climatic winter station.
Dryer and warmer than Alicante. The accommodation, however, is but indifferent.

Elgersburg.—Germany, Thuringia.
Hydrotherapeutic establishment.
Doctor: Dr. Peliziäus.
Hotels: Kurhaus.

El-Hamma.—Algiers, near Constantine.
Bicarbonate of iron water, 90° F.

Elisabethbad.—Germany, Brandenburg, near Prenzlau.
Ferruginous bicarbonate waters.
Special indications: Nervous affections.

Elisabeth-Salzbaths.—Austria-Hungary, in the town of Ofen.
Bitter waters, the oldest spring of Ofen.
Exported to the extent of over one million bottles per annum. There is also a bathing establishment.

Ellabria.—Switzerland, Canton of St. Gall.
Sulphurous waters, not used therapeutically.

Elmen.—Germany, Saxony, district of Magdeburg, near Salza.
Saline waters.
Special indications: Affections of the glands and of the mucous membrane, rheumatism.
Doctor: Dr. Kirchheim.
Hotels: Evers, Voight.

Elöpathak, or Arapatak.—Transylvania, near Kronstadt.
Ferruginous bicarbonate water, 30° F.
Doctor: Dr. Meyr.

Elorrio.—Spain, Province of Bilbao, near Durango.
Mixed sulphurous waters, 30° F.
Special indications: Skin diseases.
The site is very charming, with a very luxurious vegetation. The establishment is well-fitted and the accommodation good.
Doctor: Dr. Enriquez.

Elster.—Germany, Saxony, near the Bohemian frontier; 1,300 feet above the sea level.
Alkali-saline ferruginous waters, 25° to 30° F. A very well-appointed bathing establishment.
Special indications: Abdominal plethora, dyspepsia, and nervous affections depending on anæmia.
Mud baths, recommended in cases of rheumatic paralysis and articular disease.
Doctors: Drs. Hahn, Cramer and Flechsig.
Hotels: De Saxe, Wettiner.

Empfing.—Germany, Bavaria, near Traunstein; 1,300 feet above the sea level.
Indifferent waters; hydropathic establishment. Steam, douche, mud and pine-cone sap baths.
The establishment is good, and accommodation abundant and adequate.

Ems.—Germany, Duchy of Nassau, on the Lahn. For routes see table.
Saline alkaline and saline earthy waters, 65° to 110° F.
Season: May—October. Milk cure.
These waters have a special action on the lungs, chest, and on nervous diseases, for the treatment of which Ems is specially noted.
Bookseller: A. Pfeiffer, Colonnade 11.
Banker: L. J. Kirchberger, under the Colonnades.
Doctors: Drs. Géisse, Orth, Goltz, Vogler and Kastan.
Hotels: The *Hôtel d'Angleterre*, first-class house, good in every respect; opposite the Royal Baths; large garden belonging to the hotel. Fr. Schmitt, proprietor.

Enatbühl.—Switzerland, Canton of St. Gall; 3,000 feet above the sea level.
Alkali-saline and sulphurous waters.
The establishment is efficient, and the accommodation good.

Encausse.—France, Department of Haute Garonne, near St. Gaudens.
Sulphate of lime waters, 50° F.
Special indications: Uterine and nervous affections in general.

Engelberg.—Switzerland, Valley of Engelberg, near the Lake of Lucerne, 3,200 feet above the sea level.
Mountain air cure.
Doctors: **Drs. Daubeny, Whiteley** and **Brunner**.
Hotels: Hotel Angel, and *Hotel Müller*.

Enghien.—France, Department of Seine and Oise, near Montmorency.
Cold sulphurous and lime waters.
Special indications: Scrofula, affections of respiratory organs, herpes and rheumatism.
Bathing establishment well fitted for hydropathic medication.
Doctors: Dr. Japhet, Dr. Pozzioli.

Engistein.—Switzerland, near Berne; 1,850 feet above the sea level.
Cold ferruginous and carbonate of lime waters. Little frequented.

Enn.—France, Department of Pyrénées Orientales, near Prades.
Thermal neutral waters, 120° F.

Epidaurus.—Greece, Peloponnesus, near the ruins of the old Temple of Epidaurus.
Alkali-saline bitter waters. Strong purgative action, even in small doses.

Eppenhausen.—Germany, Westphalia.
Alkaline-earthy waters.

Eppingen.—Switzerland.
Cold sulphate of magnesia waters.

Epsom.—England, Surrey.
Sulphate of magnesian waters, rich in mineral matter, though little resorted to.
The sulphate of magnesia, which constitutes the mineral base of these waters, is usually called "Epsom Salts" in England.

Erdö-Benye.—Hungary, comitat of Zemplin, near Tokay. Rail to Szeghalom.
Sulphurous ferruginous waters; grape cure station.
Much frequented. Establishments and accommodation good.

Erlachbad.—Austria, Tyrol.
Ferruginous alum thermal waters.
Primitive establishments.

Erlau.—Hungary.
Thermal indifferent waters, 80° F.

Erlaubad.—Germany, Duchy of Baden, near Sasbach.
Thermal saline waters, 75° F.

Ermetschwyl.—Switzerland, Canton of St. Gall, near Toggenburg, 2,250 feet above sea level.
Alkaline waters, 80° F.
Special indications: Rheumatic paralysis and contractions.

Ernabrunnen.—Germany, Brunswick.
Ferruginous waters, 50° F.
Special indications: General debility, anæmia, affections of the nerve centres.

Ernsdorf.—Austria, Silesia, near Bielitz.
Whey cure station. Kumiss, pine-cone sap baths.

Erquy.—France, Department of Côtes du Nord; from Paris, viâ Lamballe.
Sea baths; sandy beach.

Escaldas (Les).—France, Department of Pyrénées Orientales.
Thermal alkaline sulphurous waters, slightly mineralised.
Doctor: Dr. Companyo.

Eschelloh.—Germany, Bavaria, near Partenkirchen.
Earthy waters.
Scarcely known even in its own immediate neighbourhood.

Escoriaza.—Spain, Province of Guipuzcoa, near Vergara.
Sulphurous earthy waters, 50° F. One ferruginous spring.
Special indications: Chronic affections of the digestive organs, bronchial and laryngeal catarrh, asthma, hyperæmia of liver and spleen, leucorrhœa, anæmia and chlorosis.
The situation is most delightful, and the gardens are very fine. The establishment and the accommodation rank amongst the best to be found in Spain. Escoriaza is consequently much frequented. The prices are, however, within the reach of everyone.
Doctor: Dr. Diez.

Escouloubre.—France, Department of Aude.
Thermal sulphurous waters.
Special indications: Rheumatism.

Eski-Schehor.—Turkey, Asia Minor.
Thermal sulphurous water.

Esparraguera and Olesa.—Spain, Catalonia, near Barcelona.
Sulphurous waters, 90° F.
Very fine promenades, and much visited. The bathing establishments and accommodation are good.

Essentuky.—Russia, Caucasus, near Pjätigorsk.
Cold alkaline and alkali sulphurous waters.
Season: May—October.
Special indications: Affections of the abdominal organs, hyperæmia of liver and spleen, arthritis, lithiasis, affections of the urinary organs, scrofula, emphysema and affections of the respiratory organs.
The bathing establishments and accommodation generally are luxurious, and fresh improvements are being made each year. No known European waters are so rich as those of Essentuky in iodine and bromine.

Estadilla.—Spain, Province of Huesca, near Monzon.
Alkaline waters, 45° F.
Special indications: Gastric affections, chronic ulcers.
Doctor: Dr. Lopez.

Étretat.—France, Department of Seine Inférieure.
Largely visited sea baths.

Eufemia (Sant').—Italy, Calabria, near Nicastro.
Saline waters, 100° F.
Very famous in antiquity, but now entirely neglected.

Euzet.—France, Department of Gard.
Sulphate of lime bituminous springs.
Special indications: Pulmonary catarrh.

Evaux.—France, Department of Creuse.
Eight sulphate of soda springs, 60° to 140° F.
Special indications: Rheumatism, paralysis, ulcers.

Evian-les-Bains.—France, Upper Savoy, on the Lake of Geneva; 1,150 feet above the sea level. For routes the see table.
Alkaline waters, and climatic air station.
Special indications: Affections of the urinary and digestive organs, chronic affections of the liver and biliary apparatus, chronic nephritis, gravel, gout, vesical catarrh, chronic cystitis, congestion of the liver, diseases of the nervous centres, insomnia.
The watering place of Evian offers to the tourist as well as to the bather an agreeable sojourn. Facing the north, it receives refreshing breezes from across the lake.

Post and Telegraph station.
Walks and excursions in great variety into the forests and mountains, &c. Excursions, by steamer (25 departures every day), to Montreux, Chillon, Vevey, Lausanne, Geneva, &c.
Doctors: Drs. Taberlet, Million, Flottard, Dantaud and Dumuz.
Hotels: The Grand Hôtel des Bains.

Evolena.—Switzerland, Canton of Tessin, 5 hours from Sitten.
Ferruginous waters.
Used only by the surrounding peasantry.

Exmouth.—England, Devonshire, 10 miles from Exeter.
Sea baths, sandy beach.
Climate: Mild
Season: June—September.
An establishment for hot and cold baths.
Special indications: Failing digestion, debility and catarrhal affections.
Doctor: Dr. Ward.
Hotel: The Beacon.

Fachingen.—Germany, Nassau, near Ems.
Cold alkaline waters; for table use only.
Special indications: Dyspepsia.
Some of the strongest alkaline waters in Germany. Largely exported.

Faenza.—Italy, Romagna.
Earthy waters.
Only used by the inhabitants. The establishment and accommodation are indifferent.

Falciano.—Italy, Tuscany, on the Ghiora.
Ferruginous muriatic waters, 60° F
Special indications: Gravel and vesical calculus.

Falkenberg.—Germany, Silesia, district of Oppeln, near Grottkau.
Ferruginous sulphurous waters.
The establishment is of a superior kind.

Falkenstein.—Germany, Thaunus, Nassau, near Kronberg.
Hydrotherapeutic station.
Special indications: Affections of the lungs.
Doctors: Drs. Meissen, Dettweiler and Lorent.
Hotel: Kurhaus.

Fållorne.—Sweden, half a mile from Wexi.
Alkaline waters.
Taken only internally at the spring; neither used in baths nor exported.

Falmouth.—England, Cornwall.
　Sea baths; sandy beach.
　Doctor: Dr. Dub.
　Hotel: The *Royal*.

Falú-Szlatina.—Austria-Hungary, district of Szigeth.
　Alkaline ferruginous and pure alkaline and saline waters.
　The country around is picturesque.

Fano.—Italy, 4 hours' rail from Bologna.
　Sea baths; sandy beach.
　Arrangements fairly good.

Farette.—France, Savoy.
　Cold ferruginous arsenical waters; a table water.

Farkas-Mezö.—Austria, Transylvania, district of Udvarhély.
　Earthy saline waters.

Farnbühl.—Switzerland, near Lucerne; 2,500 feet above the sea level.
　Air cure station.

Fécamp.—France, Department of Seine Inférieure.
　Much frequented sea baths.

Feldalfing.—Germany, Bavaria, near Munich.
　Cold sulphurous springs.

Feldberg.—Germany, Mecklenburg-Strelitz.
　Hydropathic establishment. Electro-therapeutic baths. A winter station.
　Doctor: Dr. Erfurth.

Feletekút.—Austria-Hungary.
　Sulphurous waters, containing a large quantity of sulphuretted hydrogen.
　No bathing establishment and no adequate accommodation.

Félines.—France, Department of Auvergne.
　Alkali saline waters; highly effervescent.

Felixstow.—England, Suffolk, near Harwich.
　Sea baths; firm sandy beach.
　Climate bracing.
　Season: July—September.

Fellathal.—Austria, Carinthia.
　Cold alkaline water.
　Very good establishment and accommodation.

Felsenegg.—Switzerland, Canton of Zug; 3,030 feet above sea level.
Climatic station. Very pure and equable atmosphere.

Felsö-Alap.—Austria-Hungary, comitat of Weissenburg.
Bitter waters, containing iodine. Exported since 1853. No establishment.

Felsö-Apsa.—Austria-Hungary, comitat of Mármaros, district of Szigeth.
Saline waters.
Accommodation and bathing establishments bad.

Felsö-Bajom.—Austria, Transylvania, near Baasen.
Saline waters, 70° F.
The establishment is well arranged, and the accommodation good.

Felsö-Nereszniсze.—Austria-Hungary, comitat of Mármaros, Valley of Taracz.
Ferruginous, alkaline and saline waters.

Felsö-Russbach.—Austria-Hungary, comitat of Zips, near Pudlein.
Earthy waters.
The establishment resembles a gothic castle of feudal times. Its ancient splendour is still attested by numerous antiquities. The gaseous emanations from the soil are very peculiar, and not without risk to men and animals.

Felsö-Visso.—Austria-Hungary.
Alkaline and chalybeate waters of great strength.
No establishment, and but poor accommodation.

Feredschik.—Turkey, Roumelia, Sandshak Gallipoli.
Saline waters, 70° F
The establishments are on a **grand scale**, and fitted with every luxury.

Féron.—France, Department du Nord.
Earthy waters, but little used.

Ferranche.—France, Savoy, near Chamousset.
Cold ferruginous, alkaline and magnesian waters.

Ferreira.—Spain, Province of Murcia.
Earthy ferruginous waters.
Special indications: Anæmia, dyspepsia, irregular menstruation, disorders of kidney and bladder.

Ferriere (La).—France, Department of Isère.
Cold sulphate of lime waters.

Ficoncella.—Italy, Tuscany, valley of the Paglia, near Siena.
Ferruginous alkaline waters, 105° F.
Though very much frequented in the time of the Roman Emperors, they are now scarcely visited.

Fideris.—Switzerland, Canton of the Grisons.
Cold bicarbonate of soda and chalybeate waters.

Fiestel.—Germany, Westphalia.
Sulphate of lime springs; 2 grs. of mineralisation.

Filetto.—Italy, Abruzzo, near Torano.
Earthy saline waters, 85° F.
Little used; accommodation of a very inferior kind.

Filey.—England, Yorkshire, seven miles from Scarborough.
Saline springs, deserving to be better known.
Season: June—September.
Doctor: Dr. Scrivener.

Finceschti.—Roumania, Walachia.
Sulphurous waters, 45° F.
Establishment and accommodation inferior.

Fitero.—Spain, Province of Navarre.
Saline earthy waters, 115° F.
Special indications: Rheumatism and paralysis, gout, tendinous retractions, caries, necrosis, chlorosis and anaemia.
The establishments are modern, and prices are all fixed in accordance with a tariff exhibited in the various buildings. Accommodation good.
Doctors: Dr. Gimenez de Pedro, and Dr. Coll y Amo.

Fiumorbo.—France, Corsica.
Sulphurous waters, 130° F.
Establishment and accommodation inferior.

Fläsch.—Switzerland, Argovy, near Feldkirch; 1,750 feet above the sea level.
Alkali-saline waters, 45° F.
Special indications: Gravel, vesical calculi, podagra, haemorrhoids.
The establishment is good and accommodation fair.

Fleetwood.—England, Lancashire, on the Wyre.
Sea baths; shingly beach.
Climate humid and mild.
Doctor: Dr. Orr.

Flims, or Waldhaus Flims.—Switzerland, canton of Grisons, 3 hours from Coire; 3,642 feet above the sea level. For routes see table.

Climatic mountain station; milk and whey cure; baths in the Cauma lake.

Season: June—September.

Special indications: Chronic debility and convalescence, **nervous debility,** derangements of the digestive **organs, impaired nutrition and** catarrhal **irritation** of the respiratory **organs with** phthisical tendencies.

This is a quiet and retired spot, created, as it were, by nature and aided by art, as a lounge for the summer months, far from the turmoil of the world. The scenery surrounding the establishments and lake is very beautiful, with its valleys, woods, and mountain peaks. The hill on which the establishments are erected affords a panorama of great natural beauty. It overlooks a valley clothed with pine forests, which fill the air with their balsamic odours. It is no wonder, therefore, that Flims is one of the most popular of Swiss stations. The **mean** temperature (May—September) is 55° to 60° F. Another **attraction is the** almost absolute fixity of atmospheric pressure, the mean barometric **readings being 668** $^m/_m$.

The establishment consists of three principal buildings:—The Kurhaus, with café, reading-**room,** dining, ladies' and music rooms, the Villa Belmont and the Posthouse. There is also a Swiss Chalet, all four capable of accommodating 200 visitors. Most of the rooms have balconies. The baths—wh'ch **are** floating on the lake—contain separate apartments for **ladies,** gentlemen, girls and boys, **as** also single baths, and **have complete** arrangements. There **are large and** small boats at the disposal of visitors. Temperature of **the water, 60° to 75° F.**

The excursions embrace the whole of the Grisons, and amongst others may be mentioned:—The Segnes Falls, Rhine view and Conu, Salmus, Laas, Fellers, Val Bargis and Fidaz, Martin's Loch and Six Virgins, **Segnes** glacier, Piz-Grisch and Ringelspitze, Hanz-Trons-Disentis, Waltensburgh and Brigels, Lugnetz-Vals, Piz-Mondeun, Versam and Reichenau.

English Church Service.

Post and Telegraph Office.

Doctor: A Physician constantly in attendance.

Hotel: The *Kurhaus* and its dependencies are all thoroughly comfortable and well managed. There is a dairy attached to the establishment. Guggenbühl, director.

Flinsberg.—Germany, **Silesia.**

Bicarbonate ferruginous cold waters.

Doctor: Dr. Adam.

Hotels: Thomas, *Grasser.*

Flint.—Wales.

A well-frequented sea bathing-place.

Florence.—Italy, formerly capital of, but now of the Province of Tuscany; from Paris through Mont Cenis, *viâ* Turin, Genoa, and Pisa, or *viâ* Marseilles, Nice, and Genoa, 32½ hours. 160¾ frs. 1st class.

Florence, situated in the valley of the Arno, has been called "the fairest city of the world." A great many English and American families reside here in winter and early spring.

OBJECTS OF INTEREST.—The Cathedral or Duomo, Santa Maria del Fiore, the Campanile or bell tower, the Baptistery of San Giovanni, with the doors by Ghiberti, generally called the "Doors of Paradise"; Church of St. Croce, del Carmine, San Lorenzo, Palazzo Vecchio, Uffizi gallery, Palazzo Pitti, Loggia dè Lanzi, Town Hall, Casa Buonarotti, Boboli gardens, National museum, Etruscan museum, Royal Mosaic manufactory, Grand Opera, Theatres, La Pergola, Niccolini, Nuovo, Alfieri, Pagliano, Goldoni.

EXCURSIONS: Le Cascine, the **Florentine Bois** de Boulogne or Hyde Park, the most charming drive and promenade in all Italy; Villa Demidoff in the neighbourhood; the Monte Miniato, or Viale dei Colle; a walk up the hills to Galileo's tower; Vallombrosa, Pontassieve, Pelago, Certaldo, Fiesole.

CABS: One-horse, by the drive, 80c.; by hour, 1 lire 30 c. if within the boundary; if outside, 2 lire; every ½ hour afterwards, 70 c. and 1 lire.

Post-office: Uffizi. Telegraph: Palazzo Ricardi.

Bookseller: Hrm. Loescher, 20, Via Tornabuoni.

American Dentist: Dr. H. L. Schaffner, 8, Via dei Cerretani.

Doctors: Drs. Davidson, Young, Wilson and Duffy.

Hotels: The *Hôtel d'Italie,* large first-class house for families, well situated on the Arno; has a lift and is in every way well appointed. Branch house of *Hôtel de Nice,* Nice, Bernerhof at Berne, and *Hôtel de Turin* at Turin. G. Kraft, proprietor.

The *Anglo-American Hôtel* and pension, large and small apartments, and moderate charges; good cooking. The *Hôtel du Nord,* in the centre of the town, belongs to the same proprietor, Paolo Isola.

Flue.—Switzerland, near Solothurn.
Alkaline springs.

Folkestone.—England, Kent, on the Channel coast.
A much frequented sea bathing-place; sandy beach.
Chalybeate waters.
Doctor: Dr. Tyson.

Foncirgue.—France, Ariége.
Bicarbonate of lime waters, 50° F.
Special indications: Nervous affections.

Fonfrede.—France, Department of Lot and Garonne.
Alkaline magnesian waters, 35° F.
Special indications: Gravel, affections of the kidneys, bladder, liver and diabetes; deserves to be better known.
Doctor: Dr. de Gaulejac.

Fonga.—Italy, Tuscany.
  Effervescent alkaline waters.
  Little or no accommodation.

Fonsainte.—France, Department of Cantal, near Cahors.
  Ferruginous waters.
  Mostly used in after treatment.

Fonsalade.—France, Department of Puy du Dôme, near Billon.
  Saline waters, only employed by the surrounding inhabitants.

Fonsanches.—France, Department of Gard, near Vigan, 2 hours from Nimes.
  Sulphate of soda waters, 60° F.
  Special indications: Neuralgia and rheumatism.

Fonsrouilleuse.—France, Department of Maine and Loire, near Chaumont.
  Muriatic ferruginous waters.
  Only used in the immediate neighbourhood.

Fontaccia.—Italy, Tuscany, near Monte Bicchieri.
  Sulphurous and saline waters, 65° F.

Fontaigre or Sorède.—France, Pyrénées Orientales, near Perpignan.
  Effervescent ferruginous waters.
  The hamlet is charmingly situated, while but little has been done to attract visitors.

Fonté.—Spain, Province of Zaragoza, near Chiprana.
  Sulphurous saline waters, 50° F.
  Special indications: Affections of the abdominal viscera, dyspepsia, amenorrhœa, chlorosis, hysteria, ophthalmia.
  The situation is charming. The establishment is not quite up to modern requirements. Accommodation is scanty.
  Doctor: Dr. Martinez.

Fontenelle.—France, Vendée, near Napoléonville.
  Alkaline waters similar to those of Forges-les-Bains.

Foradade.—France, Department of Pyrénées Orientales, near Tantarel.
  Bitter waters, 65° F.
  Used as a purgative, but only by the resident population.

Forbach.—Germany, Alsace-Lorraine.
  Cold saline water; 6 grs. of mineralisation.

Forceral.—France, Department of Pyrénées Orientales, near Perpignan.
Ferruginous waters, 65° F.

Fordel.—Scotland, Fifeshire.
Alkali-saline waters.

Fordignano.—Italy, Sardinia, district of Barrigado.
Saline waters, 160° F.
Renowned under the Roman Emperors, as its ruins show; the baths are now altogether neglected. The roads leading to it are, as is almost universally the case in Sardinia, in a miserable condition.

Forges-les-Bains.—France, Department Seine and Oise. For routes see table.
Cold alkaline ferruginous springs.

Forges-les-Eaux.—France, Department of Seine Inférieure. Rail from Dieppe to Paris.
Cold bicarbonate and ferruginous waters, tonic and fortifying in their action.
Special indications: Convalescence, and more especially in cases of sterility.
Forges boasts a casino.
Doctor: Dr. Caraman.
Hotels: The *Hotel des Thermes* is the only hotel which can be recommended for comfort and proximity to the baths. Miss Broderick, proprietress.

Forstegg.—Switzerland, Canton St. Gall, near Sennwald.
Effervescent sulphurous waters.
Fairly good establishment.

Fortuna.—Spain, Province of Murcia, near Cieza.
Saline waters, 125° F.
Special indications: Rheumatism and paralysis, chronic ulcers, scrofula, neuralgias.
The establishment is one of the best in Spain, and the accommodation is good. The gardens are beautiful.
Doctor: Dr. Chacel.

Fortyogo.—Austria-Hungary, Transylvania, district of Coik.
Sulphurous alkaline waters.
Used in baths only by the surrounding peasantry.

Fossino.—Italy, Tuscany, in the bed of the Valconto river.
Alkali-saline and ferruginous waters.
Very little used.

Fosso degli Ontani.—Italy, Tuscany, valley of the Fiora.
Alkali-saline waters, 65° F.

Foucaude.—France, Department of Herault, near Montpellier.
Warm bicarbonate of soda waters, 60° F.
Special indications: Sedative action, useful in certain nervous affections.

Fourchambault.—France, Department of Nièvre, near Nevers.
Bicarbonate of lime waters, cold.

Fowey.—England, Cornwall.
Sea baths on the south coast.

Frailes y La Rivera.—Spain, Province of Jaen, near Alcalá la Real and 1,140 feet above sea level.
Bitter waters, 50° to 55° F.
Special indications: Bronchial catarrhs, hyperæmia of the liver, dyspepsia, leucorrhœa, chlorosis and chorea.
The situation is lovely, and the vegetation very luxuriant. The establishment is composed of 25 houses and three large edifices, having 18 balconies. The arrangements for baths are good, and the accommodation is of a superior kind; prices are, however, moderate. Ball, concert, reading, billiard, smoking and music rooms.
Doctor: Dr. Barráca.

Frankenhausen.—Germany, Thuringia, near Sondershausen.
Saline lithic waters.
Special indications: Scrofula, rheumatism.
Doctors: Drs. Graef and Pflug.
Hotels: Nigger, Thuringia.

Frankfort-on-Main.—Germany.
Saline waters; 2 grs. of mineralisation.
Special indications: Skin diseases, but scarcely used.

Frankfurt-on-Main.—Germany, until 1866 one of the four free cities remaining of the Great Hausebund, but now annexed to Prussia; from Paris by North of France Railway, *viâ* Cologne and Mayence, or *viâ* Forbach and Mayence, 18½ hours. 85 frs. 1st class.
A very wealthy commercial town; the birthplace of the Rothschilds.
OBJECTS OF INTEREST.—The Cathedral, the Römer or Town Hall, Church of St. Paul, the Exchange, Städel Museum and Academy of Paintings, Bethman villas with Danecker's statu f Ariadne, Bethmann Museum, Goethe House, Public Gardens, Zoological, Botanical, and Palm Gardens, Old and New Cemetery; Günthersberg, near Bornheim, the seat of the Rothschilds.

Post and Telegraph Office: Zeil No. 52.
CABS: For one or two persons, 50 pfg.; three or four. 70 pfgs.; for one hour, 1 mk. 70 pfgs., and 2 mk.
Bookseller: Schottenfels, Frankfurterhof.
Dentist: Brown-Townsend, Am.
Doctors: Drs. Varrentrop, Stein and Simrock.
Hotels: The *Swan* **Hotel**. A well managed and deservedly popular house, on the Theatre Platz. The treaty of 1871 was signed in this house. Geo. Fay, proprietor.
The *Hotel d'Angleterre*, a large first-class house for families, in the best central position. J. G. Saegmüller, manager

Franzensbad.—Austria, Bohemia, near Eger; 1,900 feet above sea level. For routes see table.
Six sulphate of soda and ferruginous springs, all cold; mineralisation, 4 to 5 grs. Carbonic acid waters and mud baths.
These waters are tonic and stimulating.
Special indications: Anæmia and many uterine affections
The treatment is internal rather than external.
Doctors: Drs. Hamburger, Kallay, Klein and Sommer.
Hotels: Post, Hubner, Emperor's.

Fraserburgh.—Scotland, Aberdeenshire, 28 miles from Aberdeen.
Sea baths, sandy beach and chalybeate waters.
Doctor: Dr. Grieve.

Freiburg.—Germany, Baden, in the Breisgau; 950 feet above sea level; 34,000 inhabitants. For routes, see table.
A very old town and the ancient capital of Breisgau. There is an old Münster, a fine specimen of Gothic architecture, containing many interesting objects. Archbishop's palace, Ducal palace or Citadel, Merchant's Hall, Statue of Victory, University and Library, are well worth visiting.
OTHER OBJECTS OF INTEREST.—Ludwigshöhe, Loreto chapel, Kyffhäuser, Günthersthal and Schauinsland, Belchen, Müllheim, Badenweiler, Hoellenthal, Suggenthal, Waldkirch, Titi lake, Kaiserstuhl and Feldberg, Black Forest.
English church service at the Post-office; Rev. N. G. Lawrence.
Post and Telegraph: Close to Sommer's Hôtel.
Dentist: Von Langsdorf.
Doctors: Drs. Eschbacher, Schwerkhard and Hack.
Hotel: The *Hotel Sommer*, Zähingerhof. A handsome and well appointed hotel, in large gardens Sommer Bros., the same proprietors as those of Hôtel Sommer at Badenweiler.

Freienwalde.—Germany, Pomerania, on the Oder.
Cold ferruginous bicarbonate waters. Little frequented. Situated in a very fine valley.
Doctors: Drs. Giserius and Nath.
Hotel: Kurhaus.

Freshwater.—England, Isle of Wight.
Sea baths, pebbly beach.
Climate pure and bracing.
Doctor : Dr. Phelps.

Freyersbach.—Germany, Duchy of Baden.
Bicarbonate and ferruginous waters. Treatment by means of mud and carbonic acid gas baths. Little visited.
Special indications : Liver complaints, female diseases, disorders of the respiratory and urinary organs.

Fribourg.—Switzerland, capital of Canton of same name, from Paris, *vid* Belfort and Basle to Fribourg.
 A picturesque old town, founded in the twelfth century, and well worth staying at for a day, in order to see the various sights of interest, such as the—Temple Réforme, Convent of Ursulines, Lyceum and College, Town Hall, Cathedral of St. Nicholas, with the most celebrated organ in Europe (concerts every evening); Chancelry and Wire bridge, Church of St. Augustin, Convents of Montarge and Cordelier, Churches of Notre Dame and Capucins, Seminary, Pensionate of the Jesuits, Orphanage, Museum, Colonna (opened 1881), Suspension Bridge and Chapel of Lorette.
EXCURSIONS: Promenade of the Palatinate, Viaduct of Grandfrey, Hauterive Bourgileon, artificial lake of Perolles, fish hatching establishments, garden of the hospital, the Black Lake, Kaiseregg, Gruyères, Morat and Avenches.
Post and Telegraph Office.
Hotel : The *Grand Hôtel National.* A well-managed house, clean and comfortable, nearest to Cathedral and Suspension Bridge. Organ concert every evening at 8 o'clock. F. Bassler, proprietor.

Friedrichshafen.—Germany, Würtemberg, on the north shore of the Lake of Constance ; 1,376 feet above sea level ; 3,300 inhabitants.
A climatic summer station. Lake and Turkish baths.
The town is very old, but owing to recent additions it has acquired quite a modern aspect. The situation is very fine and picturesque, overlooking the lake and the opposite coast of Switzerland, with its mountain chains. Like all stations on the lake of Constance, it forms a good centre for excursions. There is a Kursaal, with park, on the shores of the lake. A museum, with many interesting objects illustrating the history of the lake of Constance and its primitive inhabitants. The King of Würtemberg resides here in summer; the castle and park are both open to visitors.
Amongst the favourite excursions may be mentioned Manzell, Berg, Unter, and Ober Theuringen, Lochbrücke, Trautenmühle, Waggershausen Riedle, St George, Tettnang, Ravensburg, Wolfegg and Waldburg.

Post and telegraph office at the railway station.
Doctor: Dr. Seibold.
Hotel: The *Hôtel de la Couronne*, near the landing stage of the steamers, overlooking the lake and Alps, with a fine garden; well managed house, with reasonable prices. Louis Deeg, proprietor.

Friedrichshall.—Germany, Bavaria, near Kissingen and Cobourg.
Bitter water, sulphate of soda and magnesia.
The water coming into the market under the above denomination is artificially prepared. It is less irritating, and consequently less purgative, than the waters of Pullna and Seidlitz, which allows of its being taken for a greater length of time.
Only used for exportation.

Friedsrichroda.—Germany, Thuringia, near Gotha.
Saline waters and pine-cone baths.
Very picturesque site.
Doctors: Drs. Keil and Weidner.
Hotels: De Berlin, Herzog Ernst.

Friedrich Wilhelmsbad.—Prussia, Island of Rügen, near Puttbus, in the Baltic.
A much frequented and very fashionable sea-bathing resort. Established 1816.
Season: June–September.

Fuen-Alamo.—Spain, Province of Jaen.
Sulphurous waters, 45° F.
Special indications: Cutaneous affections.
Primitive arrangements, but prices very moderate.
Doctor: Dr. Salafranca.

Fuencaliente—Spain, Province of Ciudad-Real, near Puertollano.
Ferruginous waters, 100° F.; douches, steam and mud baths.
Special indications: Rheumatism, paralysis, muscular contractions, corea, hysteria, anaemia and chlorosis.
The situation is good. The establishment is of recent construction, and the accommodation is abundant.
Doctor: Dr. Crespo y Escoriaza.

Fuente-Amargosa.—Spain, Province of Malaga,, near Vélez.
Sulphurous complex waters.
Special indications: Female complaints, affections of the genito urinary organs and disorders of the stomach.
Doctor: Dr. Ortiz.

Fuenterrabia.—Spain, Province of Guipuzcoa.
Sea baths; rocky and sandy beach.

Fuente de Piedra.—Spain, near Antequera.
Indifferent waters, 60° F.
Known under the Romans as highly efficacious in cases of calculi.

Fuente del Rosal.—Spain, Province of Cuenca, near Beteta.
Earthy waters, 65° F.
Much visited, but deficient in comfort.

Fuente del Toro.—Spain, Province of Madrid.
Cold sulphurous waters, containing much free sulphur.

Fuente Santa de Lorca.—Spain, Province of Murcia, near Lorca.
Sulphurous earthy waters.
The site is very charming, but the establishment inadequate.
Accommodation not of the best.
Doctor : Dr. Bernabeu.

Fuente Santa de Gayangos.—Spain, Province of Burgos, near Espinoza and Medina de Pomar.
Sulphurous earthy waters.
Special indications: Gravel, vesical catarrh, calculus and affections of the mucous membranes.
The establishment is old fashioned. Accommodation is abundant and moderate in price.
Doctor : Dr. Carrio y Grifol.

Fumades.—France, Department of Gard, near Alais.
Sulphate of calcium bituminous waters, 40° F.
Special indications: Skin diseases and chest affections.

Füred, or Balaton-Füred.—Hungary, comitat of Szalad.
Earthy, alkaline and ferruginous water; baths in the lake of Balaton; mud baths; milk cure.
Special indications: Anæmia, chlorosis, debility.
This town is picturesquely situated, and affords valuable therapeutic aid.
Doctor : Dr. Mangold.

Gaberneg.—Austria, Styria, near Rohitsch.
Alkali-saline ferruginous waters.
Only internally employed.

Gabian.—France, Hérault, near Beziers.
Effervescent ferruginous waters. Little visited.

Gadara.—Syria, near Lake Tiberias (Tabarich).
Hyperthermal sulphurous waters; mud baths. Much frequented by the Arabs. Primitive appliances.
Special indications: Herpes, rheumatism and elephantiasis.

Gafete.—Portugal, Province of Estremadura.
Cold effervescent sulphurous waters.

Gagliana.—Italy, Tuscany, Valley of the Arno near Figline.
Earthy ferruginous waters, 50° F.

Gaieiras.—Portugal, Province of Estremadura.
Sulphurous waters, 95° F.
Little visited.

Gainfahrn.—Austria, one hour's rail from Vienna, and close to Vöslau.
Hydropathic establishment. Much frequented.
Doctor: Dr. Friedmann.

Gais.—Switzerland, Canton of Appenzell; 3,000 feet above sea level.
Bicarbonate of iron, and carbonate of lime waters; climatic and whey cure station.

Galaxidion.—Greece, Morea.
Saline waters, 65° F.
Special indications: Spleen and liver affections.

Galera.—Spain, Province of Granada.
Cold sulphurous waters. Not much visited.

Galleraje.—Italy, Tuscany, Valley of Cecina.
One thermal sulphurous spring, 115° F.; two bicarbonate and ferruginous springs, 45° F.
Season: May and June, September and October.
Good establishment and accommodation.

Galthof.—Austria, Moravia, near Brünn.
Bitter waters, some of the mildest of their class.
The water is exclusively used for exportation, and is but little known.

Galway Spa.—Ireland, Connaught.
Muriatic ferruginous waters. Only used internally by the inhabitants.

Gamarde.—France, Department of Landes, near Dax.
Cold sulphate of lime springs.

Gandesa.—Spain, Province of Tarragona.
Thermal sulphurous waters.

Garde de Bio. (La).—France, Department of Lot.
Cold sulphate of sodium waters.
Special indications: Affections of the stomach.
Doctor: Dr. Audual.

Gardinière.—France, Department of Ain, near St. Ramberd-en-Bugey.
Sulphate of lime waters. Little used.

Garmyswyl.—Switzerland, Canton of Fribourg; 1,950 feet above sea level.
Cold sulphurous waters. Little used.

Garriga (La).—Spain, Province of Barcelona, near Granollers.
Saline waters, 100° F.; various springs.
Special indications: Chronic rheumatism, tendinous contractions, articular swellings, neuralgias, colics, vesical catarrhs and skin diseases.
The bathing establishment is good, and situated in a beautiful valley; it affords good accommodation at reasonable prices. Various interesting excursions may be made in the neighbourhood.
Doctor: Dr. Padrals.

Garris.—France, Lower Pyrénées, near Mauleon.
Sulphate of lime springs.

Garryhill Spa.—Ireland, County Carlow.
Ferruginous waters.

Gastein, or Wildbad-Gastein and Hof-Gastein.—Austria, Duchy of Salzburg; 2,900 feet above sea level. For routes see table.
Indifferent waters. Eighteen springs, of which nine only are utilised; temperature ranges from 75° to 170° F.
Although these waters are not more highly mineralised than ordinary water, they are useful in relieving nervous affections, gout and rheumatism. They are charged with electricity.
The price of lodgings is fixed by the authorities.
Doctors: Drs. Pröll and Schider.
Hotels: Straubinger, Schloss.

Gath.—Austria-Hungary, Transylvania, near Stuhlweissenburg.
Earthy waters, 50° F.

Gava.—Spain, Province of Barcelona.
Bicarbonate and ferruginous waters, 45° F.

Gaviria.—Spain, Province of Guipuzcoa.
Sulphur springs, 45° F.
Doctor: Dr. Pardo.

Gavorrano.—Italy, Tuscany, valley of the Pecora.
Earthy ferruginous waters, 90° F.
The bathing establishment is good, but the unhealthiness of the site prevents any large concourse of visitors.

Gazost.—France, Department of Hautes Pyrénées, near Lourdes.
Sulphurous sodic, bromo-iodurated waters, 50° F.
Deserve to be better known.
A very good establishment, with comfortable accommodation.

Gebangau.—Java, Dutch Indies.
Saline springs, rich in iodine and magnesia.

Gehringswalde.—Germany, Saxony, near Wolkenstein.
Earthy alkaline waters, 75° F.
Only used in baths. The establishment contains all modern appliances, and affords comfortable accommodation to visitors.

Geilnau.—Germany, Nassau.
Alkaline ferruginous table waters, analogous to those of Selters. Exported only.

Geisslingen.—Germany, Würtemberg, between Stuttgart and Ulm.
Indifferent waters.
A small establishment.

Geltschberg.—Austria, Bohemia, near Leitmeritz.
Hydropathic establishment. Climatic air station.

Gempelenbad.—Switzerland, Canton of St. Gall; 1,680 feet above the sea level.
Earthy waters.
Used only by the surrounding peasantry.

Genestelle.—France, Department of Ardèche, near Entraigues.
Effervescent ferruginous waters. Scarcely known.

Geneva.—Switzerland, on the Lake of Geneva.
Transition station. Bathers from Aix-les-Bains come here to take a few days' rest after their cure.
Milk cure.
Bookseller: Monroe, 32, Grand Quai.
Doctor: Dr. Wilkinson.
Hôtel des Bergues and National.

Genoa.—Italy, on the Bay of Genoa, in Liguria; long a Republic, but now united with the kingdom of Italy; from Paris by P.L.M., viâ Mont Cenis, or Marseilles, Nice, 25 hours. 120 frs. 1st class.
Genoa, called "La Superba," or the "Proud," is situated in the middle of the Riviera. It is a city of great antiquity, and is seen to the best advantage when entering the harbour by boat. The hills are covered with villas. A very fine harbour, protected by two moles.

OBJECTS OF INTEREST.—The Corso, Cathedral of San Lorenzo, l'Annunziata, Palazzo Brignole, Palazzo Pallavicini, Doria Torsi, Balbi, Palazzo Reale, Palazzo Doria, della Universitá with a library, Ducale; Bank of St. Georgo (the oldest bank in Europe) Exchange in Loggia dè Banchi, Academy, Public Library, Theatro Carlo Felice, Public Gardens, the Fortifications.

EXCURSIONS: Villas Doria and Pallavicini, at Pegli: Nervi and Sestri-ponente.
Post and Telegraph Office: Piazza Fontana Morose.
CABS: One-horse, 1 fr. the drive; by the hour, 1 fr. 50 c.
Doctor: Dr. Breiting, 33, Via Mamelli.
Hotels: The *Grand Hôtel de Gênes*, first-class, best position in Genoa facing the Opera House, and near the Post. Bonera Frères, proprietors; also at Nervi.

Georgenbad.—Germany, Saxony, near the Bohemian frontier.
Sulphurous acidulated waters.
The bathing establishment has been much neglected.

Gerace.—Italy, Province of Reggio di Calabria. Rail from Naples to Reggio, stop at Gerace, thence to the baths by omnibus or carriage, one hour.
Effervescent saline, ferruginous and sulphurous waters, containing arsenic; the air is pure and healthy.
Season: May—September.
Special indications: Inflammation of the respiratory and digestive organs, herpes, skin diseases; chronic, articular and muscular rheumatism; glandular obstructions, radical cure of *syphilis*.
The establishment is not such as the efficacy of the waters and the large number of visitors require.
The hotels are good and comfortable; villas and private houses may also be had for the season or by the month.

Geroldsgrün.—Germany, Bavaria.
Cold ferruginous bicarbonate waters

Gerolstein.—Germany, Rhenish Prussia, in the Eifel, near Trèves.
The waters used largely under this name in England as table waters, and exported by an English Company, are in reality those of Birresborn.

Gersau.—Switzerland, on the lake of Lucerne; 1,400 feet above sea level; a small village with 2,000 inhabitants. For routes see table to Lucerne; thence by boat in 1½ hours.
A climatic station, and also for the whey and grape cure. Inhalations, spray, pneumatic apparatus, electricity and shower baths. Bathing establishment on the lake.
Season: Spring and autumn, but open all the year.
Gersau is a quiet but charming village, situated on the southern slopes of the Righi. Its position is extremely picturesque and well sheltered. The climate is similar to Vevey and Montreux, but milder and less variable. The fact that figtrees, almonds and laurels grow well out of doors through the winter, shows how genial the climate is. Gersau is much visited by patients going to and from the Riviera and Italy, while a good many pass the whole winter here. During the summer months it is more a centre for excursions, which may be taken in endless variety, and the mere enumeration of which would occupy too much space. A large park surrounds the Villa Flora, and is open to visitors.
Post and Telegraph Office.
English Church.
Doctor: Dr. Frouler.
Hotel: The *Hotel Müller*, with its dependency the Villa

Flora, is a thoroughly well managed hotel, affording all
necessary accommodation ; 130 rooms, with all the advantages
of a Kurhaus. Pension from 7 to 10 fcs. a day.

Giengen.—Germany, Würtemberg ; 1,450 feet above sea level.
Earthy waters, of the same class as " Wild baths."
Very good establishment ; largely frequented.

Giesshübl.—Austria, Bohemia, near Carlsbad ; 1,100 feet above sea level.
Pure alkaline waters.
The water is chiefly exported. There is a whey cure and
hydropathic establishment.
Special indications : Feeble digestion and nervous stomachic
disorders, affections of the urinary organs, albuminuria, diabetes.
A very elegant and comfortable Kurhaus.
Doctor : Dr. Gastl.

Giglio.—Italy, Tuscany, Island of Giglio.
Ferruginous effervescent waters, used only as table waters.

Gigondas, or Montmirail.—France, Department of Vancluse.
Acidulated waters, 60° F.
Mild but unsteady climate.

Gigonza.—Spain, Province of Cadiz.
Cold sulphate of lime waters.

Gijon.—Spain, Gallicia, near Oscura.
Sea baths. Very fashionable.
The scenery is highly picturesque, and the excursions in the
neighbourhood very interesting.

Ginoles.—France, Department of Aude, near Limoux.
Thermal sulphate of magnesia waters, 75° F.
Mildly laxative and diuretic.

Gisi.—Switzerland, Canton Unterwalden.
Alkaline waters. Not used.

Gisland.—England, Cumberland, on the line of railway from Carlisle to
Newcastle.
Sulphurous waters, little visited.

Giunco Marino.—Italy, Tuscany, Valley of the lower Arno, district of
Lorenzana.
Alkaline-ferruginous waters, 50° F.
Only used in the immediate neighbourhood.

Glaine-Montaigut.—France, Puy de Dôme, near Clermont-Ferrand.
Cold ferruginous waters.

Glanagarin.—Ireland, Cork.
Ferruginous waters; but little used.

Gleichenberg.—Austria, Styria, near Graz. One hour by carriage from **Feldbach** station on the West Austrian Railway, and 3 hours from Spielfeld on the South Austrian line.
Alkaline, muriatic and ferruginous waters, four springs. Grape cure in autumn. Climatic air station in spring and autumn. Carbonic acid gas baths, **iron and fir-cone** baths.
Season: May—October.
Special indications: Nasal catarrh, inflammation of pharynx and larynx, chronic bronchial catarrhs, emphysema, asthma, chronic catarrhal inflammation, incipient phthisis, affections of the intestines, congestion of liver and spleen, vesical catarrh, amenorrhœa, sterility, chlorosis, scrofula, anæmia, **gout, and chronic rheumatism.**
Gleichenberg is situated in a park-like valley, offering great diversity of scenery and vegetation. The accommodation is of the best, there being more than eighty villas at the disposal of visitors. The climate is mild and equable, the mean annual temperature being 65° to 70° F. Mean atmospheric pressure, 735 $^m/_m$., Gleichenberg is consequently a convenient intermediate halting place for **invalids** passing from a southern to a northern climate.
Post and telegraph. Excursions of the most varied nature. Concerts twice daily, theatre, casino, reading rooms, tombolas, &c.
Doctors: Drs. Präsil and von Hausen.
All information may be had from the director of the baths.

Gleisweiler.—Germany, Palatinate, near Landau; 1,000 feet above sea level.
Hydro-therapeutic establishment; **koumiss, whey and grape** cures.
Doctor: Dr. Schneider.
Hotel: The *Establishment*.

Gleissen.—Germany, Prussia, near Laugsberg and Küstrin.
Saline ferruginous waters; whey cure.
A very good establishment.

Gleisslibergerbad.—Austria, Tyrol, near Botzen.
Cold sulphurous waters.
A small bathing establishment

Glengariff.—Ireland, Cork, near Bantry.
Sea baths; climate bracing but mild.

Glenn-Sulphur-Springs.—U. S. America, North Carolina, near Columbia.
Sulphate of magnesia springs.
Much frequented.

Gloucester.—England.
　Saline ferruginous waters.
　Much frequented, and possessing good bathing appliances.

Gmünd.—Austria, Carinthia, district of Villach.
　Two cold sulphurous springs.
　Only used by the surrounding peasantry.

Gmunden.—Austria, Upper Austria; 1,290 feet above sea level.
　Saline waters, 25 per cent., 2 per cent. more than at Reichenhall. Climatic station.
　The Curhaus provides saline, brine, warm, pine-cone and steam baths, and the arrangements are very efficient. The casino contains concert, card, reading and ball rooms, and is elegantly fitted up.
　The scenery around is very picturesque.

Goczalkowitz.—Germany, Silesia, near Pless.
　Very strong, bromo-iodurated saline waters, containing iron, 50° F.
　There is a very good establishment, principally for sitz baths.

Godelheim.—Germany, Westphalia, district of Höxter.
　Cold bicarbonate and ferruginous waters.
　A good establishment, but little known.

Godesberg.—Germany, Rhenish Prussia, near Bonn.
　Alkaline-saline ferruginous waters. Climatic summer station of great repute, and much frequented.
　A charming little town in a very picturesque country, near the seven mountains chain; half-hour from Bonn. Combines town with country comforts.
　Doctors: Drs. Finkelnburg, Gerber and Schwann.
　Hotels: Blinzler, Alder, and Riebes.

Golaise (La).—Switzerland, Canton Valais.
　Sulphurous and ferruginous waters.

Goldbach.—Germany, Bavaria, near Aschaffenburg.
　Alkaline ferruginous waters.

Goldberg.—Germany, Duchy of Mecklenburg-Schwerin.
　Chalybeate waters.
　An efficient bathing establishment, containing sitz, douche, spray and steam baths.

Golden Bridge.—Ireland, County Dublin.
　Strong sulphurous waters.
　Very good establishments.

Gonten.—Switzerland, Canton of Appenzell ; 2,750 feet above sea level.
Earthy ferruginous waters.
The establishments are thoroughly efficient, and afford comfortable accommodation.

Göppingen.—Germany, Würtemberg, between Stuttgart and Ulm.
Alkaline magnesia waters. Exported.
A well arranged establishment. Charmingly situated.

Görbersdorf.—Germany, Silesia, circuit of Breslau. Rail to Dittersbach ; 1,700 feet above sea level. For routes see table.
Mountain air cure station.
Special Indications: Phthisis, affections of the respiratory organs, anæmia, chlorosis, nervous asthma and emphysema.
Doctors: Dr. Brehmer and Dr. Römpler.
Hotels: The two establishments, one under the direction of Dr. Brehmer and one under Dr. Römpler.

Gori.—Russia, Caucasus, near Semu-Chwedureti, district of Gori.
Cold sulphurous waters.
Special indications: Mercurial dyscrasia, liver and spleen disorders, old sores, herpes.
No establishment, and the baths are taken in common by patients in a natural basin in the rocks.

Gortwa-Kisfalu.—Austria-Hungary, near Vargedeu.
Cold ferruginous waters.
Accommodation indifferent.

Görz or Goritz.—Austria, Illyria.
A climatic winter station. Grape cure.
Very pure air.
Season: Autumn and early spring.
Special indications: Scrofula, rhachitis, chronic catarrh, with emphysema, phthisis, nervous complaints.
Mean winter temperature, 52° F.; windless days, 71 in winter season; rainy days, 81 ; cloudless days, 53 ; mean rainfall, 841 $^m/_m$. The climate must therefore be considered as a humid one.
Doctor: Dr. Arone.
Hotels: The *Grand, De la Poste*.

Gosehwitz.—Germany, Duchy of Saxe-Weimar, near Jena.
Cold sulphate of lime waters.

Göurnay.—France, Department of Seine Inférieure, near Neufchâtel-en-Bray.
Cold ferruginous waters.
Only used in the neighbourhood as a table water.

Gourock.—Scotland, Renfrewshire, on the Clyde.
Sea baths, with good beach and healthy climate.
Doctor: Dr. Leitch.

Grabalos.—Spain, Province of Logroño.
Sulphate of lime waters.

Graena.—Spain, Province of Granada, near Guadix; 2,900 feet above sea level.
Thermal ferruginous and sulphurous waters, 100° to 105° F.
Special indications: Rheumatism, paralysis, gout, hemiplegia and paraplegia, nervous affections, corea, chlorosis, anæmia, amenorrhœa and dismenorrhœa.
Superior bathing appliances and accommodation.
Doctors: Dr. Estrada.

Gräfenberg.—Austria, Silesia.
Indifferent waters. Hydropathic establishment of great celebrity.
Doctors: Drs. Angel and Schindler
Hotels: *Crown Prince*.

Gramat.—France, Department of Lot, near Gourdon.
Cold bicarbonate of iron waters.

Gran.—Austria-Hungary, comitat of Gran.
Sulphate of magnesia or bitter waters. Mineralisation 108 gr., of which 104 are sulphate of magnesia.
One of the strongest bitter waters known. The action is rapid and very marked, even in small doses.

Granada.—Spain, the ancient capital of the Moorish Kingdom, and at present of the Province of Granada; from Paris, *viâ* Bordeaux, Irún, and Madrid, 41½ hours.
Granada is a very old town, on the confluence of the Darro and Genil, partly built in a valley and partly on the slopes of a mountain. Built in the form of an amphitheatre, it adorns the hill slopes on which the Alhambra stands.
OBJECTS OF INTEREST.—The Alhambra, Generalife, and Moorish remains; Cathedral and Capilla de los Reyes, Cartuja, Churches of various saints, Public and Private Palaces, Zacatin, Alcaicería, University, Museum, Casas Consistoriales, Court of Chancery, Lunatic Asylum, Libraries, several old Gateways, Theatre.
Post Office: Plaza de San Augustin.
Telegraph Office: Calle de la Duquesa.
Cabs: The course, 6 reals or 3 frs.; to any part of the town 12 reals or 6 frs., except Alhambra or Generalife, for which 10 reals extra are asked.
Hotels: *De la Alameda, De la Victoria, De los Siete Suelos, Washington Irving*.

Grandcamp.—France, Department of Calvados, from Paris, *vid* Isigny.
Sea baths, with good sandy beach.

Grandeyrol.—France, Department of Puy de Dôme, near Issoire.
Ferruginous bicarbonate waters.

Grandril.—France, Department of Puy de Dôme, near Ambert.
Cold bicarbonate of lime waters.

Grange-over-sands.—England, Lancashire, Morecambe Bay.
Sea baths and good beach.
Climate fresh and bracing.
Season: March—October.
Doctor: Dr. Beardsley.

Gränichenbad.—Switzerland, Canton of Argovia.
Earthy waters.
A good bathing establishment, in Swiss chalet style, with ample appliances.

Granville.—France, Department of La Manche.
Much frequented sea baths.

Grao.—Spain, near Valencia.
Sea bathing and sandy beach.
Grao shares with Carbagnal (¾ hour distant) much the same reputation in Spain, which Brighton and Hastings do in England. Accommodation very good.

Grasnawawoda.—Austria-Hungary, comitat of Neutra, one hour from Iastrabje.
Ferruginous effervescing waters. Only used internally.

Grasse.—France, Department of Alpes Maritimes, three-quarters of an hour's rail from Cannes; 1,230 feet above sea level; 11,000 inhabitants. For routes see table.
Winter station. Dry and sedative climate.
Season: October—June.
Special indications: Chest and lung complaints, laryngitis, pharyngitis, nervous affections, anæmia, chlorosis.
English Church service is held in winter in the *Grand Hôtel*.
Post Office. Boulevard du jeu de ballons.
Telegraph: Au Cours.
Doctor: Dr. Ferraud.
Hotels: The *Grand Hôtel De Grasse, de la Poste*.

Grasville-l'heure.—France, Department of Seine Inférieure, near Le Havre.
Iodurated ferruginous waters.

Gravalos.—Spain, Province of Logroño, near Calahorra.
Sulphurous earthy waters, 50° F.
Special indications: Skin diseases, leucorrhœa, prurigo, chronic bronchial catarrhs, scrofula.
The situation is picturesque, the establishment a little backward, but its accommodation cheap.
Doctor: Dr. Aguilera.

Gravesend.—England, Kent, on the mouth of the Thames.
Sea baths; good beach. Healthy air.
Season: June—September.
Doctor: Dr. Firman.

Great Yarmouth.—England, Norfolk.
Sea baths; fine sandy beach.
Hot and cold baths on the parade. Climate bracing and exciting.
Doctor: Dr. Vores.

Greifenberg.—Germany, Bavaria, 4 hours from Landsberg and 2,000 feet above sea level.
Chalybeate waters; mud baths; also a climatic station.
Special indications: Feeble digestion, anæmia, disorders of the urinary system.
The environs are very interesting, possessing numerous ruins of Roman buildings. The site is charming. The Ammersee is close by.

Greifswalde.—Germany, Pomerania.
Strongly saline and bromine waters.
Establishments and accommodation good.

Gréoulx.—France, Department of Lower Alps, near Digne.
Thermal bromo-iodurated and sulphate of lime waters. The various springs remain much as they were in the time of the Romans.
Special indications: Scrofulous affections, and general constitutional weaknesses, atonic sores, blenorrhœa.
In proper hands, Gréoulx would soon become a first-class bathing and climatic station.
Doctor: Dr. Joubert.

Gries.—Austria, Tyrol, **close to** Botzen.
Climatic winter **station, in a well** sheltered position.
There are very few days here during winter when an invalid cannot take outdoor exercise between 11 and 2 o'clock. Much patronised by Austrians and Germans.
The accommodation is good, but amusements are few.
Doctor: Dr. Mayrhofer.
Hotel: Belle Vue.

Griesbach. Germany, Duchy of Baden, **in the valley of the Rench,** on the road from Strasbourg to Stuttgard.
Two chalybeate springs, **one of** 60° F. Mineralisation, 3 gr.
A much frequented station, providing also the milk cure.
Doctors: Drs. Jaegerschmid (speaks English) and Haberer.
Hotels: Kurhaus.

Griesbad.—Germany, Bavaria, near Ulm.
Earthy ferruginous waters.
A good establishment.

Grindelwald.—Switzerland, Bernese Oberland; 3,460 feet above sea level; 3 hours **from** Interlaken.
A very important climatic mountain station, especially suitable in cases of anæmia, hydræmia, chlorosis and rhachitis.
Goats,' cows' and asses' milk. Baths of every description. The scenery all around is unsurpassed, and **walks** and excursions of great variety abound. Latterly a mineral spring has been discovered, but has **not** yet been analysed.
The *Bear* (I. Boss, **proprietor**) is the hotel where English comforts are most studied. Arrangements made **for** families.

Grinneaux.—France, Department of Puy de Dôme, near Riome.
Bicarbonate of iron waters, 60° F.
These waters are used for incrustation more than for medical purposes.

Grodeck.—Austria, Galicia, district of Lemberg.
Sulphurous waters.
Special indications: Rheumatism and gout.

Gross-Albertshofen.—Germany, Bavaria, near Sulzbach.
Cold sulphate of magnesia water.

Grosskarben.—Germany, Bavaria, on the Nidda.
Earthy saline waters, very effervescent.
A good establishment, **with** abundant accommodation, reasonable **in price.**

Gross-Schlagendorf.—Austria-Hungary, comitat of Zips; 3,025 feet above sea level.
Alkali saline waters.
A very efficient establishment, with comfortable and cheap accommodation.

Grosswardein.—Austria-Hungary, near Hájos.
Sulphate of lime waters, 90° to 110° F.
Sulphurous indications in general; a much frequented station.
Special indications: Rheumatism and old wounds.
Two very elegant establishments, with concert, ball, reading, billiard and smoking rooms.

Gross-Wunitz.—Austria, Bohemia, near Püllna.
Sulphate of soda and magnesia water, 22 grs. of mineralisation; purgative action.

Grotta del Cane.—Italy, Naples, near Agnano.
Natural steam baths.
A grotto with gaseous emanations, much visited as a natural curiosity.

Grüben.—Germany, Silesia.
Sulphur ferruginous waters, very slightly mineralised.
Has a small bathing establishment.

Grull.—Germany, Westphalia.
Saline springs.
Special indications: Scrofulous affections

Grund.—Germany, Brunswick.
Pine-cone, steam baths, and inhalations.
In the midst of the magnificent pine forests of the Harz.
Special indications: Convalescence after severe illnesses of all kinds, affections of the respiratory organs.
Doctor: Dr. Brockmann.
Hotel: Town Hall Hotel.

Grundhofen.—Germany, Saxe-Weimar, near Salzungen.
Effervescent ferruginous waters.
In a charming valley. Its elegant and comfortable establishment attracts many visitors.

Guagno.—France, Island of Corsica.
Two sulphate of soda springs, 95° to 125° F.
A thermal military establishment.
Special indications: Rheumatism in its chronic forms.

Guarda-vieja.—Spain, Province of Almeria, near Berja.
Thermal sulphate and saline waters, mineralisation 44 grs.
Special indications: Rheumatism, herpes, glandular swellings.
The establishment is romantically situated, is good and affords comfortable accommodation at low prices.
Doctor: Dr. Serrano.

Guernsey.—England, an island in the Channel, off the coast of Normandy.
Sea baths much visited.

Guesalivar or **Sta. Agueda de Mondragon.**—Spain, Province of Guipuzcoa.
Cold sulphurous waters.
The bathing establishment is one of the best appointed in Spain, with good accommodation.

Guillon.—France, Department of Doubs, near Baume-les-Dames.
Cold saline springs, slightly mineralised; have a strong smell of sulphuretted hydrogen. This water frequently induces erysipelas.
Special indications: Rebellious neuralgias and syphilis.
The bathing establishment is well fitted up, and has ample accommodation.
Doctor: Dr. Brauley.

Guitera.—Corsica.
Undetermined waters, 110° F.

Guitiriz.—Spain, near Lugo.
Sulphurous waters, 50° F.
Special indications: Herpes.

Günthersbad.—Germany, Thuringia, near Erfurt.
Cold sulphate of soda and saline water.

Gurnigel.—Switzerland, Canton of Berne; 3,700 feet above sea level.
Sulphate of lime and ferruginous waters.
Climatic station.
Very pure air. The establishments are thoroughly well arranged with all modern accommodation.

Gustafsberg.—Sweden, near Uddevalla.
Sea baths and ferruginous waters; mud baths.
A much frequented station.

Gyrenbad.—Switzerland, Canton of Zürich; 2,320 feet above sea level.
Effervescent earthy waters.
An establishment with sitz, swimming and steam baths. The water is seldom used internally. Climate severe, but healthy.
Special indications: Very efficient in rheumatism and abdominal obstructions.

Gythium.—Greece, south coast of Laconia.
Cold sulphurous and alkaline waters.

Gyüzy.—Austria-Hungary, comitat of Hónth.
Alkali-**saline** and sulphurous waters, 60° to 70° F.
Special indications : Affections of the eyes, gout, rheumatism, chronic skin diseases.

Häbernbad.—Switzerland, Canton of Berne, near Hutwyl.
Cold sulphurous waters; whey cure. Primitive arrangements.

Hackelthal.—Germany, Bavaria, half-an-hour from Haag.
Alkaline waters.

Hafkreuz.—Germany, Hollstein.
Sea baths; very fine sandy beach. Much visited by the inhabitants of Lübeck and Hamburg.

Hague, the, or s'Gravenhage.—Present capital of Holland; from Paris, *via* Brussels, Antwerp, Rotterdam.
The residence of Court and Government, founded A.D. 1250, is one mile from the sea. Has been called the "largest village in Europe," owing to its having never had any fortifications.
OBJECTS OF INTEREST.—Royal Museum, Naval Museum, Town Hall, Buitenhof and Binnenhof, Library, Royal Palace, Prince of Orange's Palace, Great Church, Opera-house, Palace in the Wood.
Post and Telegraph Office: Close to the Great Church and Town Hall.
Cabs: One or two persons, 60 c. or 1s., three to four persons, 85 c.; by hour, 1 florin; to Scheveningen and back, 2 florins.
Doctor : J. H. Blom Coster.
Hotels : Paulez, Oude Doelen, de l'Europe, de Belle Vue.

Haiti or St. Domingo.—An island in the Antilles.
Several sulphurous thermal springs.

Haj-Stubna.—Hungary, comitat of Thuraz on the Stubna.
Thermal sulphate of soda waters, 105° F.
Laxative action.

Haldenstein.—Switzerland, Grisons, near Chur.
Alkaline waters; little used.

Hall.—Austria, near Linz; 1,157 feet above sea level.
Cold saline and bromo-iodurated waters, mineralisation 17 grs.
Special indications : Scrofula, goitre.
Doctors : Drs. Körbl and Rabl.

Hall.—Austrian Tyrol, near Innspruck, 1,610 feet above sea level.
Cold saline waters.
The bathing establishment is at Baum-Kirchen.
Special indications : Scrofula.
Doctor : Dr. Katzer.
Only private dwellings; there is no hotel.

Hall.—Germany, Würtemberg, near Stuttgart; 670 feet above sea level.
Cold saline waters.
Special indications: Scrofula.
A much frequented station.
Doctors: Drs. Bilfinger, Dürr and Haucisen.
Hotels: Bath, Lamb.

Halle.—Germany, Prussian Province of Saxony, district of Merseburg.
Cold saline waters.
There is a very good establishment here, and the University is one of the best in Germany.

Hallein.—Austria, near Salzburg on the Salzach.
Saline waters. Climatic station in summer.
Very picturesquely situated, and a good centre for many interesting excursions.

Halsbrücke.—Germany, Saxony, near Freiberg.
Sulphurous and chalybeate waters.

Hambach.—Germany, Rhenish Prussia, 10 hours from Trèves.
Alkaline ferruginous waters.
These springs are also called the Birkenfeld water. Much frequented in the 18th century. Establishment destroyed in 1792, but rebuilt in 1845. Visitors have increased in number every year since then.

Hammam Berda.—Algiers, between Bone and Constantine.
Bicarbonate of lime waters, 70° F.

Hammam de Gabes.—Tunis.
Thermal waters not yet analysed.

Hammam el enf.—Regency of Tunis, near Tunis.
Thermal waters.

Hammam-Melouan.—Algiers, near Rovigo, about 20 miles from the capital, Algiers.
Saline waters, 103° F., strongly mineralised, 26gr. of chloride of sodium.
Special indications: Very efficient in rheumatism and abdominal obstructions.

Hammam-Meskutin.—Algiers, Province of Constantine, 12 miles west of Guelma.
Hyperthermal waters, 203° F., saline, sulphate of lime and arsenical springs.
Military bathing establishment.
Special indications: Paralysis, paludal cachexia, herpes, syphilis, old wounds.

Hammam R' Irha.—Algiers. Railway from Algiers to Bou-Medfa, 4 hours; thence to Hammam R' Irha by carriage, 8 miles.
One thermal sulphate of lime spring, 133° F., and one bicarbonate of iron, 45° F.
Season: Spring and autumn months.
Special indications: Articular pains and stiffness, muscular retractions, anchylosis, rheumatism and anæmia.
It has a military hospital.
Doctors: Dr. Savill.
Hotels: The Establishment. Mr. A. A. Dufour, proprietor.

Hampstead.—England, a suburb of London.
A place visited more for the beauty of its situation than for the medicinal properties of its waters.

Hanau.—Germany, Hesse-Nassau, near Frankfort-on-the-Main.
Alkali-saline and ferruginous waters.
The establishment is tastefully built, and contains sitz, douche and steam baths. It is surrounded by a large park.

Hanover.—Germany, former capital of Kingdom of same name. Since 1866 annexed to Prussia, and now capital of the Province of Hanover. From **Paris**, *viâ* Erquelines, Venlo, and Lehrte, or *viâ* Cologne, 15½ hours. 87 frs. 50 c. 1st class.
Beautifully situated on the banks of the river Leine, an affluent of the Weser.
Very much visited by English and American families, both for the beauty of the town and its environs, as also because the German language is spoken here with greater purity than in any other part of Germany.
OBJECTS OF INTEREST.—Royal Palace, Opera-house, Mint, Viceroy's Palace and Arsenal, Polytechnic School, Royal Church or Chapel, Museum, Collection of Pictures of George V., Town Hall in Old Town; Royal Theatre; Royal Library, Aquarium, Odeon and Tivoli.
EXCURSIONS: To Herrenhausen, Galleries, Orangeries, Berggarten, Museum of antiquities, Mausoleum, Eilenrode and Zoological Gardens, and numerous others.
CABS: One or two persons, by the course, 60 pfgs., in the environs of the town, 72 pfgs.; by hour, 1 m., 20 pfgs.
Dentist: F. L. Heyne.
Doctors: Drs. Becker, Dyes, Cumme and Eysell.
Hotels: Royal, de Russie.

Hapsal.—Russia, Esthonia, on the coast of the Baltic Sea.
Sea and mud baths, with temperature of about 60° F.
Celebrated for its mud baths.

Hardeck.—Germany, Bavaria, Palatinate.
Cold sulphurous and ferruginous waters.

Harkanyi.—Hungary, comitat of Baranya, near Siklós.
Alkaline sulphurous waters, 130° F.
A small establishment.
Season: May–October.
Special indications: Gout, rheumatism, feeble circulation, hæmorrhoids, lymphatism, scrofula, affections of the female organs, mercurial dyscrasia, atonic sores.

The vegetation is very luxuriant, and the site picturesque. The near neighbourhood of Siklós and Fünfkirchen, with all the advantages offered by their size, &c., the mild climate and elegant bathing establishments make Harkanyi a very favourite Hungarian bathing station. The environs are full of most interesting antiquities. Accommodation is abundant and good, and prices moderate.

Harrodsburg.—U. S. America, Kentucky.
A much-frequented thermal watering-place.

Harrogate.—England, Yorkshire. About a hundred trains pass daily through Harrowgate.
Its waters are athermal. They contain sulphuretted hydrogen, chloride of sodium and iron.
Special indications: Chlorosis, anæmia and herpes.

Harrowgate is a very fashionable watering-place, with every accommodation for visitors.

The old sulphur spring has a mineral percentage of 137 grs., and is highly charged with carbonic acid.
Doctors: Drs. Ford and Oliver.
Hotels: Queen's, Crown.

Harsfalva.—Austria-Hungary, comitat of Bereg; rail to Munkácz.
Chalybeate waters; hydropathic establishment and climatic station.

The establishments and accommodation are up to all modern requirements. The environs are very charming. Large forests, with fishing and hunting.

Hartfell.—Scotland, Dumfriesshire.
Chalybeate waters.
Special indications: Chronic ulcers, anæmia, chlorosis, skin diseases, gout, rheumatism, paralysis.

Hartlepool.—England, Durham.
Sea baths.

Harwich and Dovercourt.—England, Essex, on the North Sea.
Sea baths; sandy beach. Chalybeate spring at Dovercourt, with a good bathing establishment.
Climate mild and healthy.
Season: June–September.
Special indications: General debility, anæmia, scrofula.
Doctors: Drs. Cook and Dalton.
Hotel: The Pier.

Harzburg.—Germany, Duchy of Brunswick, in the forest of the Harz.
Saline waters, mineralisation 6 grammes of chloride of sodium.
Whey cure, pine-cone sap baths, climatic station in summer.
Very charming locality, surrounded by pine forests.
Doctor : Dr. Franke.
Hotels : *Kurhaus Iuliushall, Löhr, London.*

Hassan-Pacha-Palenka.—Servia, near Lemendria.
Ferruginous waters, strongly impregnated with carbonic acid.

Hastings and St. Leonard's.—England, Sussex.
Sea baths and winter station; sandy beach. A favourite rendezvous of fashionable English society. Very mild and salubrious climate.
Season : September to March.
Special indications : Dyspepsia, bronchitis, nervous irritation and pulmonary complaints.
Doctors : Drs. Brodribb, Croft and Greenhill.
Hotels : The *Queen's*, The *Albion*.

Hauterive.—See Vichy.

Havre.—France, Department of Seine Inférieure, at the mouth of the Seine, 142 m. from Paris, and 105 sea miles from Southampton. Steamers of the L. & S. W. Railway (London, Waterloo Station) ply three times a week between Southampton and Le Hâvre, 6½ hours; 28 fcs. 1st class. Steamboats continuously ply between Le Hâvre, Honfleur, Trouville and Deauville.
Sea baths. The beach being stony, bathers require shoes.
Season : 1st June to 1st October.
A very thriving commercial town, containing two theatres, chamber of commerce, several interesting churches, exchange, town hall, aquarium, port and docks. After Marseilles, it is the largest seaport in France.
English Church : Rue Mexico.
Post and Telegraph Office : Boulev. de Strasbourg.
Doctors : Dr. Farral, 170, Boulev. de Strasbourg.
Hotels : The *Hotel, Casino* and *Baths of Frascati* are situated close to the sea-shore. Large first-rate establishments, band, various amusements every day. The hotel cannot be too highly recommended for its excellent management and comfort. Theoph. Fötsch, proprietor.

Hayling-Island.—England, Hants, on the South coast.
Sea baths, fine sandy beach. Climate mild and bracing.
No particular season; open all the year round.
Special indications : Convalescence.
The vegetation of the island is luxurious.
Doctor : Dr. Aldersey.
Hotel : The *Royal*.

Hechingen.—Germany, Hohenzollern, between Würtemberg and Baden, near Siegmaringen; 1,570 feet above sea level.

Sulphurous calci-iodurated waters.

Special indications: Skin diseases.

The bathing establishment is situated in a park, and is thoroughly well fitted up.

Heckinghausen.—Germany, Rhenish Prussia.

Earthy ferruginous and sulphurous waters.

Heidelberg.—Germany, Duchy of Baden. For routes see table.

Climatic station

Most pleasantly situated on the banks of the Neckar, and in what is certainly one of the most beautiful localities in Germany. It owes its celebrity chiefly to its university, which, after that of Prague, is the oldest in Germany

The castle of Heidelberg was formerly the residence of the electors, but is now in ruins; in its cellars is the celebrated "Heidelberger Fass." The whey cure station, Church of Holy Ghost, St. Peter, university, with an archæological institute, zoological and botanical gardens, zoological museum and mineral collection, are close by.

Excursions to the Königsstuhl, Wolfsbrunnen, Schwetzingen, and Philosophen weg.

CABS: One hour, 1 m., 20 pfgs.; by course, one person, 50 pfgs.; two persons, 90 pfgs.

Dentist: Dr Middelkamp, Grabengasse 14.

Bookseller: Chas. Schmidt, Grabengasse.

Doctors: Dr Erb, Dr. Arnold, Dr. Czerny.

Hotels: The *Hotel Prince Charles*, a first-rate popular house, with view of the castle; railway tickets can be obtained and luggage registered in the hotel. Excellent cuisine and wines of their own growth. Sommer & Ellmer, proprietors. In connection with *Hôtel du Pavillon* at Cannes.

The *Grand Hôtel*, a large well-appointed house, with first-rate accommodation; near the railway station. E. Thoma, proprietor.

The *Castle Hotel*, the only hotel in the castle grounds; air, grape, and whey cures. H. Albert, proprietor.

Miss Sutton's Pension, 49, Anlage, first-class comfortable house, quietly and well situated on the chief promenade; has the advantage of a large and shady garden.

Heiden.—Switzerland, Canton of Appenzell, 1 hour from Rorschach, on the Lake of Constance.

Ferruginous and sulphurous waters; but chiefly known as a climatic station for summer.

The climate is mild and healthy in winter.

Doctor: Dr. Altherr.

Heilbrunn.—Germany, Bavaria, near Munich.
 The Adelaide spring supplies cold iodo-bromurated water.
 Special indications: Scrofula, chlorosis.
 Doctor: Dr. Petrequin.

Heiligekreuzbad.—Austria, Tyrol, near Hall.
 Alkaline, **saline**, sulphurous waters, very mild, and little used.

Heiligenstadt.—Austria, Lower Austria, half-hour from Vienna.
 Ferruginous waters, but frequented more as a summer villeggiatura.

Heilstein.—Prussia, District of Aix-la-Chapelle, 1½ hour from Gmünd, in the Eifel.
 Earthy alkaline waters. Exported.

Heinrichbrunnen.—Germany, Silesia, District of Neisse.
 Cold ferruginous waters.

Heinrichsbad.—Switzerland, Canton of Appenzell, near St. Gall, and Herisau; 2,410 feet above sea level.
 Cold ferruginous bicarbonate waters.
 Goats' milk cure.
 Elegant Curhaus, with all modern requirements.

Helensburgh.—Scotland, on the Clyde, eight miles from Dumbarton.
 Sea baths, sandy beach; no machines.
 Doctor: Dr. Finlay.

Helenskilde.—Sweden, Zealand, Westerbotten, near Fredericksborg.
 Alkaline ferruginous waters.
 Only used in the immediate vicinity.

Heligoland.—Island of, at the mouth of the Elb, belonging to England.
 Much frequented sea bathing place.
 Season: June 15th to September.
 Doctor: Dr. Zimmermann.
 Hotels: London, Queen of England.

Hellopia.—Greece, Epirus, on the Kimara.
 Saline, sulphurous thermal waters, mentioned by Homer and Hesiod.
 Special indications: Skin diseases.
 In the vicinity was the oracle of Dodona.

Helmstädt.—Germany, Brunswick.
 Saline ferruginous waters.
 A bathing establishment for **sitz, steam, shower** and **dust baths.**

Helsingfors.—Russia, 5 hours from Reval.
Seabaths; sandy beach.
A luxurious bathing establishment, situated in a fine garden.

Helwân-les-Bains.—Egypt, near Cairo, on the Nile.
Saline, sulphurous thermal springs, 110° F.
Single bath, swimming baths, inhalations, and hot sand baths.
Special indications: Pulmonary diseases, phthisis, gout, rheumatism.
Winter station, mean temperature, 60° F. in the shade; mean moisture, 67%; rainy days in winter season, 11.
Doctor: Dr. Engel.
Hotels and pensions.

Hennebon.—France, Department of Morbihan.
Sea baths, little frequented.

Heppingen.—Germany, Rhenish Prussia, District of Ahrweiler.
Mild, alkaline earthy waters, containing a little iron.

Herculesbad.—*See* Mehadia.

Heringsdorf.—Germany, Prussia, on the island of Usedom, in the Baltic, 22 miles from Stettin; 150 feet above sea level.
Sea baths, excellent beach of very fine sand.
Much frequented; mild and steady atmosphere.
Season: June—September.

Herlein.—Hungary, near Kaschau, also called Rank-Herlein.
The beauty of the site, rather than the cold chalybeate spring, attracts visitors.
An establishment recently re-built, together with the new hotels adjoining, offer every comfort, and are well kept.
Doctor: Dr. Thomas.

Hermannsbad.—Germany, Saxony, near Lausigk.
Ferruginous and sulphurous waters. Not used therapeutically.

Hermannsbad.—*See* Muskau.

Hermida (La).—Spain, Province of Santander, near St. Vicente.
Thermal saline waters, 145° F.
Special Indications: Rheumatism and paralysis.
Doctor: Dr. Andres.

Hermione.—Greece, Argolia.
Bitter waters, mostly exported to Turkey.

Hermonville.—France, Department of Marne, near Rheims.
Bituminous sulphurous waters.
Special indications: Rheumatism, asthma, gout.

Herne Bay.—England, Kent.
Sea baths, good beach. Climate mild.
Season: July—September.
A very much frequented resort. Two establishments.
Doctor: Dr. Bowes.
Hotel: The Brunswick.

Hernösand.—Sweden, Angermanland.
Ferruginous and effervescent waters.

Herrscha.—Roumania, on the Transylvanian frontier.
Very potent sulphurous waters; used only in the neighbourhood. No visitors.

Herse (La).—France, Department of Orne, near Montagne.
Cold ferruginous bicarbonate waters, very mild.
Special indications: Debility of the stomach, leucorrhœa.

Hervideros del Emperador (Los.)—Spain, near Ciudad-Real.
Bicarbonate of lime waters, 60° F.

Hervideros de Fontillesca.—Spain, near Ciudad-Real.
Ferruginous waters, 45° F.

Hervideros de Fuen Santa (Los.)—Spain, Province of Ciudad-Real, near Pozuelo de Calatrava.
Ferruginous bicarbonate waters, 50° F., very rich in carbonic acid, which escapes from the soil in great quantities.
Season: June 10th to September 15th.
The number of visitors during the season is over 6,000.
Special indications: Rheumatism and herpes, old syphilis, hysteria, epilepsy, chorea, neuralgia, leucorrhœa.
The establishment is elegant, and very clean, a quality not always found in Spanish hotels.
One of the most frequented of Spanish baths. Excellent establishments. The accommodation is abundant and efficient.
Doctor: Dr. Fernandez.

Hervideros de Villar del Pozo (Los.)—Spain, near Ciudad-Real.
Ferruginous bicarbonate waters, 65° F.
Special indications: Rheumatism and herpes.

Heselwangen.—Germany, Würtemberg, district of Black Forest.
Cold sulphurous waters, little known.

Heucheloup.—France, Vosges, near Mirecourt.
Ferruginous iodo-arsenical waters.

Heustrich-Bad.—Switzerland, in the Bernese Oberland ; 2,050 feet above sea level.
Cold sulphate of sodium waters, rich in nitrogen. Establishment pretty good.
Special indications : Catarrhs and nervous affections.

Hevitz.—Austria-Hungary, comitat of Zala.
Indifferent waters, 75° to 85° F.
The springs form a lake. The establishment is well arranged and surrounded by a park. The accommodation is ample and comfortable.

Heyst.—Belgium, on the North Sea, near Ostend and Blankenbergh.
A rising sea bathing place, in which every effort is made to amuse and accommodate visitors.

Hildegarde-Brunnen.—Hungary, near Ofen.
Sulphate of soda and magnesia waters, 16 grammes of mineralisation; purgative.

Hilversum.—Holland, near Amsterdam.
Artificial mineral waters.
The establishment is elegant, and is every year increasing in favour. Scrupulous cleanliness is the rule.

Hing-tchon.—China, North of Pekin.
Thermal waters not yet analysed.

Hinnewieder.—Germany, Silesia.
Bicarbonate ferruginous cold waters; a much frequented establishment.

Hofgeismar.—Germany, Province of Hesse-Cassel, on the route from Cassel to Paderborn.
Cold alkali-saline chalybeate waters.
A small but very good establishment.

Hohenberg.—Germany, Bavaria, near Wunsiedel.
Bicarbonate ferruginous table water.

Höhenstadt.—Germany, Bavaria, near Passau.
Cold indifferent and slightly sulphurous waters; mineral muds.
A very good establishment, with every luxury and comfort.

Hohenstein.—Germany, Saxony, near Chemnitz.
Earthy ferruginous waters; whey cure and hydropathic establishment.
The Kurhaus is well arranged, and the accommodation abundant and good.

Holbeck.—England, Yorkshire, near Leeds.
Alkaline sulphurous waters.

Holkham.—England, Norfolk, near New Walsingham.
Sea baths; satisfactory beach.

Holyhead.—Wales, Anglesea.
Sea baths; good beach.
Doctor: Dr. Hughes.

Holywell.—England, Lancashire, near Cartmell.
Muriatic ferruginous waters.
Special indications: Anaemia, obstructions, skin diseases.

Holywood.—Ireland, Downshire.
Sea baths; sandy beach and chalybeate waters at Ballymahon and Cultra.
The air is very fine.
Doctor: Dr. Smith.

Holzhausen.—Germany, Prussia.
Cold sulphate of lime waters.
Special indications: Rheumatic neuralgias.

Homburg-les-Bains.—Germany, in the Province of Hesse Nassau; 10 miles north-west of Frankfort-on-the-Maine. For routes see table.
Four springs of saline, ferruginous and acidulated waters.
The mineral matter consists principally of chloride of sodium, and is present to the amount of 6, 10, 13 and 19 grammes per cent.
Special indications: Dyspepsia, scrofula, anaemia.
The thermal bathing establishment is comfortable and luxurious.
Doctors: Drs. Weber, Hœber, Zurbuch, Deetz, Lewis and Haase (specialist for eye and ear affections).
Hotels: Ritter's Park Hotel, combined with Café-Restaurant: first-class house, the only one on the promenade and park, close to Kurhaus, and mineral springs. Splendid view of Taunus mountain chain. Chas. Ritter proprietor, for many years manager of the Kurhaus-Restaurant.
The *Hôtel de Russie,* a large first-class well managed house, near the English church. F. A. Laydig, proprietor.
The *Hôtel d'Europe,* first-class family house, opposite the Kursaal; moderate charges. Georg Schmidt, proprietor.

Home-Varaville (Le).—France, Department of Calvados, from Paris *vid* Cobourg.
Sea baths.

Homok.—Austria-Hungary, comitat of Szathmar.
Ferruginous and **sulphurous** waters. Only used by inhabitants as a specific against **gout**.

Homorod.—Hungary, Transylvania.
Ferruginous bicarbonate waters, 30° F., tonic action.

Honfleur.—France, Department of Calvados.
Sea baths on the mouth of the Seine.

Horcajo de Lucena.—Spain, Province of Cordova, near Lucena.
Saline waters of 75° F.
Special indications: Skin diseases, leucorrhœa and hyperæmia of liver.
The establishment might be improved. The accommodation is abundant and moderate in price.
Doctor: Dr. **Urdapilleta**.

Horley-Green.—England, Oxfordshire.
Sulphate of iron waters.

Hornsea.—England, Yorkshire.
Sea baths, shingly beach.
Climate **mild** and bracing
Doctor: Dr. **Hodson**.

Horn.—Switzerland, Thurgau, 2 hours from St. Gallen.
Ferruginous and sulphurous waters.
Baths in the Lake of Constance and the whey cure.

Houches, Les.—France, Upper Savoy
Cold ferruginous, alkaline, magnesian and carbonate waters.
Special indications Chlorosis; **acts** on the digestive organs.

Houlgate.—France, Department of Calvados, *via* Dives.
Sea baths.

Hourdel.—France, Department of Somme.
Sea baths; retired and quiet.

Hovehampton.—England, Sussex.
Sea baths.

Hovingham.—England, Yorkshire.
Alkaline springs.

Howth.—Ireland, Dublin.
Sea baths, sandy **beach**.
Climate mild.
Doctor: D. **Wright**.

Hozumezö.—Austria-Hungary, comitat of Sarós.
　Cold sulphurous and ferruginous waters.
　Only used by the surrounding inhabitants.

Hradiszko.—Austria-Hungary, comitat of Sarós, near Zeben.
　Cold sulphurous waters.
　A small bathing establishment.

Hubertusbrunnen.—Germany, Prussia, Province of **Saxony**; 800 feet above sea level.
　Saline springs.
　Doctors: Dr. Schläger.

Huelva.—Spain, capital of the Province of same name, on the Atlantic coast, and on the junction of the Rios Tinto and Odiel, 9,500 inhabitants. For routes see table.
　Winter station and sea baths.
　Season: For baths the autumn months, and generally from November till May.

The climate is delightfully steady, mild, and invigorating, and as a consequence, the station has been coming into favour of late. The bay is a fine expanse of water, and the town is well sheltered from northerly winds. It is an old place, having been built by the Romans.

There are as yet no statistics concerning the atmosphere, rainfall, &c., but the number of cloudless days in the year is large. Some interesting old buildings exist in the town.
　Post and telegraph office.
　There is a resident English physician.
　Hotel: The *Columbus* Hotel is owned and managed by an English Company, and is supplied by produce from their own farm. The prices are moderate and the accommodation excellent. A. Adrion, manager.

Humera.—Spain, Province of Madrid.
　Ferruginous bicarbonate waters, 50° F.

Hunyadi-János.—Hungary, near Budapesth.
　Cold sulphate and saline waters.
　Purgative action.

Hunstanton.—England, Norfolk, near Lynn.
　Ferruginous water and sea baths.
　Sandy beach.
　Climate bracing and fresh.
　Doctor: Dr. Whitty.

Hüttersbach.—Germany, Baden.
　Saline waters.

Hyères.—France, Department of Var, near Toulon. For routes see table.

A winter station and sea baths.

Winter season, November—June; summer, or sea bathing season, May—October

Special indications: Diseases of the chest and lungs, scrofula, affections of the larynx, diabetes, gout and rheumatism.

The **town proper is three miles** from the sea, and is built on the southern **slope of the Maurettes chain** of hills, here about 700 ft. high. North of the Gapeau, a small river east of the town of Hyères, the chain of Maures mountains protects the town against north-east and north winds. To the south and south-east of Hyères are the three islands of Porquerolles, Port Gros, and Levant, also called the Hesperides, or Golden Islands. These protect the roadstead of Hyères, and **form a safe harbour.** The situation of this winter station has been frequently wrongly stated to be on these islands. The mean winter temperature at noon in the sun is 55° F., and in spring from 65 to 68° F. Approaching Hyères from La Pauline on a calm day, the traveller will be astonished **at the rapidity with** which the temperature rises, while the **vegetation assumes a** more thoroughly tropical character, until **the culminating point is** reached in the **date palms on the Place des Palmiers.** The mistral, that great enemy **of Provence, is little felt here.** The valley of Costebelle **is well sheltered against this** scorching wind, and is also more protected and **favoured as** regards vegetation.

The mean barometric pressure is 762 to 766 mm., and the mean humidity **of the air is 58 per cent.** in the year. Forty-five rainy **days, from October to end of** April, have been noted; **snow falls very rarely, indeed not more** than two or three **times in a century.** Fogs occur sometimes in autumn and spring, but are of short duration. The health of the community is very good, and **scrofula and phthisis** scarcely exist. The water is good. Drainage has been improved considerably of late, but much still remains **to be** done in this respect.

As everywhere on the Riviera, so in Hyères, the hour of sunset is a dangerous one, and patients have to be careful not to remain out of doors. The **climate of Hyères, on the** whole, may **be called a** dry and stimulating one, and well-suited to patients **who desire** to avoid the damp and cold of a northern winter, and **who do not** depend on free exercise in the open air as their principal treatment.

The town has **12,000 inhabitants,** and the excursions are very varied and beautiful. · A special feature is the three avenues planted with palm-trees, whence the name "Hyères les Palmiers." A large casino, with theatre, reading rooms, concert hall, &c., exists on the Boulevard Napoleon, where the syndicate has its offices.

English church, post and telegraph on the Route National, near the *Hôtel des Iles d'Or*.

Bookseller: T. Hébrard.

Doctors: Drs. Griffith, Biden, Cessens and Marquez.

Dentist: Dr. Harwood.

Bankers: R. C. Corbett & Co.

The Hôtels **Peyron** and *de l'Ermitage*, healthily and beautifully situated in the pine forest. First-class house; telegraph in the Hotel. A. Peyron, proprietor.

The *Grand Hôtel des Palmiers*, excellent, first-class 200 rooms, lift, and all modern improvements; most sheltered position; Lawn-tennis and gymnasium; *Grand Hôtel du Parc*. Same proprietor, A. Wattebled.

The *Hôtel d'Europe* is now under new management; it has been entirely refurnished and greatly improved; it is centrally situated; the cooking is good, and prices moderate. Marius Arnoux, proprietor.

Hypate.—Greece, Peloponnesus.

Saline sulphurous waters, 50° F., mildly purgative action.

Special indications: Leprosy, elephantiasis, syphilis and mercurial dyscrasia.

Hythe.—England, Kent, on the Channel coast.

Sea baths much frequented, and good bathing appliances.

Ibenmoos.—Switzerland, Canton of Lucerne; 1,800 feet above sea level.

Cold earthy waters. Charming site.

Internal and external application.

The establishment is well arranged, and the accommodation satisfactory.

Iberg.—Switzerland, Canton Schwyz, near Schwyz; 3,050 feet above sea level.

Sulphurous magnesian waters.

Exported in stone bottles to the surrounding districts.

Ibero.—Spain, province of Pamplona, near Echauri; 1,400 feet above sea level.

Alkaline waters, 60° F.

Special indications: Gastralgia and dyspepsia, hyperæmia of liver and spleen, plethora abdominalis, chronic metritis, calculus and gravel, vesical catarrh.

The situation is beautiful and the vegetation luxuriant. The gardens are extensive and enclose a small lake for boating. The establishment is completely fitted up in modern style; the accommodation is ample and moderate in price.

Doctor: Dr. Cervera.

Iceland.—Island of, belonging to Denmark.
Intermitting springs, known under the name of "Geysers." The hottest springs and the largest in volume in the world.

Ikaria.—Greece, **island in** the Ægean **Sea.**
Very strong saline waters, 120° to 125° F.
Notwithstanding a lack of accommodation and an unsatisfactory climate, **the place is** frequented by the inhabitants **of** Samos and Asia Minor.

Ildjak.—Bosnia.
Saline waters.

Ilfracombe.—England, North Devonshire, at the entrance of the Bristol Channel.
Sea baths, much frequented. Air bracing and fresh.
Doctor: Dr. Stonehame.
Hotel: The *Ilfracombe Hotel*.

Ilkeston.—England, Derbyshire.
Sulphate of iron waters.

Ilkley.—**England,** Yorkshire, on the Wharfe.
Climatic station in summer, with hydropathic establishment at Ben-Rhydding.
Doctors: Dr. Leeson, Dr. **Scott.**

Ilmenau.—Germany, Thüringia, Duchy of Saxe Anhalt; 1,620 feet above sea level.
Waters various; hydro-therapeutic establishment; and owing **to the** surrounding pine forests, a climatic station.
Special indications: Affections of the eyes and nervous and heart diseases.
Doctor: Dr. Hassenstein.
Hotels: *Kurhaus*, **Sun**, *Lion*.

Imnau.—Germany, Principality of Hohenzollern; 1,260 feet above sea level.
Ferruginous bicarbonate waters, 22° F. Exported. Pine-cone-sap baths.
Special indications: Affections of the chest.
Much frequented station.—Good establishment.

Imola.—Italy, Province of Bologna; 45 m. from Bologna by rail, quarter of an hour carriage from the station; 12,000 inhabitants.
Ferruginous, sulphurous saline waters. The establishment, erected in 1830, is efficient, and is situated in a very charming valley. The hotels afford every accommodation and comfort.
Doctors: Drs. Lesi and Ferretti.

Inchaurte.—Spain, Province of Guipuzcoa.
Cold sulphurous waters.

Innichen.—Austria, Tyrol, valley of the Puster.
Saline sulphurous and saline ferruginous waters. Little frequented.
Special indications: Gout, scrofula, blenorrhœa, amenorrhœa, hysteria, hypochondriasis, colic.

Inowrazlaw.—Germany, Prussian Province of Posen.
Saline waters.
Special indications: Scrofula, skin diseases, rheumatism and female affections.
Doctors: Drs. Mannheim, Rakowski and Winkler.
Hotels: Kurhaus, Villa Weiss.

Inselbad.—Germany, Westphalia, near Paderborn.
Saline and saline chalybeate waters, 45° F. Inhalations, whey cure.
Special indications: Tuberculosis, chronic bronchial catarrhs, pleuritic exudations.
Doctor: Dr. Brüggelmann.
Hotels: Kurhaus, Bathhouse.

Instow.—England, Devonshire.
Sea baths.

Interlaken.—Switzerland, Canton of Berne, between the Lakes of Brienz and Thune.
Climatic station in summer. Whey cure.
It possesses only artificial baths, and is visited almost entirely on account of its beautiful envirous. Very fine scenery, and pure healthy air.
English service in the church at the Nunnery, also at Beatenberg.
Doctors: Drs. Zürcher and Strasser.
Hotel: The *Rugen Hotel*, Jungfraublick, a first-class, well appointed house; magnificently situated on the Rugen. J. Oesch-Muller, proprietor.
The *Grand Hotel Victoria*, a magnificent first-class family hotel, the best as regards situation in Interlaken; 400 beds, lift. Special arrangements made for a prolonged stay. Ed. Ruchti, proprietor.

Inverleithen.—Scotland, Peebles, on the Tweed.
Saline waters.

Iósza.—Austria-Hungary, comitat Ung, rail to Nagy-Mihálz.
Alkali-saline waters. Establishment good.

Irno (Valley of).—Italy, Province of Naples; from Naples to Salerno by rail, 1½ hours, thence by carriage half-an-hour.

Alkali-saline waters; not thoroughly analysed (but superior to Vichy, and assimilating more to the Karlsbad waters), 59° F.

Season: June—September.

Special indications: Chronic catarrh of the stomach, splenic and hepatic complaints, uric acid diathesis, catarrh of the urinary organs, gout and chronic rheumatism.

The site is beautiful, but the management of the establishment is scarcely on a par with the efficacy of the waters.

Ischia.—Italy, island in the Bay of Naples.

Alkali-saline waters, 145° F. Thermal establishment of Casamicciola. Climatic station for spring and autumn.

Special indications: Uterine affections, rheumatism, but more especially diseases of the bones, sores, gout and paralysis.

An island celebrated for its rich vegetation and the beauty of its sky. It is much visited, too, on account of the abundance of its mineral springs. On its small surface there are no fewer than 22 springs, all well known and all of which were in use under the Romans. The principal are St. Restituta, Cappone Fontana (similar to cold Carslbad), and Gurgitello. (See articles on above). The earthquake of July, 1883, has wrought so much havoc amongst the bathing establishments, that it is unlikely they will ever be rebuilt.

Doctors: Drs. Manzi and Storer.

Ischl.—Austria, Salzkammergut, on both banks of the Traun, on the confluence of the Traun and Ischl, 1,600 feet above sea level; 9,000 inhabitants. For routes see table.

Saline and cold sulphurous waters; climatic summer station; whey cure; saline, steam, hot and cold brine and sulphurous baths; mud, malt, pine-cone sap, and wave baths; inhalation; medical gymnastics; large swimming basin.

Season: May—October; the pleasantest months here are September and October.

Special indications: Lymphatism, incipient goitre, affections of the mucous membrane of the stomach and intestines, chronic vomiting, diarrhœa from debility, dropsy, chronic affections of the respiratory organs, chronic skin diseases, Nervous Affections, disorders of ovaries and uterus, and sterility.

Ischl is charmingly situated in the centre of three valleys, and surrounded by high mountains, off-sets of the Tyrolese Alps. The country around offers an ever-varying change of picturesque beauty. The vegetation is very luxurious, especially in spots exposed to the sun's rays. The air is soft and pleasant; nor does Ischl suffer from the defect common to most valleys, namely, a very limited period of sunshine daily, with a proportionately high temperature, which at times may become well-nigh unbearable. The thermometer at midsummer seldom rises above 86° F., and the change in temperature from morning to evening is scarcely perceptible. Large Government salt-works are carried on here, and owing to this circumstance, the air is always impregnated with saline particles, which render it very invigorating. The surrounding mountains are clothed with pine forests, the emanations from which contribute in a large degree to the health-giving qualities of the air.

The chief promenade is the Esplanade, where visitors congregate, and which is thus the centre of "cure-life." The streets and roads are admirably kept, and the population is very civil and attentive. Beautiful villas and gardens surround Ischl for a wide range. The price of baths, &c., is posted up in all the establishments. Rudolph's Garden is the centre of visitor-life in the morning, when the band plays from seven till half-past eight. Around it are the different bathing establishments. The hotels, villas, and houses with furnished apartments, are prettily situated, mostly in their own gardens.

The *Curhaus* surpasses most establishments of its kind even in the best watering-places along the Rhine, both in architectural beauty, and in the elegance and comfort of its arrangements. A very extensive garden surrounds it: its prominent features are:—The terrace, where daily concerts are held, a large concert and ball-room, reading-room, gaming and smoking-rooms. The view from the terrace is beautiful. The theatre is in the town on the Kreuzplatz. Operas and comedies.

Society at Ischl is composed of the highest aristocracy. The Imperial family passes several months here every year, in the castle.

The *Post and Telegraph Office* is in the centre of the town, near the church. From here the various diligences start.

The excursions are very varied and too numerous to name in detail. Fishing is an amusement much indulged in, and amongst the fish caught are, salmo frio, salmo salvelinus, Salmo Schiefermulleri, and Salmo Thymallus. Permission to fish is obtainable from Mr. Koch, mayor of the town.

English church : Hôtel Kaiserin Elisabeth.
Exchange and Bank : Gottwald, Kreuzplatz.
House and Estate Agent : Heuschober, Crown Hôtel.
Bookseller : Manhart, Pfarrstrasse.
Chemist : Krupitz, Landstrasse.
Doctors : Drs. Kaan, Fürstenberg, Heinemann, Stieger and Kottowitz.
Hotels : The *Hôtel Kaiserin Elisabeth* is splendidly situated. The proprietor owns the trout fishing. F. Koch, Proprietor. Crown, Victoria, Post.

Islington.—England, a district of London.
Ferruginous waters of great repute in the eighteenth century; at present not used.

Ismailia.—Egypt.
Sea baths.

Isola Bona.—Italy, near Porto Maurizio.
Cold sulphurous waters.

Ivanda.—Austria, on the frontier of Servia and Wallachia, near Temesvar.
Bitter waters, mineralisation 17 gramms., 12 gramms. of which are sulphate of sodium. It contains no sulphate of magnesia, and is laxative, mild in action, and therefore suitable for prolonged and gentle treatment.

Ivanyi.—Austria-Hungary, comitat of Bereg; half-an-hour from Mumkácz.
Earthy ferruginous water.
A small bathing establishment.

Ivonicz.—Austria, Galicia, near Krosmo.
Two saline bromo-iodurated, a bituminous and a ferruginous spring. This water contains 0·017 iodide of sodium.
Special indications: Goitre, articular disease, and herpes.
The establishments are efficient and the environs charming.

Jabalcuz.—Spain, Province of Jaen, and near this town.
Sulphurous, magnesian or bitter waters, 80° F.
Special indications: Nervous complaints, epilepsy and hysteria, rheumatism, gout and the uric acid diathesis, and general paralysis.
The establishment is not good. Patients can be lodged at the baths; but prefer to stay, as a rule, in Jaen, whence Jabalcuz is only half-an-hour's drive. Prices are moderate. The situation is mountainous.
Doctor: Dr. Lacort.

Jacintos (de los).—Spain, New Castilia, near Toledo.
Ferruginous alkaline waters, only exported.
Special indications: Difficult and irregular menstruation, chlorosis.

Jacobfalva.—Austria, Transylvania.
Cold bicarbonate ferruginous waters, mineralisation 5 gramms.
A small establishment, with limited accommodation.

Jaen.—Spain, Capital of Province of Jaen.
Thermal sulphate of magnesia springs, 85° to 90° F.
Special indications: Rheumatism, paralysis.

Jahodnika.—Austria-Hungary, comitat of Turocz.
Highly effervescent ferruginous waters. Only used locally.

Jalleyrac.—France, Department of Cantal.
Cold bicarbonate of iron waters.

Jallova.—Turkey, Asia Minor, near Brussa.
   Thermal sulphurous springs.
   A very good establishment and **an interesting place to stay at.**

Jamnicza.—Austria, Croatia, near Agram.
   Cold **alkaline and** sulphurous ferruginous waters, **16 gramms.** of mineralisation, of which 6 are carbonate of sodium and 3 sulphate of sodium.
   Special indications: **Anæmia.**
   A much frequented establishment.

Janischek.—Russia, Lithuania.
   Sulphurous waters.
   Only used in the neighbourhood.

Jano.—Italy, Modena, near Scandiano.
   Cold sulphurous waters.
   Only used in baths by the surrounding peasantry.

Jaraba.—Spain, Province of Saragoza, near Ateca.
   Thermal **sul**phurous waters, 85° F.
   Special indications: Rheumatism, urinary affections, calculus gravel, chronic metritis, difficult **m**enstrua**tion.**
   The **site is romantic.** There are three establishments **equally good.** The accommodation is ample and moderate in price.
   Doctor: **Dr. de** Gregorio.

Jaróslaw.—**Austria,** Galicia, district of Przemysl.
   Effervescen**t** ferruginous waters, used in bathing.

Jaxtfeld.—Germany, Würtemberg, near Heilbronn; 450 feet above sea level.
   Saline waters, mineralisation 113 grammes., of which 108 are chloride of sodium. Inhalation.
   A good establishment, **with** comfortable accommodation.

Jelen.—Austria-Hungary, comitat of Gömór, 1½ hour from Rimn-Szombath.
   Effervescent ferruginous waters, used only in the environs as table water.

Jenatz.—Switzerland, Canton of Grisons.
   Cold ferruginous and mixed bicarbonate waters of 60° F.
   Whey cure.
   The establishment is fairly well arranged, and the accommodation good.

Joanette.—France, Department of Maine and Loire, near Angers.
   Saline ferruginous **springs,** one sulphurous.
   A small establishment.

Job.—France, Department of Puy de Dôme.
Slightly mineralised waters, used only as table water.

Jobsbad, or Wiesenbad.—Germany, Saxony, near Annaberg; 1,370 feet above sea level.
Earthy alkaline waters.
A fairly good establishment. Scenery grand and charming.

Johannisbad.—Austria, Bohemia, near Melnik.
Ferruginous waters.

Johannisbad.—Austria, Bohemia, near Pardubitz; 2,000 feet above sea level.
Alkali-ferruginous thermal water, 70° F.; climatic station.
Special indications: Nervous affections, hysteria, neuralgia, gastralgia.
Very charming site and mild air.
Doctors: Drs. Kopf and Pauer.
Hotel: Kurhaus.

Johannisberg.—Germany, Nassau, near Geisenheim on the Rhine.
Hydropathic establishment with electrotherapy. Grape cure.
Very elegant accommodation.

Johnstown.—Ireland.
Ferruginous spring. Much frequented.

Jood.—Hungary, comitat of Mármaros, Isa Valley.
Sulphate of soda and magnesia waters, with purgative action.

Jordansbad.—Germany, Würtemberg, near Biberach; 1,660 feet above sea level.
Effervescent ferruginous waters; establishments, with shower, spray, douche, and sitz baths. The situation is retired and pretty.

Jorullo.—South America, Venezuela.
Thermal sulphurous springs, 155° F.

Jose.—France, Department of Puy de Dôme.
Bicarbonate of soda water; only used as table waters.

Josephsbad.—Austria, Bohemia, near Tetschen.
Ferruginous waters.

Jouene.—France, Jura, near Dole.
Saline waters.

Juliushall.—*See* Harzburg.

ungbrunnen.—Germany, Würtemberg, near Rottweil.
Saline waters, whey cure.
A good establishment.

unqueiro.—Portugal, on the Atlantic, close to Lisbon.
Sea baths, sandy beach.
Very good establishments, and much frequented.

urowla.—Austria, Galicia, district of Sánok.
Cold saline waters.
Poor arrangements.

Kabolapolyána.—Austria-Hungary, comitat of Marmáros, district of Rahó.
Highly effervescent ferruginous waters.
Thoroughly efficient bathing establishments with all modern requirements.
The position of the town is picturesque.

Kács.—Austria-Hungary, district of Mezö-Kovesd, near Miskolcs.
Indifferent waters, 65° F.
A very good establishment, with good accommodation.

Kaiapha—Greece, east coast of Morea, Elis, near Navarino.
Thermal sulphurous waters.
Special indications: Herpes, psoriasis, elephantiasis.
These waters are mentioned by Strabo. They are very efficient, but the severity of the climate makes residence scarcely advisable.

Kaissariani.—Greece, Attica, at the foot of Hymettus.
Indifferent waters; climate very good in summer.
There are many ruins in the neighbourhood.
Ovid speaks of these waters in the following lines

"Est prope purpureos colles florentis hymetti,
Fous sacer et viridi cespite mollis humus."

Kalauria or Poros.—Greece, island in the Ægean Sea.
Slightly sulphurous waters, used in urinary and intestinal disorders.

Kalimaneste.—Roumania.
Saline sulphurous waters, 55° F.
A small establishment.

Kaltenleutgeben.—Austria, Lower Austria, near Vienna.
Hydropathic establishment.
Very mild air and picturesque site.
Doctor: Dr. von Winternitz.

Kammietz-Podolsk.—Russia, Government of Podolia.
    Sulphurous waters.
    A much-frequented station.

Kammer.—Austria, near Linz and Salzburg.
    Climatic summer station; baths in the Attersee; whey cure.
    Very charmingly situated, pure air and equable climate; good trout fishing and interesting excursions for tourists. The establishment is excellent, and accommodation comfortable and moderate.

Kanitz.—Germany, Bavaria, near Werdenfels; 2,500 feet above sea level.
    Alkaline iodurated waters.
    Whey cure.
    A good establishment.

Karlsbad.—Austria, Bohemia; 1,150 feet above sea level. For routes see table.
    There are twelve springs, of which the four principal ones are the Sprudel, 170° F., the Neubrunnen, 145°, the Mülbrunnen, 130°, and the Theresienbrunnen, 125° F.
    Special indications: Constipation, liver and bilious complaints, plethora, obesity, **gout**, gravel, diabetes, &c.
    Doctors: Drs. Seegen, Kraus, Pleschner, von Hochberger, Kafka, Preiss, London, Neubauer and Sehnee.
    Bookseller: Hans Feller, Alte Wiese.
    Hotels: National, Hanover, de Russie.

Karlshafen.—Germany, Prussia, near Minden.
    Saline waters.
    A good establishment with satisfactory accommodation. Climate very mild and agreeable.

Karlsruhe.—Germany, Baden, between Karlsruhe and Durlach.
    Cold effervescent ferruginous waters; little used.

Karpfen.—Austria-Hungary, comitat of Sohl, on the Kropona.
    Sulphurous magnesian waters.
    A small establishment, used only by the inhabitants.

Karytena.—Greece, Peloponnesus.
    Cold sulphurous waters.
    Like the majority of Greek springs, these were used in ancient times, but are now neglected.

Kaschin.—Russia, Government of Tver, near Moskau, on the Masletka.
    Very cold ferruginous waters.
    Primitive arrangements, used only by residents.

Kastanowka.—Russia, Government of Kiew, near Schpoli.
    Sulphurous waters, little used.

Kastenloch.—Switzerland, Canton of Appenzell.
Alkaline waters, little used.

Katharinenbad.—Russia, Caucasia, on the Terek.
Alkali-saline waters, 175° to 180° F.
Primitive arrangements and accommodation.

Katharsion.—Greece, Island of Lesbos.
Saline waters, 75° to 80° F.; they contain a good deal of sulphur. The place is called, in consequence, the "ill-savoured pool."
The baths enjoy a high reputation, and are much frequented.
The ruins of the old Grecian baths have been partially restored, and afford visitors satisfactory accommodation.

Katwyk.—Holland, one hour from Leyden.
Sea baths and sandy beach; very quiet place, though much frequented. The simple and tranquil life act very favourably on invalids.

Kaudenbach.—Germany, Rhenish Prussia, near Bertrich.
Earthy ferruginous waters, 70° F.; used in drinking, as auxiliary to those of Bertrich.

Keked.—Austria-Hungary, comitat of Abauj, near Kaschau.
Cold sulphurous waters.
Special indications: Chronic catarrhs and muscular contractions.

Kellberg.—Germany, Bavaria, 1½ hours from Passau; 1,200 feet above sea level.
Ferruginous and slightly effervescing waters.
The establishment is well arranged, and the accommodation comfortable and moderate in price.

Keménd.—Austria-Hungary, comitat of Hunyad.
Earthy waters, 45° F.

Kemmern.—Russia, near Riga.
Very strong sulphurous waters, containing a large quantity of sulphuretted hydrogen.
The bathing establishment is elegant and the accommodation satisfactory.

Kerö.—Austria, Transylvania, district of Dées.
Cold saline sulphurous waters.
Good establishment, with tolerable accommodation.

Keruly.—Austria, Transylvania, district of Udvarhély, 4 hours from Lövete.
Effervescent alkaline ferruginous waters. Used only by the surrounding peasantry.

Keswick.—England, Cumberland, Lake District.
Summer holiday resort, with fishing and boating.
Doctor: Dr. Knight.

Keuchreæ.—Greece, on the Isthmus, south of Corinth.
Saline waters, 70° F.
Known in the days of Pausanias as the Helena baths.

Kiel.—Germany, Province of Holstein. Important maritime port on the Baltic.
Sea baths; fine white sandy beach.
Season: June–September.
The establishment is very elegant and much frequented.

Kierling.—Austria, Province of Lower Austria, near Klosterneuburg.
Climatic station, but better known through its whey cure establishments.
Very fine climate and luxuriant vegetation.
The establishment is comfortable.

Kilburn.—England, near Hampstead, in Middlesex.
Bitter waters.
Special indications: Similar to those of Friedrichshall.

Kilkee.—Ireland, Donegal.
Sea baths; sandy beach; chalybeate springs.
Doctor: Dr. Griffin.

Kilkenny College Spa.—Ireland, on the River Store.
Chalybeate waters.

Killymard.—Ireland, Donegal.
Sulphurous springs, entirely neglected.

Kilrush.—Ireland, Clare.
Sea baths; sandy beach; chalybeate waters.
Doctor: Dr. Counihan.

Kimpolung.—Roumania, near the Austrian frontier.
Saline sulphurous waters.
Primitive establishment, with indifferent accommodation.
The site is pleasant and pretty.

Kinsale.—Ireland, County Cork, on the Bandon river.
Sea baths, fine sandy beach, chalybeate waters near Castle-in-Park.
Doctor: Dr. Bishopp.

Kiralyi.—Austria-Hungary, comitat of Gömör, near Tornallya.
Sulphurous earthy **waters**, 60° to 75° F. Only used in the environs.

Kiralymezö.—Austria-Hungary, comitat of Marmáros, district of Tecső.
Ferruginous iodo-bromurated saline waters.
The spring is the property of the Government, and is almost neglected.

Kirchberg.—Germany, Bavaria, near Reichenhall; 1,520 feet above sea level.
Earthy alkaline saline waters, whey cure, and aromatic herb extract baths.
Much frequented. The situation is charming.

Kirchbrunnen.—Germany, Würtemberg, near Heilbronn.
Bitter waters.
Formerly much frequented, and the water exported, but now abandoned.

Kirchheim.—Germany, Würtemberg, near Teck.
Cold sulphurous waters.
Special indications: Chronic, rheumatic and arthritic complaints, chronic skin diseases.
There is a good establishment.

Kirchleerau.—Switzerland, Canton of Aargau, 3 hours from Aarau.
Earthy waters.
Used only by the peasantry of the neighbourhood.

Kirkilissa.—Turkey, Bulgaria, on the road from Adrianople to Constantinople.
Sulphurous water, 80° F.
The bathing arrangements are the same as in all Oriental baths.

Kirstenpils.—Denmark, near Copenhagen, in the Zoological Gardens.
Indifferent waters.
Special indications: Paralysis and rheumatism.

Kis-Czeg.—Austria, Transylvania.
Alkali-saline waters; four springs.
Purgative and very efficacious action. A cold Karlsbad.

Kis-Kalan.—Austria, Transylvania, 2 hours from Hunyad.
Mixed bicarbonate waters, 75° F.

Kis-Sáros.—Austria-Hungary, comitat of Sáros, near Eperiés.
Alkaline ferruginous waters.
A very good establishment, with comfortable accommodation.
Very picturesquely situated.
Doctor: Dr. Jácz.

Kissingen.—Germany, Bavaria, near Würzburg, on the right bank of the Saale; 602 feet above the sea level. For routes see table.

Cold saline waters, **strongly mineralised.** Three principal springs; the Rakoczy, the Pandur, and the Maxbrunnen.

Season: May **15th—September 30th.**

Number of visitors: About 30,000 annually.

Special indications: Tonic and excitant, they are admirably suited to abdominal and hæmorrhoidal congestions.

Prince Bismarck passes about two months here every year.

Bookseller: Ph. Hailmann.

Doctors: Drs. Gaetschenberger, Sotier, Stöhr, Scherpfert, Diruf, and Welsh.

Hotels: The *Hotel Victoria,* facing the baths, Kurhaus and gardens. The hotel is now well managed by the new proprietor, W. Todt. It has recently been refurnished; large shady garden. Omnibus **meets all trains.**

The *Kaiserhof,* first-class, well recommended, next to the Rakoczy spring and bathing establishment. Omnibus at station; **carriage in the house.** In May and September **reduced prices.** L. Waltner, proprietor.

Kislovodsk.—Russia, Caucasus; 2,375 feet **above the sea level.**

Alkali-saline earthy waters, 45° F.

A special feature of this water is **its action on** the kidneys. It is very efficacious **in all vesical diseases.** Caution must be used in drinking it.

Klaussen.—Austria, Styria, near Graz.

Ferruginous bicarbonate waters. Composition almost identical with those of Spa; 0·098 of iron.

Special indications: **Catarrhal affections, hypochondriasis, hysteria, anæmia.**

Klein-Chocholna.—Austria-Hungary, comitat of Trenczin and near this town.

Alkali-saline ferruginous waters.

Only used by the peasantry in cases of intermittent fever.

Kleinengstingen.—Germany, Würtemberg, near Marbach; 2,100 feet above sea level.

Indifferent waters; little used.

Kleinern.—Germany, Principality of Waldeck-Pyrmont.

Bicarbonate of magnesia waters. Complementary to those of Wildungen.

Klein-Schirma.—Germany, Saxony, between Freiberg and Chemnitz.

Mineral and mud bathing establishment.

Klemutzion.—Greece, Morea, Province of Elis, facing Zante.
Sulphurous waters, 70° to 95° F.
Well known in ancient times.

Kliening.—Austria, Carinthia, District of Klagenfurt.
Earthy alkaline and ferruginous waters.
Special indications: Debility, anæmia, difficult menstruation.

Klokocs.—Austria-Hungary, comitat of Sohl.
Effervescent earthy waters, little used.

Klosters.—Switzerland, Grisons; 4,000 feet above the sea level.
Cold sulphurous waters.
Indifferent establishment.

Klutschewsk.—Russia, in the Ural mountains, Government of Perm.
Sulphurous waters.

Klütz.—Germany, Mecklenburg, near Wismar.
Sea baths, sandy beach, and good arrangements.

Knaresborough.—England, Yorkshire.
Sulphate of lime waters.

Knutwyl.—Switzerland, Canton of Lucerne; 2,110 ft. above the sea level.
Sulphate of lime and iron waters. Very mild and pure air.
Good establishment and tolerable accommodation.

Kobersdorf.—Austria-Hungary, comitat of Oedenburg, near Rabnitz.
Alkali-saline ferruginous waters.
Good establishment, with comfortable accommodation at moderate rates.

Kochel.—Germany, Bavaria, on the Lake of Kochel.
Alkaline waters, climatic air station; very mild and equable air.
The establishment is luxurious, and accommodation ample.

Kondrau.—Germany, Bavaria, near Ratisbon; 1,000 feet above sea level.
Saline waters.
Special indications: Catarrh of the vesical organs and gravel.

Königsborn.—Germany, Westphalia, near Unna.
Cold saline waters.
Very efficient establishment.

Königsbrunn.—Germany, Saxony, near Königstein, 1½ hour from Dresden.
Hydropathic establishment.
Very fine site; place thoroughly well appointed.

Königsdorff-Jastrzembs.—Germany, Silesia, near Ratibor; 800 feet above sea level.
Bromo-iodurated saline waters, very strongly iodurated. 65° F.
Special indications : Chronic rheumatism, paralysis, syphilis, scrofula, affections of the uterus, difficult menstruation, caries, necrosis, skin diseases, goitre.
Doctor : Dr. Schenk.
Hotels : *Königsdorff, Hohenzollern.*

Königstein.—Germany, Nassau in the Taunus.
Hydro-therapeutic establishment.
Doctors : Drs. Pingler and Thewalt.
Hotels : *Pfaff, Amsterdam.*

Königswarth.—Austria, Bohemia, near Marienbad; 2,155 feet above sea level.
Bicarbonate of iron waters.
Special indications : Anæmia, rhachitis, scrofula, chronic paralysis.
The establishment is thoroughly fitted up and the site is charming.
Doctor : Dr. Kohn.
Hotel : *Kurhaus.*

Konopkowka.—Austria, Galicia, district of Tarnopol.
Cold sulphate of lime waters.
Special indications : Herpes, rheumatism.
Only used in baths; the establishment is efficient.

Kopenhagen.—Denmark, capital of.
Sea baths, sandy beach.

Kornwestheim.—Germany, Würtemberg, half-hour from Ludwigsburg.
Cold sulphurous waters.
The Kurhaus combines bathing establishment and hotel.

Korond.—Austria-Hungary, Transylvania.
Saline waters.
No establishment, and but primitive accommodation.

Koroud.—Austria, Transylvania, near Parajd.
Effervescent earthy waters.
There is a bathing establishment, and several buildings affording good accommodation.

Korsow.—Austria, Galicia.
Ferruginous bicarbonate waters.

Korytnica.—Austria-Hungary, in the Carpathians, near Neusohl.
Saline ferruginous waters, and climatic station.
Korytnica is magnificently situated. The vegetation is very luxuriant, the air mild and equable, and the place very fashionable.
Doctor: Dr. Vogel.

Kösen.—Germany, Saxony.
Saline bituminous waters, 40 grammes of mineral principles, of which 33 grammes are chloride of sodium; the salts are exported.
Special indications: Scrofula.
Very good establishments, with abundant and fashionable amusements.
Doctors: Drs. Wahn, Knorr and Rosenberger.
Hotels: Zum Muthigen Ritter, Kurzhalz.

Kosia.—Roumania.
Sulphurous-saline waters, 60° F.
Very little used.

Kostendjil.—Turkey.
Sulphurous waters.
Special indications: Skin diseases, gout, paralysis and intestinal disorders.

Kostreiniz.—Austria, Styria.
Carbonate of soda water; 10 grammes of mineralisation, of which six are carbonate of sodium.

Köstritz.—Germany, Principality of Reuss.
Saline, pine-cone sap, dry, warm, and sand baths.
Special indications: Rheumatism, anchylosis.
Doctor: Dr. Sturm.
Hotel: Kurhaus.

Kötschenowa.—Russia, near Moscow.
Effervescent ferruginous waters.
Good establishments, long known.

Kovaszna.—Austria, Transylvania, near Kronstadt.
Cold alkaline-muriatic springs.
Primitive establishment, with accommodation of an indifferent kind.

Krähenbad.—Germany, Würtemberg; 1,100 feet above sea level.
Alkaline-earthy waters.
There is a primitive establishment.

Krankenheil.—Germany, Bavaria, near Tölz; 3,452 feet above the sea level, on the river Isar.
Cold iodo-bromurated saline waters.
Special indications: Scrofula, inflammation of the ovaries, hyperæmia of the uterus, indurated breasts, inflammation of the prostate gland and urinary organs, goitre.
Kursaal and Trinkhalle; much frequented establishments.
Doctors: Drs. Höffer and Jungmayer.
Hotel: *Kurhaus.* Chas. Loder, proprietor.

Krapina Töplitz.—Austria, Croatia, comitat of Warasdin.
Two thermal carbonate of lime springs, 100° to 115° F.
Special indications: Nervous affections, eczema, prurigo, lichen, psoriasis.
Excellent establishment and accommodation. The site is beautiful and the climate mild and healthy.
Doctors: Drs. Von Aigner and Rak.
Hotel: *Kurhaus.*

Kreuth.—Bavaria, Germany; 2,910 feet above the sea level.
Bitter waters, whey cure, climatic station.
A well-appointed establishment; climate mild, yet damp.
Doctor: Dr. May.
Hotel: *Kuranstalt.*

Kreuzen.—Austria, Province of Upper Austria, near Grein; 1,360 feet above the sea level.
Hydropathic establishment. Climatic station, surrounded by pine forests.
Complete establishment.
Doctors: Dr. Krischke, Urbaschiek, Fleishanderl.

Kreuznach.—Germany, Rhenish Prussia, on the Nahe. For routes see table.
Bromo-iodurated saline waters.
Season: From 1st May to 1st October; open all the year round.
Special indications: Scrofulous affections, diseases of the ears, eyes, respiratory organs, bones and joints, all female and skin diseases, and in chronic affections generally.
The excursions in the environs are very varied and beautiful. English Church.
Post-office. Telegraph.
Bookseller: Rhein. Schmidthals and George Barth.
Doctors: Drs. Engelmann, Stabel and Strahl.
Hotels: The *Kurhaus Hotel.* G. Simpson, proprietor.
The *Hôtel Royal* and *d'Angleterre,* well situated, close to the *Kurhaus,* 100 bed and 20 sitting-rooms. 20 new baths and a magnificent new dining hall added for the summer of 1881. J. C. Heitz, proprietor.

Krevenish.—Ireland, County Fermanagh.
Sulphurous waters, with only a local reputation.

Kronberg.—Germany, Nassau.
Saline and chalybeate waters. Whey cure.
There is a very efficient establishment.

Kronthal.—Germany, Nassau; 500 feet above sea level. For routes see table.
Alkali-saline **waters, 40° F.; exported**.
Special indications: **Nervous** diseases and affections of the mucous membranes, bronchial catarrh.
Good establishment.

Krumbach.—Germany, Bavaria, between Ulm and Memmingen.
Carbonate of lime waters.
Special indications: **Nerve** affections.
Very little frequented.

Krynica.—Austria, Galicia, in the Carpathian mountain chain; 2,200 feet above sea level.
Gaseous alkaline ferruginous waters; climatic station; whey cure.
The establishments and the accommodation are thoroughly good.
Doctors: Drs. Mars, Zdun and Zieliniewsky.
Hotels: Seyfert, Krakau.

Krzessow.—Austria, Galicia.
Ferruginous and sulphurous waters; mineral muds.
Primitive arrangements, and the place is but little frequented.

Kugelbad.—Austria, Bohemia, ½-hour from Prague.
Alkaline ferruginous waters.
Good establishment.

Kunda.—Russia, Esthland, near Makholm.
Cold sulphurous waters.

Kungara.—Russia, Caucasus, near Zelesnowodsk.
Alkali-saline waters, 80° F.
Very primitive arrangements, although the waters are highly efficacious.

Kunzendorf.—Germany, Silesia, district of Neustadt.
Ferruginous and sulphurous waters.
Tolerable establishment.

Kuppis.—Russia, Finland, 2 miles from Abo.
Alkaline ferruginous waters.

Kythmos.—Greece, Island in the Ægean Sea.
Two thermal saline springs, from 100 to 130° F.
Special indications: Rheumatism, gout, and elephantiasis.

Labassère.—France, Department Hautes Pyrénées, near Bagnères de Bigorre.
Cold sulphurous soda waters.
Special indications: Pulmonary, laryngeal and bronchial affections.
This spring is now the property of the Company, who hold the Concession of Bagnères de Bigorre.
Doctor: Dr. Lacoste.

Labestz-Biscaya.—France, Department of Basses Pyrénées, near Mauleon.
One sulphate of lime and one cold ferruginous spring.

La Caldare.—Austria-Hungary, comitat of Mármaros, near Kösep-Vissó, valley of Vinului.
Alkaline waters. Used only in the environs.

Lacanne.—France, Tarn, near Castres.
Thermal sulphate of lime waters.
Special indications: Herpes, rheumatism, and gravel.
Doctor: Dr. Strehalano.

La Cava dei Tirefii.—Italy, near Naples.
Climatic summer station.
Season: June—August.

Lacvillers.—France, Department of Doubs, Arrondissement of Pontallier.
Cold ferruginous waters.

Laemnoli.—Switzerland, Canton St. Gall.
Cold sulphurous springs.

Laer.—Germany, Hanover, near Iburg.
Muriatic waters, 11 grammes of chloride of sodium.

Lago d' Averno.—Italy, near Pienza.
Ferruginous sulphurous waters. They form a lake which exhales a disagreeable odour.
Special indications: Chronic skin diseases, old sores and anchylosis.

Laifour.—France, Department of Ardennes, near Mezières.
Cold ferruginous waters. Exported.
Special indications: Leucorrhœa and vesical catarrhs.

Lake District.—England, Lancashire and Westmoreland.
Comprises several summer stations and holiday resorts, forming centres of excursions, and of boating and fishing on these picturesque lakes.
Season: June—September.

Lake of Constance.—*See* Bregenz, Friedrichshafen, Lindau, Rohrschach.

Lalliaz.—Switzerland, Canton of Vaud, near Vevey; 1,760 feet above sea level; a fine Alpine situation.
Sulphate of lime waters.
Special indications: Herpes, dyspepsia, atonic and discrasic sores.
The establishment is very good, and the accommodation comfortable.

Lama.—Italy, Department of Siena, near Poggibonsi.
Earthy magnesian waters, little used.

La Malou.—France, Hérault. For routes see table.
Thermal and cold polymetallic waters, 75 to 115°. Alkaline, ferruginous, brine and arsenical waters; gaseous and natural steam baths.
Special indications: Rheumatism, neuralgia, affections of the nerve centres and locomotor ataxy.
The climate is warm, and changes of temperature rapid. The establishments are thoroughly well fitted up.
Doctor: Dr. Cros.

Lamotte.—France, Department of Isère; 1,450 feet above the sea level. For routes see table.
Thermal saline springs, 145° F.; 6 grs. of mineralisation, of which 3 grs. are chloride of sodium. Action excitant and tonic.
Special indications: Rheumatism, uterine affections, scrofula and white swelling.
The bathing establishment is in an old castle, with good airy rooms.
Doctor: Dr. Gubian.

Lamscheid.—Germany, Prussia, **Lower Rhine,** near Coblentz.
Bicarbonate of iron waters, also known under the name of "Leiningen acidulous waters."
Special indications: Debility, anæmia, chlorosis, difficult menstruation.

Lanaskede.—Sweden, district of **Iönköping.**
Cold sulphate of iron water.

Landeck.—Germany, Silesia, near Glatz; 1,460 feet above the sea level.
Six alkaline-saline and sulphurous springs, with a varying temperature of 45° to 75° F.
Special indications: Rheumatism, affections of the larynx, nervous affections and female complaints.
A well-appointed establishment.
Doctors: Drs. Joseph, Wehse, **Langner and Ostrowitz.**
Hotels: Bismarck, Carlshof, **Schlössel.**

Landetta.—Spain, in the vicinity of Cuenca.
Bicarbonate of lime waters, 45° F.

Landskron.—Germany, Rhenish Prussia, district of Ahrweiler, near Neuenahr.
Alkaline magnesian waters.
No establishment.

Langassa.—Greece, near Thessalonika.
Sulphurous thermal waters, much used in antiquity, but now neglected.
No establishments and no accommodation. Especially efficacious in arthritic complaints.

Langenberg.—Germany, Principality of Reuss.
Hydropathic establishment. Very mild and healthy air.

Langenbrücken.—Germany, Baden, between Bruchsal and Heidelberg; 440 feet above sea level.
Saline sulphurous waters, 35° F.
Special indications: Catarrhal affections of the respiratory organs and of the bladder, rheumatism, gout.
Doctor: Dr. Ziegelmeyer.
Hotel: Kurhaus.

Langensalza.—Germany, Thuringia, Province of Saxony.
Cold sulphate of lime bituminous waters.
The establishment is excellent.
Doctor: Dr. Baumbach.
Hotel: Kurhaus.

Langenschwalbach.—*See* Schwalbach.

Langrune.—France, Department of Calvados, near Caen.
Sea baths; very quiet. The sea water here is highly charged with iodine.
Season: May—September.

Lanjarron.—Spain, Alpajurras, Province of Granada; 3,300 feet above the sea level.
Alkali-saline waters, superior to those of Vichy; climate mild and bracing—45° to 80° F.; also a ferruginous effervescent spring.
Season: June—September.
Special indications: Gravel, disease of the kidneys and bladder, gout, rheumatism, obesity, feeble digestion, chlorosis and anæmia.
The establishment is antiquated, and the accommodation not of the best.
Doctor: Dr. Valenzuela.

Largs.—Scotland, Ayrshire, near Glasgow, railway station of Wemyss Bay.
Sea baths, gravelly **beach**, climate **mild** and **steady**.
A good bathing establishment.
Season: June—September.
Doctor: Dr. Caskie.

Larivière sous Aigremont.—France, Department of Haute Marne, near Bourbonne les Bains.
Cold effervescent **ferruginous waters.**

Laroche-Posay.—France, Department of Vienne.
Sulphuro-ferruginous springs.

La Salute.—Italy, **near** the Lago Maggiore.
An intermediate **station**, which deserves to be better **known.**

Lasserre.—France, Lot and **Gard, near Fronsecas.**
Cold effervescent **earthy waters, slightly laxative.**
Auxiliary to **the waters of Luchon.**

Laubbach.—Germany, Rhenish Prussia, half-hour from Coblenz. For routes see table.
Indifferent waters. Hydropathic establishment **and summer station of first** importance. Inhalations. **Milk, buttermilk and grape cure.** Air pure and bracy.
Season: May—October, but open all the year round.
Special indications: Morbus Brightii as a speciality, affections of the respiratory organs and of the nervous system, constitutional diseases and circulatory disorders, anchylosis, contractions and **articular swellings.**

This establishment, situated in the most beautiful portion of the Rhine valley above Coblentz, ranks as one of the most important of its kind. It was founded in 1840, and has subsequently been adorned and perfected both by nature and art. The surrounding woods, which belong to the town of Coblentz, **secure perfect** quietude, and at the same time render the **air** bracing. **There is absolutely** no dust.

The balneal **appliances** in the various building are well worth the attention **of medical men.** There are Turkish, Roman, Irish and Russian **steam baths; douches of all** kinds and shower baths. The methods **of** treatment embrace **diet cure,** hydropathy, thermotherapy, together with gymnastics, massage **(kneeding), electricity** and inhalation. Moreover, the long-standing experience **of the** director and proprietor ensures success in cases where other similar establishments have failed. Bright's **disease** has been made an especial study, and the cures obtained **in this** malady are attracting more and more the attention **of the** medical profession and of **the** public, especially of **those who** have long resided in India, &c.

The environs of Bad-Laubbach are very lovely, and excursions are numerous. The immediate circuit of the Kurhaus is laid out in gardens, and the accommodation is abundant, comfortable and moderate. The banks of the Rhine, from the town of Coblentz to this establishment are laid out in gardens, in which concerts are given every day in summer.

The excursions embrace: Ehrenbreitstein, the Karthause, the Oelberg, Andernach and Laacher Sea, Neuwied Castle, the Moselle and its many old castles; up the Lahn to Ems, and up the Rhine as far as Bieberich and Bingen. All of them are interesting, and well worth making.

Post and Telegraph in the establishment.

Doctors: Dr. Averbeck, the proprietor, and several assistants.

Hotels: The establishment, which contains above 100 rooms, thoroughly comfortable, entirely refurnished, and greatly improved; drainage perfect. Address the Director.

Lauchstadt.—Germany, Prussian Province of Saxony, near Halle and Merseburg.

Earthy saline ferruginous waters.

Special indications: Neuroses, rhachitis, chorœa, debility, anæmia, disorders of the urinary organs.

A fairly well-appointed establishment.

Laugenberg.—Russia, Caucasus, near Pjàtigorsk.

Alkali-saline ferruginous waters.

The site is picturesque. The establishment is primitive, and little has been done to bring these very excellent waters into prominent notice.

Laurion.—Greece, Attica, near Keratix, six hours from Athens.

Thermal saline waters, little used.

Lausanne.—Switzerland, Canton Vaud, 61 m. from Berne, an irregular picturesque town; 30,000 inhabitants; 1,600 feet above the sea level.

Climatic station; bracing air and grape cure.

Season: Summer and autumn.

Magnificently situated on the slopes of the Swiss Jura, dominating Lake Leman, from whose shores it is 1½ mile distant. Air mild and especially suitable to convalescents and weak and anæmic persons. Winter less cold and summer less hot than in other stations in the same latitude. The annual mortality ranges between 21 and 22 per thousand, the average of other European towns being 24 to 30 per 1000. Lausanne, with Geneva, has the lowest mortality of any Swiss town. Spring and rock water of finest quality; the supply amounts to 575 quarts for each person daily.

Numerous educational establishments for young gentlemen and ladies. University studies can be followed up at the academy. Conservatory of music.

Very **varied**, well-shaded promenades and **excursions**; boating on the lake.

English, Scotch, Presbyterian, Catholic and Wesleyan church.

Post and Telegraph Office: Place du Grand Pont.

House and Estate Agent: Amadée de la Harpe, who gives all information gratis.

Bookseller: Th. Roussy, Rue du Bourg 7.

Doctors: Drs. Dufour and Recordon, oculists; Dr. de Cérenville, chest and nervous diseases; Drs. Rouge and Dupont, surgeons; Dr. Challand, cerebral affections; Drs. Larguier and Rogivue; Dr. Eug. Sécrétan, ear diseases; Drs. de la Harpe, and Joël; Drs. Heer & Rapin, ladies' doctors.

Hotels: At Ouchy the *Hôtel Beau Rivage* is well named, for it is one of the best appointed and finest hotels in Switzerland; it stands in its own grounds, near the landing place of the steamers, distant about 10 minutes by pneumatic railway from the centre of the town. A. Martin-Rufenacht, manager.

The *Hôtel Gibbon*, a first-class old family hotel. Best situation in the town. E. Ritter, proprietor.

The *Hôtel du Faucon* is recommended as a very comfortable first-class house. Excellent kitchen. A. Raach, proprietor.

The *Pension Chevallier—Beau Séjour*, situated in a garden overlooking the lake. Avenue de la Gare. Very comfortable and homelike house; winter garden and billiard rooms. Madame Chevallier, proprietress.

The *Pension Campart—Avenue du Théâtre*, a first-class quiet family house; well managed. F. Campart, proprietress.

ausigk.—Germany, Saxony, 3 miles from Grimma.
Sulphate of iron waters.
The establishment is good.

auterberg.—Germany, in the Harz, near Nordhausen.
Mild waters and hydrotherapeutic establishment.
Doctor: Dr. Ritscher.
Hotels: Kurhaus, Crown.

avardens.—France, Department of Gers.
Bicarbonate of lime, 45° F.

avey.—Switzerland, Canton of Vaud, near Bex.
Thermal mixed sulphurous waters, 125° F.
Special indications: Rheumatism, herpes, scrofula.
Good climate. Well-appointed establishment.

eamington.—England, Warwickshire, three hours' rail from London.
Saline, sulphuretted, and chalybeate waters, containing iodine and bromine. Winter health resort. Eleven springs, very effervescent.
Season: May to October; but open all the year.

Special indications: Dyspepsia, hepatic colics, portal or uterine congestions, **liver complaints**, disorders of the kidneys, gout, sciatica, blood **poison**, **skin** diseases, rheumatism, anaemia, chlorosis, scrofula, **rhachitis**, strumous, lymphatic or other glandular enlargements, **tabes mesenterica, phthisis**, tubercular and wasting diseases, chorea, uterine tumours.

The waters of Leamington, neglected **since the** days of Dr. Jephson, **are**, owing to the **complex character** of their composition, very efficacious, and **can safely compete with the** best of their kind on the Continent. The **long abandonment is partly due** to fashion, partly to the **apathy of the Councilmen of the town**, who have **not** marched with the time. The bathing **appliances in** the establishment **are not quite even** now what they ought to be to successfully compete with the rival stations. As, however, the wave of fashion has again taken a patriotic turn, and English watering-places **see their long-dormant advantages resorted to** again, Leamington, it **is to be expected**, will not remain behindhand in entering the arena **of competition for** public favour it **so amply deserves** through the excellency of its waters.

The climate is very equable, the annual mean temperature being 48° F.; mean rainfall, 27 inches, and death rate, 12 per thousand. The town is well built, well drained **and** clean. The amusements are very varied and more numerous than 50 **years ago**, the zenith of its fame. Those who are fond of natural beauty **and scenery** will find in its environs an ample field for most interesting excursions.

Doctors: Drs. Eardley Wilmot, Haynes, Holmer & Thursfield

Hotels: The *Regent Hotel*, first-rate well-known old family house, situated near the baths and **Jephson Gardens**; the best position in **Leamington**. The management, and all the appointments of this house, **will** be found to meet the **requirements of all visitors. Address, L. Bishop**, proprietor.

The *Manor House Hotel*, situated in its **own** extensive pleasure grounds near the railway station. It has recently been **entirely** renovated; it is bright and homelike, and replete with every modern **comfort. First-rate wines and** cooking. **Address, Jesse Percival, proprietor.**

Lebetzoba.—Greece, Morea, Laconia.
Sulphurous waters.
Much used in antiquity. For modern requirements the arrangements are rather of a primitive kind.

Leccia.—Italy, Tuscany, near Siena.
Thermal alkaline-magnesian bicarbonate waters, 95° F.
Primitive arrangements and accommodation.

Ledesma.—Spain, 3 hours from Salamanca, 2,000 feet above sea level.
Thermal sulphurous waters, 120° F.
Special indications: Rheumatism and paralysis, muscular and tendinous retractions, neuralgia, **skin** diseases, sciatica,

necrosis, old wounds, **chronic** ulcers and sores, asthma, **bronchitis,** chorea.

A much frequented **station.** Good **establishment, dating from** the days of the Moors. One of the best in Spain, at **moderate prices.**

Very **good accommodation in** the new baths.

There **are two large swimming baths,** and other **appliances, which are all up to modern requirements. In this respect, the Ledesma** establishment may be called the best in **Spain.** The site is romantic, **and the establishments are surrounded by** well-kept gardens **and a large park. Excursions abound in** every direction, **and residence is made very agreeable.**

Doctor: Dr. Garcia Lopez.

ee.—England, **Devonshire,** near Ilfracombe.

Sea baths; **sandy** beach, chiefly frequented **as a** summer station. Climate bracing.

Season: June—October.

eghorn.—Italy, **Tuscany, at** the mouth of the Calambrone, connected **by a canal with the Arno.**

Sea baths; also a good winter station.

The climate is almost the same as at Pisa, only **a** little more **moist and exhilarating.**

The drinking water is excellent, and the houses being **all of more or less** recent **construction,** epidemics, such as **occur in other Italian towns,** are exceedingly **rare.**

The existing **meteorological observations by** Schouw **are** untrustworthy. **It is well known that invalids from** the north with **delicate chests have remained here for** years, with very favourable results.

Post: Piazza **Carlo Alberto.**
Telegraph: **Piazza Cavour.**
Bankers: Macquay, Hooker **& Co.**
Chemist: Galligo, via **della** Madonna.
Doctors: Dr. **Schintz, Dr.** Taddei.
Hotel: The *Anglo-American* **is a first-class** house, facing the gardens **and sea baths, combining** comfort and moderate prices. **Leop.** Foccacci, proprietor.

égué St. Brieuc.—France, **Department of Côtes du Nord.**

Sea bath, sandy beach, climate mild yet bracing.

eipzig.—Germany, **second city of** the **kingdom of Saxony; from Paris** via Forbach, **Frankfurt, 27 hours.** 134 **frs. 1st class.**

One of the most busily **commercial cities of** Europe. Chiefly interesting during **the great fairs, of which** three are held; one at Michaelmas, one on New Year's **Day,** and one after Easter. It is the centre of the German **book**-trade, which has an

Exchange of its own. Every work, whether book, brochure, or pamphlet, edited in any part of the globe, can be procured here.

OBJECTS OF INTEREST.—Kingshall, the University, Buchhändler Börse, Paulinum with the University Library; Town Hall, Theatre, Museum.

EXCURSIONS: To residence of Baron Speck, Pleissenburg and Battle Field.

CABS: One to four persons from ½ to 1½ marks by time, 1¼ to 2 marks by the hour.

Post Office: Augustus Platz.

Dentist: P. G. Kneisel.

Doctors: Drs. Leukardt, Schildbach and Thiersch.

Hotels: *Hauffe, Rome, de Prusse, de Russie, de Barière.*

Leissigen.—Switzerland, District of Interlaken.
Cold sulphate of lime springs.

Lemnos.—Greece, Island in the Ægean Sea.
Saline sulphurous thermal waters.
There is only a small grotto, admitting two persons at a time, to bathe in.

Lendershausen.—Germany, Bavaria, near Hofheim.
Effervescent ferruginous waters.
No establishment.

Lepanto.—Greece, in the Monarchy of Akarnania-Ætolia.
Sulphurous and bitter waters.
Season: May—August.
Much frequented; possesses a well-arranged establishment.

Le Prese.—Switzerland, on the Lake of Poschiavo; 3,215 feet above the sea level.
Indifferent waters; a climatic station.
The establishment is efficient, and the accommodation cheap and good.

Lès.—Spain, near the French frontier and Luchon.
Sulphate of soda waters, 80° and 45° F. Only frequented by those living in the environs.

Lesbos.—Island in the Grecian Archipelago, also called Mitylen.
Saline waters, 125° F.; about half-hour from the town of Lesbos. Sulphurous thermal waters near the port of Lebanus.
The baths are taken by both sexes in common; much freedom prevails. The establishments leave room for improvement, while the accommodation is far from what it ought to be.

Lesina.—Austria, Dalmatia, an island in the Adriatic, on the coast of Dalmatia.
Winter station. Grape cure, with sea baths in autumn.
Though one of the best of winter stations, the arrangements for properly receiving patients are still wanting.

Les Salins.—France, Department of Puy du Dôme.
Saline waters.

Leszina.—Austria, Croatia, near Carlstadt.
Sulphate of soda water; 5 grs. of mineral principles, of which 2 grs. are sulphate of sodium. Highly charged with carbonic acid gas.

Letantus.—Greece, island of Euboea.
Saline waters, 130° to 140° F.
No establishment and but little accommodation; in fact, the waters are used only by the inhabitants of the neighbourhood.

Leutstetten.—Germany, Bavaria, district of Starenberg.
Indifferent waters.
Very well arranged establishments.
The scenery is charming and the climate mild.

Levane.—Italy, Tuscany.
Alkaline ferruginous waters, 35° F.
Special indications: Rhachitis, scrofula.
Primitive arrangements, but the waters enjoy a great reputation.

Levico.—Austria, Trentino, near Trent, valley of the Sugana; 1,000 feet above sea level.
Effervescent ferruginous waters, containing arsenic. Grape cure.
Season: May—October.
Special indications: Anæmia, chlorosis, incipient affections of the spinal chord, nervous diseases, irregular menstruation.
The newly-erected bathing establishment has been fitted up with all modern requirements. The stay here is very agreeable, and the scenery and vegetation fine.
Excursions very varied. Balls, concerts, and other amusements during the season, which is sometimes prolonged through November.

Lichtenthal.—Germany, Baden, near Baden-Baden.
Cold ferruginous waters.
Two establishments, which are fitted up with all modern requirements and elegance.
Special indications: Debility, scrofula, anæmia, chlorosis, menorrhagia.

Lidja.—Turkey, Anatolia.
Thermal springs, 120° F.; known in ancient Greece as the baths of Agamemnon.

Liebenstein.—Germany, Thuringia, Saxe-Meiningen; 1,000 feet above the sea level.

Cold ferruginous bicarbonate waters: 0·0775 of carbonate of iron. Climatic air station, hydropathic establishment, whey cure.

Very fashionable resort. Air mild and equable, charming and very picturesque situation. One of the most *recherché* resorts of German society. Excursions in endless variety.

Doctors : Drs. Döbner and Hesse.
Hotels : Müller, *Martius.*

Liebenzell.—Germany, Würtemberg, near Stuttgart; 1,100 feet above sea level.

Light alkali saline waters, 60° F. Climatic station. The establishment is small but elegant.

Doctor : Dr. Trion.
Hotel : Kurhaus.

Liebwerda.—Austria, Bohemia, near Görlitz; 1,220 feet above sea level.

Alkali-saline, ferruginous, and effervescent waters. Exported. Whey cure.

The establishment is thoroughly fitted up, and accommodation comfortable.

Doctor : Dr. Plunert.
Hotels : Kurhaus, *Black Eagle.*

Lienzmuhl.—Austria, Carinthia, district of Klagenfurt, 1½ hours from Wolfsberg.

Alkaline-magnesian ferruginous waters, highly effervescent.

Lierganes.—Spain, near Santander. Diligence from Boó.

Sulphate of lime springs, 50° F.

Special indications : Skin diseases, intestinal derangements, laryngeal and bronchial catarrhs, uterine disorders, vesical catarrhs, rheumatism and scrofula.

The town is situated in a very narrow valley; vegetation is ever green. The establishment is modern, and the accommodation good. There is a kitchen for the use of those who may wish to prepare their own meals, as indeed is customary in Spanish watering places.

Doctor : Dr. Salazar.

Limmer.—Germany, Hanover, near Hanover.

Cold saline sulphurous waters.

Special indications : Chronic mercurial dyscrasia, paralysis, gout, rheumatism, and skin diseases.

Limpach.—Switzerland, near Berne; 1,200 feet above sea level.

Indifferent water.

Special indications : Neuroses.

The establishment is thoroughly fitted up, and the accommodation reasonable in price.

Lindau.—Germany, Bavaria, a small island in the Lake of Constance, on the north-eastern shore; 5,600 inhabitants; 1,230 feet above sea level. For routes see table.

Climatic station. Very mild and equable air. Milk and whey cure; baths in the lake. Hechelmann's bathing establishment has Turkish, Russian, sulphurous, pine-cone sap, brine baths, and is otherwise well fitted up.

The island of Lindau is in one of the best sheltered and most healthy positions on the lake. Epidemics have never occurred here. Its situation is picturesque, overlooking the lake and the whole panorama of East Switzerland. Its environs will bear comparison for scenery with the best parts of Switzerland. The town of Lindau is very old, and contains many interesting antiquities. It is connected with the main land by the railway bridge and an old wooden bridge. There are sulphurous waters in Schachenbad, half-hour from Lindau.

The island of Lindau has been much beautified of late years. The quays and walls of the town have been converted into promenades and planted with trees. Parks and gardens at the Landthor, Schützengarten, Reichsplatz, and Seehafen are all admirably kept. Steamboats ply frequently to the different places of interest on the lake. Concerts, etc., during the season. Good fishing in the lake and the various streams. Reading room at the Harmony Assembly Rooms. Illuminations on the lake. Post and Telegraph office at the station.

Chemists: The Angel. The Deer.
Dentist: Dr. Pfirsch.
Carriages: Mr. Seutter.
Doctors: Drs. Volk, Mayer, Otto, Kimmerle and Bever.
Hotels: The *Hotel de Bavière*, a first-class house. C. Ludwig, Proprietor.
Hotel Crown, Rentemann, Helvetia, and *Lindau.*

Lindenholzhausen.—Germany, Nassau; 452 feet above sea level.
Magnesian saline-ferruginous waters.
Very little frequented.

Lintzi.—Greece, Peloponnesus.
Thermal sulphurous waters, 90° F.
Special indications: Rheumatism, skin diseases, eczema.
Fairly well frequented.

Lion-sur-Mer. France, Calvados, near Caen.
Sea baths, sandy beach.
The air is impregnated to a large extent with iodine.
Season: May—October.
The climate is said to be especially beneficial in skin diseases.

Lipari.—Italy, capital of Liparian Islands.

Alkaline waters, 110° F., but more frequented for its natural steam baths. (The *Secca di Bagno*, described by Diodsrus Siculus.)

Special indications: Gout, rheumatism, and chronic skin diseases.

Lipetsk.—Russia, Government of Tambov.

Effervescent ferruginous waters.

The establishment is good, but the place is visited only by the inhabitants of the neighbourhood.

Lipik.—Austria, Slavonia; 560 feet above the sea level.

Bicarbonate, soda, and iodurated water, 150° to 155° F.; iodide of calcium, 0·029; resolvent action.

Very steady and equable climate.

Special indications: Scrofula and syphilis, periostitis, tophi, psoriasis, cachexia.

The new establishment is thoroughly fitted up and very elegant. The rooms are high and airy, and covered verandahs connect them with the baths, thus avoiding any prejudicial influences from the weather.

The gardens are tastefully laid out, and there is plenty of amusement.

Doctor: Dr. Kern.

Lippspringe.—Germany, Westphalia.

Effervescent ferruginous waters, 50° F. Climatic station.

This station possesses a chamber for inhalation, for catarrhal and phthisical patients.

The town is quiet, and well suited to certain classes of patients.

Doctors: Drs. Frey, Roden and von Brunn.

Hotel: Kurhaus.

Lisbon.—Portugal, capital of.

Thermal sulphurous spring, 70° F.

Sea baths and climatic winter station.

Several very good bathing establishments.

Formerly much frequented by Englishmen as a winter station; it has been somewhat neglected of late, probably owing to its changeable climate. Its mean winter temperature is 55° F. Mean humidity of the air, 79 per cent. Rainy days in the year, 111, in the winter season, 39·7.

Snow and fogs are unknown.

The drinking water is excellent, and a stay in town very interesting.

Lisdoonvarna.—Ireland, County Clare, near Limerick.

A much frequented sulphurous spring, sea baths.

Season: June—October.

Littlehampton.—England, Sussex, near Worthing and Bognor.
   Sea baths, sandy beach, climate dry and mild.
   Season: May—September.
   Doctor: Dr. Vines.

Little Haven and Broadhaven.—Wales, Pembrokeshire, or St. Bride's Bay.
   Sea baths, sandy beach, climate mild yet bracing.
   Season: April—October.

Llandridod (Wye).—Wales, Radnorshire.
   Ferruginous and saline sulphurous thermal waters.
   Special indications: Rheumatism, skin diseases and affections of the eyes.
   Doctor: Dr. Davies.

Llandudno.—Wales, Carnarvonshire.
   Sea baths, much visited; sandy beach, climate dry and mild
   Season: July—September.
   Doctor: Dr. Nicol.

Llandwrtd.—Wales, Breconshire.
   Strongly sulphurous and ferruginous waters.
   Special indications: Scrofula, scurvy, skin diseases, ulcers, and nephritic affections.
   Doctor: Dr. Tombs.

Llanfairfechan (Conway).—Wales, Carnarvonshire.
   Sea baths, sandy beach.
   Season: July—September.

Llangranog.—Wales, Cardiganshire.
   Sea baths. Summer and holiday resort. Climate healthy and cool.

Llanstephan and Ferryside.—Wales, Carmarthenshire.
   Sea baths. Extensive sandy beach. Climate mild.
   Season: May—October.
   Doctor: Dr. Phelan.

Llo.—France, Department of Pyrénées Orientales.
   Sulphate of sodium, 70° F.
   Enjoys a very high reputation in the surrounding country.

Löbenstein.—Germany, Thuringia, Principality of Reuss.
   Chalybeate waters, 82° to 90° F. Summer station.
   Special indications: Anæmia, chlorosis, anchylosis, rheumatism, gout.
   Doctor: Dr. Aschenbach.
   Hotels: Kurhaus, Bavière.

Locarno.—Switzerland, on the Lago Maggiore.
   Intermediate spring station. Grape cure in autumn. More resorted to for the beauty of its site. Terminus of the Gothard Railway and Lake steamers.
   Hotel: The *Grand Hôtel*, Locarno, in a magnificent position overlooking the lake; 200 rooms open all the year. G. Seyschab & Co., proprietors.

Lochbachbad.—Switzerland, Canton Berne; 1,810 feet above sea level.
   Alkali-saline ferruginous waters.
   The establishment is thoroughly well fitted up, and the accommodation abundant and reasonable.

Lodova.—Spain, Province of Navarre.
   Cold ferruginous waters.

Loeches.—Spain, Province of Madrid, near Alcalá
   Cold sulphurous water.
   Special indications: Scrofula, skin diseases, hyperæmia of liver and spleen, plethora abdominalis, leucorrhœa, chronic syphilis.
   The establishment is of recent date and situated in a large garden. The accommodation is moderate and prices low.
   Doctor: Dr. Lucientes y Pueyo.

Loka.—Sweden, near Philipstad, on the Lake of Lersjö.
   Indifferent waters; mud baths.
   The establishment is ample and surrounded by a beautiful park. The scenery in the neighbourhood is very picturesque.

Longbranch.—United States.
   Sea baths on the Atlantic; very fashionable; much frequented.
   Season: June—September

Lons-le-Saulnier.—France, Department of Jura.
   Cold saline water.
   Special indications: Rheumatism, skin diseases, scrofula.

Looe, East.—England, Cornwall, near Liskeard.
   Summer health resort; air pure and mild
   Doctor: Dr. Kerswill.

Loreto.—Italy, Romagna, Province of Forli, 2 hours' carriage from Forli.
   Saline waters.
   Special indications. Inertia of stomach and intestines, obstructions of liver and spleen.
   No bathing establishment.

Los Baños.—Philippines, Island of Luçon.
   Sulphurous waters, 200° F.
   Very much visited by the inhabitants of Manila.

Losdorf.—Switzerland, Canton of Argovia; 2,200 feet above sea level; between Olten and Aarau.

Cold sulphate of lime waters.

Laxative action.

The establishments are very efficient, and the sojourn pleasant.

Louèche-les-Bains.—Switzerland, Canton of Valais; 4,642 feet above the sea level, at the foot of the Gemmi. For routes see table.

Earthy waters, containing arsenic and iron, 125° F., climatic mountain air station; whey and milk cure. Baths and drinking.

Season: May **15th** till September **30th**.

Number of visitors: Above 5,000 annually.

Special indications: Convalescence, anæmia, chlorosis, rhachitis, scrofula, chronic skin diseases, rheumatism, gout, catarrh of the mucous membranes, female complaints and muscular contractions.

Loèuche-les-Bains is situated in the midst of a grand Alpine country, 3 hours from the railway station (Louèche Susten). The road is very good, and passes through a magnificent country. Vegetation is very luxuriant.

The establishments are complete, there being arrangements for douches, electric baths, hydropathy, and shampooing.

Surrounding the town there are extensive larch and pine woods, which render the air very bracing. The excursions are charming and varied. Among them are to Torrenthorn, one of the finest views in Switzerland, Gemmi, Wiedstrubel, Altels, Balmkorn; communication by road with Bernese Oberland, Belalp Æggischhorn and Valley of Zermatt.

Doctors: Dr. Brunner, Dr. Mengis, Dr. Rey, Dr. de Werra.

Hotels: Hotel des Alpes, Hotel de France, des Freres Brunner, Maison Blanche, l'Union.

Loujo.—Spain, at the mouth of the Arosa, Province of Pontevedra.

Bitter saline water, 60 to 120° F.

Special indications: Rheumatism, contractions, paralysis, hemiplegia, scrofula, skin diseases and gunshot wounds.

The establishment is situated on an island in the Atlantic, half-an-hour from the shore. Vegetation is abundant and the accommodation cheap. The bathing arrangements are very inadequate.

Doctor: Dr. Avila.

Lowestoft.—England, Suffolk, 10 miles from Yarmouth.

Sea baths; firm sandy beach mixed with shingles. Climate bracing.

Season: June—September.

Doctor: Dr. Ray.

Lu.—Italy, Piedmont, near Alessandria, district of San Salvatore-Monferrato.
Saline, sulphurous, and iodurated waters; mineralisation 3 grammes. Deserves to be better known. Mud baths.
The establishment is primitive.

Lubien.—Austria, Galicia, near Lemberg.
Cold sulphate of lime waters. Inhalation.
Special indications: Rheumatism and herpes, laryngitis.
Modest arrangements.

Lucainena de las Torres.—Spain, Province of Almeria, near Sorbas.
Carbonate of lime, 50° F.
Special indications: Herpes, dyspepsia, scrofula, chlorosis, amenorrhœa and dismenorrhœa, vesical catarrh, hyperæmia of liver and leucorrhœa.
The establishment and accommodation are inadequate.
Doctor: Dr. Vecino Villar.

Lucan.—Ireland, county Dublin, on the Liffey.
Alkaline sulphurous waters.
Special indications: Biliousness, scurvy, gout, rheumatism.
Doctor: Dr. Watson.

Lucca.—Italy, Tuscany.
Thermal sulphate of lime waters, 70 to 125° F. Ten springs, very abundant.
The baths are about 13 miles from the town, at the foot of the Apennine mountains.
Fine climate, beautiful site, and agreeable sojourn.
Special indications: Herpes, rheumatism and gout.
Doctors: Drs. Gason and Giorgi.

Lucerne.—Switzerland, capital of Canton of same name; from Paris by East of France Railway, *viâ* Basle.
Is charmingly situated on the bank of the Lake of the Four Cantons, and is the terminus of the Gotthard Railway on this side of the Alps. A very old town, with walls; divided into two parts by the Reuss; opposite the Riga Mountains, and at the foot of the Pilatus.
OBJECTS OF INTEREST.—Bridges over the Reuss, Water-tower, Cathedral, Town Hall, with a Museum of Antiquities; Arsenal, College or Lyceum, and a Glacier Garden; Stauffer's Museum, Organ in Cathedral; Lion Monument, by Thorwaldsen.
EXCURSIONS on the lake to the Rigi, Stanstadt, Bürgenstock, Engelberg, Seeligsberg, Mount Pilatus, Beggenried, Brunnen, Schönegg, and many others. To the Bernese Oberland, *viâ* the Brünig Pass to Brienz, Meyringen, Grindelwald, and Interlaken.
CABS: Half hour, one or two persons, 80 c.; three or four persons, 1. fr. 20 c.; by hour, 2 or 3 frs.
Post and Telegraph Office: Near Schweizerhof.

Bookseller: Prell, Kramgasse.
Doctors: Drs. Stöcker, 315, Kapellplatz, Nager and Steiger.
Hotels: The *Schweizerhof* and *Lucernerhof*. Large first-class houses in the best part of Lucerne; open all the year round. Hauser Bros., proprietors.

Luchon (Bagnères de).—France, Department of Haute Garonne, near the Spanish frontier. 1,990 feet above sea level. For routes see table.

There are more than 50 different springs, rich in hydrosulphuric acid; also ferruginous waters and one alkaline spring.

Season: June–September.

Special indications: Rheumatism, scrofula, affections of the bronchi, skin diseases, and chronic sores.

Luchon is one of the most frequented of bathing stations, and affords every comfort and luxury.

Doctors: Drs. Garrigou, Kuhnemann, Laspales, Verdalle and Lambron.

Hotels: The *Hotel d'Angleterre*, a quiet and comfortable family hotel, centrally situated in a large garden. De Vries, proprietor.

The *Hotel Richelieu*, first-class, facing the bathing establishment and park. There are two villas to let for families, one containing 16 rooms elegantly furnished, Louis Estrade, proprietor.

Lucsky.—Hungary, comitat of Liptau, near Rosenberg; 1,000 feet above sea level.

Ferruginous waters, 80° F. Climatic station.

Season: June–September.

The site is most picturesque, and the surrounding vegetation consists chiefly of pine forests. The establishments are fitted with all modern requirements, but more should be done for the accommodation, seeing the waters attract so many visitors.

The springs yield very abundantly.

Luc-sur-Mer.—France, Department of Calvados.
Sea baths; sandy beach.

Ludwigsbrunnen.—Germany, Hessen, near Schwalheim.
Table water, similar to seltzer.

Special indications: Gravel, biliary calculi and urinary disorders.

Lugano.—Switzerland, Canton of Tessin, on the Lake of Lugano.

Intermediate station, with fine climate, beautiful situation on the lake. Grape cure.

Season: Spring and autumn.

Mean humidity of air in winter, 72°; 48 to 50 rainy days, 31 windy days. Snow frequent.

Doctor: Dr. Cornilo.

Lugo.—Spain, Province of Lugo, on the right bank of the Miño; 1,930 feet above the sea level.

Sulphurous saline waters, 80° to 105° F.

Special indications: Rheumatism and herpes, scrofula, tendinous contractions.

The spring is about 3 k. south of the town.

The establishment is large, and situated in an extensive park. It is not, however, up to modern requirements. Accommodation is plentiful, and cheap in proportion.

Doctor: Dr. Carretero.

Luhatschowitz.—Austria, Moravia; 600 feet above sea level.

Alkali-muriatic, iodide and bromine waters. One of the strongest springs known.

Special indications: Affections of the respiratory and digestive organs and urinary disorders.

The establishment is thoroughly fitted up. Parks, gardens and covered walks of great extent give Luhatschöwitz the appearance of a station of first importance. The accommodation is excellent, and moderate in price.

Doctors: Drs. Gallus and Küchler.

Hotel: Kurhaus.

Luhi.—Austria-Hungary, comitat of Bereg, near Munkács.

Alkali-saline waters, containing iron.

Special indications: catarrhal affections.

There is no bathing establishment, but the waters are much exported.

Luisenbad.—Germany, Pomerania, near Polzin.

Chalybeate waters; mud baths.

Special indications: Rheumatism, anchylosis, articular affections.

Doctors: Drs. Deets, Bechert and Lehman.

Hotel Kurhaus.

Lukhas.—Greece, four hours from Libadia.

Saline waters.

The establishment is in the convent of St. Lukhas, where good accommodation can be had.

Lulworth West.—England, Dorsetshire, near Worl.

Sea baths frequented all the year round.

Climate bracing, in winter warm.

Lund.—Sweden, Province of Gothland, district of Ludsjö, near Löddekö.

Effervescent indifferent waters.

Establishment good, but the accommodation is indifferent.

Lüneburg.—Germany, Hanover, on the left bank of the Ilmenau.

Cold saline waters; 254 grammes of mineralisation, of which 246 are chloride of sodium.

The brine being so excessively strong, is mixed with common water for the baths.

Lutraki.—Greece, on the isthmus of Corinth.
 Very good waters. 150° F. Only used in baths and by the inhabitants of the neighbourhood. The springs are mentioned by Xenophon.

Luxburg.—Switzerland, Canton Thurgau, near Salmsach, on the Lake of Constance; 1,200 feet above sea level.
 Sulphurous alkaline ferruginous waters.
 Ample establishments.

Luxeuil.—France, Department of Haute Saône, 1,300 feet above sea level. For routes see table.
 Saline, ferruginous and magnesian waters. Eleven springs, of from 70° to 125° F.
 These waters act first as excitants, then as sedatives.
 Special indications: Neuralgia, rheumatism, paralysis, gastralgia.

Luxor.—Upper Egypt, on the right bank of the Nile, amid the ruins of Thebes.
 Winter station; warm and very dry climate.
 Season: January--March.
 Doctor: Dr. Maclean.
 Hotel: Luxor Hotel, pension £1 per day in season. Thos. Cook & Son, proprietors.

Lygourio.—Greece, Argolis, near Nauplia and Epidaurus.
 Muriatic bitter thermal waters, much frequented in antiquity, but now entirely abandoned.

Lyme Regis.—England, Dorsetshire, near Bridport.
 Sea baths; firm sandy beach.
 Season: July--August.
 Doctor: Dr. Skinner.

Lymington.—England, Hampshire. South Western Railway.
 Sea baths; sandy beach.

Lynmouth (Lynton and).—England, on the west coast, near Ilfracombe; 430 feet above sea level.
 Frequented sea baths; sandy beach.
 Mild but bracing climate.
 Season: July--August.
 Doctor: Dr. Hartley.

Lyons.—France, at the confluence of the Saone and Rhone; from Paris by P.L.M. Railway, 9 hours. 63 fr. 1st class.
 The first industrial city in France, and for silk manufactures, in the world. The best view of the city is obtained from the observatory of Notre Dame.
 OBJECTS OF INTEREST.—Public squares and Fountains, **City Library**, Botanical Garden, Cathedral, **Church of St. Jean**, Museum, Hall of Sculpture, Hospital, Commercial School, Exchange at the Palace of Commerce, **Law Courts**, Theatre, Park de la Tête d'Or.

EXCURSIONS to Mont Ceindre, Roche Cardou, Ile Barbe and Croix Rousse, and many others.
Post-office: Place Bellecour.
Telegraph-office: Place de Lyon, 53.
CABS: 1½ fr. the course, 2 fr. first hour.
    Bookseller: V. Cantal, 9, Rue Bourbon.
    Hotel: The *Grand Hôtel de Lyons*. A first-class house in every way, with the best position in the town. M. Kussler, Director.

Lytham.—England, Lancashire, near Blackpool.
    Sea baths; sandy beach; climate very mild. Season all the year.
    Doctor: Dr. Garstang.

Mablethorpe.—England, Lincolnshire, near Sutton.
    Sea baths; firm sandy beach.
    Season: June—September.
    Doctor: Dr. Campbell.

Macerato.—Italy, Tuscany, district of Siena, on the Merse.
    Alkaline sulphurous waters, 150° F.
    Season: June—September.
    Special indications: Gout, rheumatism, paralysis, skin diseases.
    The situation is very fine. The establishment has only two piscinæ, and the baths are taken in common. Frequented mostly by the neighbouring inhabitants and by those of the Maremma.

Macon.—France, Department of Saône and Loire.
    Cold ferruginous bicarbonate waters.
    OBJECTS OF INTEREST.—Town Hall, Prefecture, Cathedral, Law Courts, Statue of Lamartine, Old Towers of St. Vincent.
    A good place in which to lay in a stock of Burgundy, as Macon is the centre of the wine trade.
    Hotels: The *Hôtel de l'Europe* is a first class house for families; five minutes from the station; a good stopping place between the Riviera and Paris.

Macviller.—Germany, Alsace-Lorraine.
    Cold saline waters; little used.

Mád.—Austria-Hungary, comitat of Zemplén, near Szerencz.
    Sulphurous, ferruginous and aluminious waters.
    A small bathing establishment.

Madagascar.—An Island in the Indian Ocean, on the East of Africa.
    Two thermal springs, not yet analysed.

Madeira.—One of the Canary Islands, a Portuguese possession, **north of Teneriffe.** By steamer from Lisbon in from 50 to 54 hours. **For routes** see table.

Winter station; select and pleasant society.

Owing partly to the sea journey, partly to vague and malicious reports, Madeira is not visited as much as it deserves to be by patients suffering from pulmonary complaints. The climate is delightful, and there is absolutely no winter. Complete absence of dust, an ever luxuriant vegetation, no bad drainage, a pleasant humidity, and a mild, equable and pure atmosphere, are the principal advantages which Funchal, the capital, offers. These, though not entirely absent in many of the now fashionable winter stations on the Riviera, are in some of them very imperfectly enjoyed. Madeira has suffered much from false statements. English visitors decreased in number greatly from 1870 to 1880, but their numbers have been on the increase since the latter year.

The town of Funchal, except in the quarter close to the sea shore, resembles a large garden. The view of it obtained from the sea is very beautiful, the town being built in the form of an amphitheatre. It is completely sheltered against north, north-east and north-west winds by mountains which rise to a height of 6,000 feet. During the whole year both northern and tropical fruits ripen. Wet days are rare, and invalids can consequently take out-door exercise during some hours every day. The winds and storms, which are of such frequent occurrence on the Riviera during the spring months, and which prevent not only invalids, but even healthy persons, from taking out-door exercise, are unknown in Madeira.

Mean temperature, 68° F. The mean barometric pressure is 762 $^m/_{in}$; the mean humidity of atmosphere 70%. Weather statistics are —170 clear days, with bright sun; 110 with a cloudy sky, and 85 rainy days. The thermometer stands at an average 10° to 15° higher than on the Riviera in winter, and proportionably lower in summer. Death-rate 29 per thousand.

Locomotion in the town and its environs, and on excursions (which are numerous) is effected by means of palankins, hammocks or sledges, drawn by oxen, mules or horses.

Amusements are amply provided for, and although a stay in Madeira may be called on the whole quiet, patients will find ample recompense in a delightful climate and scenery.

The slight difference (viz. 15° to 20°) between the mean of the coldest and hottest months, shows how equable the climate of Funchal is.

The mosquitoes which annoy on the Riviera as early in the year as April, are entirely absent. Indeed Madeira, especially its southern slopes, might with justice be called " *The Sanatorium* **for** *Phthisical patients.*"

English Church Service.
Bankers: Krohn Bros. & Co.

Chemist: Dos dois Amigos.
Doctors: Drs. Goldschmidt, Grubham, Larisa and Vieira.
Hotels: *Miles Carmo Hotel*, well situated on the Place St. Laurent, close to the baths.

Madonna a Papiano.—Italy, Tuscany, near Caprese, valley of the Teverino.
Alkaline-ferruginous waters.
A small establishment.

Madonna die tre Fiumi.—Italy, Tuscany.
Cold bicarbonate of soda waters.
Special indications: Calculous affections.

Madrid.—Capital of Spain, on the left bank of the Manzanares. From Paris, *viâ* Bordeaux, Irun, and Saragozza, 41½ hours. 184½ fr. 1st class.
Situated on the slope of a sandy plateau surrounded by limestone hills. The climate, owing to the barrenness of the environs, is disagreeable and unsuited to persons with delicate throats and chests. Since the commencement of Don Alfonso's reign, the city has begun to make some progress.
OBJECTS OF INTEREST.—The Puerta del Sol, the Alcala Street, the Prado, the Casino, Royal palace, Artillery museum, Topographical museum, Naval and Natural History museums, Royal Picture Gallery, several private ones, National Library, and numerous private libraries, open daily. Several churches of little or no interest. The only one worth visiting is the Atocha, with the Virgin's image. Town Hall, Panaderia, Audiencia, Casa de los Consejos, Congress House, Senado (House of Lords), the different Ministeries, Palaces of Duke of Alba, of Villahermoza, Marquis of Casariera, Duke of Medina Celi and Marquis of Salamanca, Buen Retiro, Casino de la Reina, Botanical Gardens, which have been turned into Zoological; Campos Eliseos, Italian Opera, various Theatres, Circus, and arena for Bull and Cock-fights.
EXCURSIONS: Casa de Campo La Manelon; Casino de la Reina; El prado Alameda, las Carabancheles; and many others of minor importance.
Post-office: Calle de San Ricardo 5.
Telegraph: Ground floor of Home Office.
CABS: One horse, one to two persons, 2 fr. the course; two horses, 4 fr.; by hour, one horse, 4 fr.; and two horses, 6 fr.
Doctors: Drs. Hysem, Nuñez, Kispert, Jelly, Carretero and Toca.
Hotels: *Des Princes, de la Paix, de Paris, de Russie.*

Magnac.—France, Department of Cantal.
Cold ferruginous bicarbonate waters.

Magyar-Szent-Lazlo.—Hungary.
Sulphurous waters.

Mala.—Spain, Province of Granada.
Sulphate of magnesia waters, 50° to 55° F

Malaga.—Spain, on the Mediterranean coast.
Winter station. Fine mild, dry and warm climate.
Mean winter temperature, 55° F.
If more were done to make visitors **comfortable, Malaga would soon become a serious competitor to the stations on the Riviera.**
Doctor: Dr. Bundsen.

Malahá.—Spain, Province of Granáda, near Santafé.
Sulphurous saline waters, 50° to 85° F., containing iron.
Special indications: Skin diseases and rheumatism, dysmenorrhœa, chlorosis, anæmia, hysteria.
The site **is charming, the establishment modern and efficient. The gardens are** good and **accommodation reasonable.**
Doctor: Dr. Doiz.

Malahide.—Ireland, county of Dublin.
Sea baths; good sandy beach
Doctor: Dr. Stainstreet.

Mallow.—Ireland, **county of Cork. Rail from Dublin to Cork. Thermal springs from 145 to 180° F.**

Malmedy.—Germany, Rhenish Prussia, near the **Belgian** frontier.
Alkaline-magnesian ferruginous waters.
Although the waters are **very** strongly **chalybeate, they are** little used.

Malnas.—Austria-Hungary, District of Haromszék.
Sulphurous waters, 65° F.
Special indications: Articular contractions.

Maloja.—**Switzerland,** Grisons, Upper **Engadine, near St.** Moritz; 5,688 **feet above the sea level. For** routes **see table.**
Climatic mountain and winter station; chalybeate effer**vescent** waters; whey and milk cure; grape cure; hydropa**thy.**
Season: July—September, but open all the year.
Special indications: Anæmia, chlorosis, **debility, dyspepsia,** rhachi**tis, affections** of the lungs and respiratory organ**s, and** nervous **system.**
The **situation of the plateau of** Maloja is very fine, **the** panoramic **view from it being** one not readily **forgotten.** Maloja **is the** highest **inhabitable** point in Europe. **Vegetation is of course** scanty at such an altitude. **Cultivation is limited** to pasture, beyond **which** comes a forest of **larches and red** pine. The flora **is** very varied. In the valley **the air is dry,** pure and clear. The **heat in** summer rarely exceeds 60° **F.,** while the winter mean is from 7° to 9° F. below freezing point. Maloja enjoys in winter more sunshine, to the extent of one hour daily, than either St. Moritz, Davos or Wiesen.

The ground in front of the Kurhaus is laid out as a park, sloping towards the lake. At the back there is a large garden with pavilions, skating rink, croquet and lawn tennis grounds. The ventilation of the building is based upon the most recent scientific principles. It is heated by calorifères, which dispense with the necessity of stoves or double windows. Ozone will be introduced by a new method. A band plays daily, and there is a theatre, ball, concert, ladies', reading, billiard and smoking room.

Excursions are of endless variety—too numerous, indeed, to mention. Sailing and rowing boats will be placed on the lake, and regattas are in contemplation. Carriages, horses, and mules are kept by the hotel.

Post and Telegraph Office in the *Kurhaus.*
English Church Service.
Bank and Exchange Office with the Director.
Doctors: Dr. Tucker Wise and Dr. Duwer.

Hotel: The *Kurhaus Hotel* combines every modern comfort with the lowest possible charges; lighted by electricity; 400 beds; two lifts; the best cooking and wines. For all particulars, address the director.

Malta.—Island in the Mediterranean, between Sicily and Tunis. Winter station; **sea baths.**
Doctors: Drs. J. Sammret, Stilon and Schembri.

Malvern.—England, Worcestershire, 129 miles from London, 3 hours by rail; 5,800 inhabitants; 502 feet above sea level.

The waters are slightly alkaline and earthy, **and** range from 40° to 42° F., but Malvern is chiefly celebrated for its brine and saline baths at the Imperial Brine and Hydropathic Establishment and climatic summer and winter station; inhalation room.

Season: May to October.

The number of visitors is above 50,000 annually.

Special indications: Acute and chronic rheumatic gout and rheumatism, and their various sequelæ as a speciality, scrofulous disorders, neuralgia, sciatica, lumbago, &c.; swellings of the joints, contractions and anchylosis, paralysis, morbus Brightii, bronchial affections and chest complaints, skin diseases, old wounds.

The town of Malvern is beautifully situated on the slopes of the Malvern Hills, which rise to the height of 1,500 feet above the sea level. These hills are traversed by roads and paths easy of ascent. From their summit a succession of the grandest and most picturesque views in the United Kingdom embracing 12 different counties, can be obtained.

The town consists of Great and Little Malvern, which are about 3 miles apart. The picturesqueness of the site and surroundings of Malvern have been celebrated in prose and verse by Byron, Walpole, Southey, Browning and Macaulay.

The Malvern air is dry and bracing, the climate tonic and stimulating, and the temperature equable. The drinking water is of absolute purity and softness. The death rate is the lowest in the kingdom, being only 8·85 per cent. The sanitary arrangements are based on the best scientific principles. Debilitated constitutions and overworked literary and business men, after a residence here of only a few weeks, often return completely restored to health.

The arrangements for treatment are very complete. The Imperial Baths and Sanatorium contain each and every appliance which modern balneology requires, and can fairly hold their own with the best establishments on the Continent. Indeed, they are superior to many of the most famous. This establishment is divided into two departments, one for gentlemen and one for ladies. It is thoroughly heated in every room and corridor by hot water pipes. It contains—besides the large swimming bath, which is one of the best in England—Turkish and Russian vapour baths, sitz, wave, medicated and electric baths, douches of all kinds, plunge baths, douches for rain and needle sprays, ascending, descending and horizontal shower douches; shampooing in all its branches under medical supervision. The ventilation, etc., of the bath establishment will be found perfect. The Droitwich saline water and brine has been brought to Malvern at a considerable expense, and constitutes one of the most valuable therapeutic additions to the establishment. The mechanical arrangements for its application are very numerous and complete.

Its dry and highly oxygenated atmosphere is an important recommendation in favour of Malvern as a winter resort. To see the country, however, at its best, spring should be chosen.

The envirous afford an ample field for beautiful excursions, more perhaps than any other portion of the kingdom. A few of the most important may be mentioned here:—The Beacon, Malvern Abbey and Avecote Monastery, Eastnor and Bransil Castles, Ledbury, Westwood Park, Claines, Kenlip, Upton, British Camp, and Cowley Park.

Doctor: Dr. Roden, M.D., F.R.C.S.

Hotels: The Imperial Hotel, Baths and Sanatorium, exceedingly well managed and appointed; close to the Great Western Railway Station, which is reached by a covered archway. The ventilation and sanitary arrangements are very superior. Stabling. Large garden and park. Apply to the Manager.

Mannheim.—Germany, Baden, on the confluence of the Neckar and the Rhine, on the left bank of the latter, from Paris by E. of F. Railway, *via* Forbach and Bingen, or by N. of F., *vid* Cologne-Mayence, 17½ hours. 74 fr. 1st. class. 1½ hours from Dürkheim by carriage, by rail ½ hour. A high class and comfortable steamer now plies between Rotterdam and Mannheim direct, thus avoiding the frequent changes, which are so tiresome to invalids.

A comparatively modern town of great commercial importance. It is built like Philadelphia, in regular squares.

OBJECTS OF INTEREST.—The Castle, Picture-gallery, Cabinet of natural history, Collection of engravings and plaster casts, the Theatre, the Schillerplatz with statue, Church of the Jesuits, &c.

Cabs: Course, one or two persons, 60 pfgs.; by hour, 1 fr. 20 cts.
Post and Telegraph-office: At the **railway station**.
Dentist: C. H. Glöckler.
Doctors: Drs. **Mermagen** (throat) **Stephani** and Zeroni.
Hotels: The *Hotel Palatin*, a large first-class very comfortable **house for** families. Freytag & Kramer, proprietors.

Mara.—Africa, Abyssinia.
Bitter waters.

Marat.—France, Department of **Puy de Dôme.**
Cold gaseous waters.

Marbella.—Spain, Province of **Granada.**
Sulphate of lime **waters,** 60° F.

Marching.—Germany, **Bavaria**; 1,160 feet above sea level.
Sulphurous waters.
Little **known or used.**

Marcols.—France, Department of Ardèche.
Bicarbonate of soda and ferruginous waters. **For table use** only.

Margate.—England, **Kent, at** the mouth of the **Thames.**
Sea baths.
Firm sandy beach. Much frequented.
Season: April—September.
Doctor: Dr. Chambers.

Mariabrunnenbad.—Germany, Bavaria, half-an-hour from Munich, 1,500 feet above sea level.
Effervescent earthy waters.
The bathing establishment is very efficient, **and** affords **at** the same time good accommodation.

Maria dell' Aquila.—Italy, Tuscany, valley of the Fiora.
Alkaline magnesian waters, 85° F.
Little used, and no establishments.

Maria in Bagno.—Italy, Tuscany.
Sulphurous waters, 90° F.
The establishment is well **arranged, and** also affords comfortable accommodation.

Marienbad.—Austria, Bohemia, near Carlsbad; 1,920 feet above sea level.
Alkali-saline and ferruginous waters, 17° to 25° F. Mud baths.
Climatic station in summer.

The same indications as the waters of Carlsbad.
The establishments are excellent and the situation beautiful.
Doctors: Drs. Guetz, Kiseh, **Ott, Opitz, Von** Heilbrunn and Ingrish.
Hotels: **English**, *Weimar, Leipzig.*

Marienfels.—Germany, Nassau, near Schwalbach.
Cold saline waters.

Marienlyst or Marienlust.—Denmark, near Elsinore, in the circuit of Fredericksborg.
Very quiet and fashionable **sea** baths.
A very picturesque site, **with** lovely view of opposite Swedish coast.
Close by is a castle belonging **to the** King of Denmark.
Hotel: Bath Hotel.

Markammer.—Germany, Bavaria, near **Neustadt.**
Grape cure station.
Season: **September—October.**
Good accommodation.

Marlioz.—France, Savoy, close to Aix-les-Bains; **600 feet above** sea level.
Sulpho-sodic and bromo-iodurated waters, **30° F.**
Bathing, drinking, and inhalation.
Special indications: Skin diseases, struma, and **affections of** the larynx.

Marmolejo.—Spain, Province **of Jaen.**
Bicarbonate of magnesia and ferruginou**s waters, 50°** F.
The situation is fine, but little has **been done** to attract **visitors.** Primitive **arrangements.**
Doctor: Dr. Oton.

Marsala.—Italy, Sicily.
Sea **baths** and saline spring.

Marseilles.—France, Department of Bouches du Rhône.
Sea **baths**; sandy and shingly beach. Climate fine.
The bathing establishments are very good, and are fitted up **with** every luxury.
Bookseller: P. Perrard, 3, Rue Noilles.
Doctor: Dr. Seaux.
Hotels: The *Terminus Hôtel*, entrance from the platform **of the railway station; baths and every accom**modation, **such as is found in** first-class hotels.
The *Hôtel Louvre and de la Paix*, large, first-class house in every respect, the only one of the large hotels which faces south; very popular. P. Neuschwander & Co., proprietors.

Marstrand.—Sweden, two hours from Gothenburg.
Sea baths; the most frequented **in** Sweden.
Season: **June—September.**

Martigné-Briand.—France, Department of Maine and Loire.
Cold ferruginous bicarbonate waters.

Martigny.—Switzerland, **Canton** Valais, 41 m. **from** Lausanne; 5,000 feet above sea level.
Starting point for the Tête Noir, 4 hours; and Col de Balme, 5 hours; leading to Chamounix, 8 hours; to the Gorges du **Durnau**, 1 hour; and numberless other excursions; the Grand St. Bernard, 8 hours.
Hotel: The *Hôtel Clerc* is recommended as comfortable **in all** respects. 1. Clerc, proprietor.

Martigny-lez-Lamarche.—France, Vosges.
Cold sulphate of lime waters.

Martinique.—An island in the **West** Indies, one of the Carribees.
Several thermal **bicarbonate** of iron springs. Generally **used in cases** of rheumatism **and old wounds.**

Martos.—Spain, Province **of Jaen,** near Granada.
Sulphurous waters, **45°** F.
Special indications: Inveterate skin diseases, leprosy and elefantiasis, leucorrhœa, herpes and scrofula.
The establishment is **very** old fashioned, **but the** waters **are** very efficient, and **consequently** there **is a large** concourse of visitors. Accommodation is plentiful and cheap.
Doctor: Dr. Cerdó y Oliver.

Martres de Veyr (Les).—France, Department **of Puy de** Dôme.
Bicarbonate of soda ferruginous water, **50° to** 85° F. Mildly purgative.

Masino.—Italy, Lombardy, **6 hours** from **Chiavenna**; 3,270 feet above sea level.
Saline waters, **100°** F.
Special indications: **Nervous** affections.
The establishment is **very** antiquated, and the visitors have considerably decreased in number of late years.

Massa.—Italy, near the Bay of Genoa.
Sea baths; sandy beach.
Only frequented by the inhabitants of Genoa and its environs.

Mastineez.—Hungary.
Ferruginous bicarbonate **waters,** 30° F.
Not used.

Matlock.—England, Derbyshire, **on the** Derwent.
Bicarbonate of lime water, **66°** to 68° F.
In a very charming position; a well-appointed establishment, much frequented. Cool and agreeable climate.
Special indications: Muscular and acute rheumatism, **dyspepsia,** biliary and glandular **affections.**
Doctor: Dr. Holland.

Mattigbad.—Austria, Upper Austria, near Moos; 1,200 feet above sea level.
Ferruginous waters, climatic station in **summer**.
Very sheltered position, mild and pure air.
The establishment is thoroughly **well fitted up, and affords ample and** comfortable accommodation.

Mauer.—Austria, ¾-hour from **Vienna.**
Effervescent ferruginous waters.
Special indications: Debility, nervous complaints, hemicrania, rhachitis, anæmia, chlorosis, menstrual derangements.

Mayence, or Mainz.—Germany, on the left bank of the Rhine, **at the** confluence **of** the river Maine. From Paris by North of France, *viâ* Cologne, **or** by East of France Railway, *viâ* Forbach and Bingerbrück, 17¼ hours. 75½fr. 1st class.
The largest town in the Grand Duchy of **Hesse**-Darmstadt. **It was annexed to Prussia in 1866.** Very **strongly** protected **by the** castle opposite. Of great antiquity, having been **founded by** Drusus, 14 B.C.; **was seat of one of the electors.**
It is the favourite stopping place for travellers **going to any of the neighbouring watering places,** as **also on the route to** Vienna, **Switzerland** and the Bavarian Highlands.
OBJECTS OF INTEREST.—The Palace, Museum, Library, Town Gardens **or** Neue-Anlage **outside the walls,** Schillerplatz **and** Cemetery, Cathedral, **Citadel, Theatre, Tower of Drusus.**
EXCURSIONS: To Wiesbaden, **Biebrich, Bingerbrück,** and **various spots on the Rhine.**
CABS: ¼ hour one horse, 50 pfgs.; ½ hour, 1 mark.
Dentist: Dr. König.
Doctors: Drs. Hellwig, Wittmann and Dörr.
Hotels: The *Hôtel de Hollande*, a first-class house of European reputation; opposite the landing place of the Rhine steamers, **and near the** railway station; very comfortable, **with** moderate charges. Patronised by the highest aristocracy. Ferd. Büdingen, proprietor.

Mayres.—France, Department of Ardèche.
Sulphate of lime, table water.

Mazel (Le).—France, Department of Lozère.
Cold mineral water.

Mecina-Burbaron.—Spain, Province of **Granada.**
Ferruginous **bicarbonate** waters, 35° F.

Medagues.—France, Puy de Dôme.
Cold bicarbonate of lime waters.
Special indications: Dyspepsia, gravel, chlorosis, and intermittent fevers.

Medewi.—Sweden, Government of Linköping.
Effervescent ferruginous waters. One of the strongest of Swedish ferruginous waters.
Special indications: Debility, nervous disorders, anæmia, chlorosis.
The establishment is thoroughly well fitted up, and the accommodation is comfortable, though somewhat expensive.

Mehadia, also **Herculesbad.**—Austria-Hungary, Province of the Danube.
Alkali-saline sulphurous waters, 80° to 130° F. The volume of water is larger than that of any known spring in Europe.
Special indications: Affections of the respiratory organs, skin diseases, syphilis, icterus, dropsy, mercurial dyscrasia.
Most frequented bathing place in Hungary.
Doctors: Drs. Chlorin, Munk, Miesko and Pacsn.
Hotel: Kurhaus.

Meidling.—Austria, on the route from Vienna to Trieste.
Sulphurous waters.
A well-appointed establishment, with theatre, concert hall, and comfortable accommodation.

Meinberg.—Germany, Lippe-Detmold, near Pyrmont; 638 feet above sea-level.
Mixed sulphurous waters and mud baths.
Special indications: Rheumatic affections, female complaints and scrofula.
Very well-appointed establishment.
Doctor: Dr. Caspari.
Hotels: Red House and Kurhaus.

Melcombe Regis.—England, county of Dorset, on the South coast, near Weymouth.
Sea baths; firm sandy beach.
The beach is very steep and the surf heavy.
There is a spring slightly sulphurous here; the waters are mixed with warm sea water.

Melksham.—England, county of Wilts, near Salisbury.
Saline and ferruginous waters. Artificially prepared with carbonic acid they are exported for table use.

Melos.—Greece, one of the Cyclades in the Ægean sea.
Ferruginous waters, 80° F.
Much renowned in antiquity; at present the available accommodation is very primitive.

Meltingen.—Switzerland, Canton Soleure.
Cold ferruginous bicarbonate waters. These waters are tonic.
A much frequented and well-appointed establishment.

Mentone.—France, Department of Alpes Maritimes, **near** the Italian frontier. For routes see table.

Winter station of first importance, and sea baths.

Winter Season: November—June.

Summer, or sea-bathing Season: May—October.

Special indications: All forms of chest diseases, consumption, bronchitis, chronic laryngitis; nearly every form of rheumatism; all cachectic diseases, and general debility; it also suits admirably very young or very old persons.

Mentone lies at the foot of the Maritime Alps, on the strip of shore to which the name of Riviera, or Cornice, has been given, and which extends from Nice to Genoa. It is a small town of 6,000 inhabitants. Latitude, 43° N., longitude, 11° E.

The coast line of mountains here rises to an elevation of 4,000 feet or more, and, receding, forms a sheltered amphitheatre, well protected from the north, north-east, and north-west winds. The scenery is more picturesque than in any other portion of the Riviera. The mingling of mountain and sea, and olive-clad hills, intersected by innumerable charming valleys, is indescribably lovely, and must be seen to be realised. The flora is varied, and spring or summer reigns even in winter. The watershed of the country—the mountain masses—are formed of secondary white limestone. But tertiary sand and gravel enter into the soil and diversify vegetation in a very charming manner.

The climate is milder than in any other portion of the north shore of the Mediterranean, as shown by the presence of forests of lemon trees which creep up the hillsides to a considerable elevation. Nowhere on the northern Mediterranean shore does the lemon-tree grow with equal luxuriance. It is a tropical tree which cannot bear more than three or four degrees of frost without perishing. The fruit is destroyed by even one or two degrees. At Mentone lemon growing is the chief agricultural industry.

The most trustworthy meteorological data respecting the winter climate of Mentone are those given by Dr. Henry Bennet, in his well-known work: "Winter and Spring on the Shores and Islands of the Mediterranean," 5th edition. They are founded on fifteen years' personal observation; for November his mean temperatures are min. $49.2°$, max. $60.1°$; December, min. $44.2°$, max. $55.1°$; for January, min. $42.8°$, max. $53°$; for February, min. $43.5°$, max. $55.7°$; March, min. $45.3°$, max. $59.3°$; April, min. $50.8°$, max. $66.3°$. The mean minimum for the six winter months is thus for fifteen years, $45.9°$; the mean maximum, 58.2. The combined mean for the six months during fifteen years is $52°$ F. During this period, observations were daily made of the hygrometric state of the air with wet and dry bulb thermometers, and a mean of 5.1 was obtained. This corresponds to 68 % of dryness—saturation being 100°, with a temperature of about 58°, and implies moderate dryness in the air. In the winter there are no sea or land fogs, although sea fogs are occasionally observed in May and June. The average number of days on which rain falls—little or much—is, according to Mr. Brea, about 81 in the year. The amount of rainfall is about 24 inches. (*Farina*.)

The climatic conditions above described render Mentone an exceptionally valuable winter resort in all diseases and in all conditions of health in which a dry, sunny, mild winter climate is required. There is a peculiar condition connected with the climate of Mentone, which Dr. Henry Bennet was the first to point out. The moisture-laden sea breeze, which is drawn in daily by the sun's heat in cloudless weather, is arrested by the amphitheatre of mountains, which it cannot pass. The colder atmosphere of the elevated regions condenses its moisture, and so forms a haze over the landscape from about 11 a.m. These condensed vapours soften the atmosphere in an exceptional manner, and no doubt explain the fact of so many invalids improving, or even getting well at Mentone, while other regions of the north Mediterranean coast and the Riviera disagree with them. The rapid changes of temperature common to Nice, Cannes and Hyères, are less noticed here, temperature being more equable.

The drainage has improved of late, owing to the united efforts of the medical men and the municipal authorities.

There is a casino, with theatre, reading rooms, &c.

English church.

Post and Telegraph office, Rue St. Michael.

American Dentist: A. Preterre.

Bookseller: Bertrant, Queyrot & Co.

Bankers: Credit Lyonnais, Avenue Victor Emmanuel.

House & Estate Agent & Provision Merchant: T. Willoughby, Rue St. Michel.

English Chemist: P. Bezos, 27, Rue St. Michael.

Doctors: Drs. Henry Bennet, Marriott, Daremberg, Fitz-Henry, Siordet, Cazenave de la Roche, and Morin.

Hotels: The *Hôtel des Iles Britanniques*; exceptional position, well recommended. Rosnoblet, proprietor.

The *Hôtel des Anglais*, full south in the East Bay, well sheltered, near the Chalet des Rosiers (residence of Queen Victoria in 1882). Em. Arbogast, proprietor.

The *Hôtel Victoria*, in the centre of the town, fine garden, and well sheltered. Sèmeria and Leubner, proprietors.

Meran.—Austria, Tyrolean Alps, near the Italian frontier, 1,100 feet above sea level. For routes see table.

Climatic station in winter. Very fine and pure air. Pneumatic baths. Whey cure, grape cure.

Season: October—April.
Grape cure Season : September—October.
Mean winter temperature, 40° F.; spring and autumn, 55° F.; summer, 70° F. The greatest proportion of days in winter are sunny and cloudless.

The climatic residences in winter are Ober and Untermais. The latter is beautifully situated on the left bank of the river Passer, and extends, with its castles, villas, gardens, parks and old ruins, over a rising slope. Obermais has made rapid strides of late, and its fine villas and large hotels offer very comfortable accommodation.

Gratsch is a quarter which has come into existence only within the last few years. It is gaining rapidly in favour, as its situation is the best and its climate exceptionally mild. Invalids as a rule prefer to stay at one of the above-mentioned villages.

The situation is a charming one. The scenery is picturesque, and suggestive of excursions. The promenades in the town and suburbs afford ample space for long walks. The streets are well kept and life in the town is agreeable.

The "Winteranlage" is the principal promenade for visitors and patients during winter; between 11 and 3 o'clock the Foreign Colony meets here. The vegetation is evergreen, composed mainly of sub-tropical plants.

The "**Curhaus**" is a very fine building, and contains besides assembly-rooms, concert, ball and billiard rooms, a café and a bathing establishment.

Chemist: Hallersche.
Bookseller: Pötzelberger.
Bank and Exchange Office: Biedermann Bros.
Post and Telegraph Office: 36, Rennweg.
Doctors: Dr. von Kaan and Dr. Settari.
Hotels: Comte de Meran, Sun.

Mercatale.—Italy, Tuscany, valley of the Arno.
Ferruginous sulphurous waters, 45° F.
The establishment is very primitive, though the waters enjoy a great reputation in the neighbourhood.

Merens.—France, Department of Ariège.
Sulphate of soda water, 90° to 110° F.

Mergentheim.—Germany, Würtemberg, 8 hours from Würzburg.
Bromo-iodurated bitter waters, 30° F.
Slightly **purgative.**
The establishment is elegant and efficient.
Theatre, assembly and reading-rooms, and fine promenades.
Doctors: Dr. Lindemann, Dr. Höring and Dr. Krauss.
Hotels: Strauss and Rose.

Merligen.—Switzerland, Canton Berne, on the Lake of Thun. By boat from Interlaken and Thun.

Climatic station, remarkable for its mild climate. The situation of Merligen is one of the most beautiful in this part of Switzerland. The scenery is highly picturesque, and rendered still more attractive by the luxuriance of the vegetation. Numerous excursions can be taken. Merligen, in a word, is a very attractive spot.

The *Hôtel Beatus* will be found a **good comfortable house**, with good kitchen and moderate **prices**. Ph. Hofmann, proprietor.

Mers.—France, Department of Somme.
Sea baths; sandy beach.
A small establishment.

Messina.—Italy, Sicily, 45 minutes by steamer from Reggio. Pullman cars are now attached to all express trains from Naples. For routes see table.

A winter station **and sea baths.**

Much exposed to winds. It is mentioned here only because it is selected occasionally as a winter residence, though it has actually one of the severest climates of any town in Sicily. On the average there are 12 snowy days in winter.

Bookseller: G. Welbatus.

Hotel: The *Hôtel Victoria* is the best house in the town. William Möller, proprietor.

Meta.—Italy, between Castellamare and Sorrento.
Climatic station in summer.
Used by inhabitants of Naples as a summer residence. The whole country is one vast garden.

Methana.—Greece, near Kato-Muska, Argolis.
Sulphurous waters, 85° F.
The arrangements are on a system of taking the baths in common.
Accommodation primitive.

Metyline.—Asiatic Turkey.
Sulphate of soda water, 80° to 100° F.
Special indications: Rheumatism and skin disease.

Mezières.—France, Department of Ardennes.
Cold saline waters.

Michelstadt.—Germany, Hesse, near Erbach, in the Odenwald; 800 feet above sea level.
Summer station, with hydropathic establishment.
Doctor: Dr. Spies.

Middelkerke.—Belgium, West Flanders, near Ostend.
Sea baths, with fine beach.
Is rising in importance.

Miemo.—Italy, Tuscany, in the Maremma-Volterana.
Alkaline **sulphurous** waters, 80° F.
Primitive arrangements and accommodation. Almost exclusively used by the surrounding peasantry.

Miers.—France, Department of Lot.
Cold sulphate of soda waters.
Special indications: Hepatic affections, constipation and hæmorrhoids.
Doctor: Dr. Fraisse.

Milan.—Italy, Capital of Lombardy. From Paris by P.L.M. through Mont Cenis, 27 hours. 116¾ fr. 1st class.
The town is nearly circular in form, and surrounded by walls; erected for the most part by the Spaniards. The space between the walls and canal is laid out in gardens and well planted with trees. Like all old cities, Milan is very irregularly laid out, but is, for all that, one of the most interesting in Europe.

OBJECTS OF INTEREST.—The Cathedral, the Amphitheatre, the **various** Old Gates, the Church of Sant' Ambrosio, the Military hospital, the Gallery of Victor Emanuel, Palazzo di Corte, Brera, Observatory, Pinacoteca, Ambrosian Library, Museum, Theatre della Scala, Theatre Royal, Hospital, Civic museum, Town hall, several private Palaces, Loggia degli Osii, City hall, Mint, Judicial and government Palaces, Customhouse, Treasury, Public gardens, the Castle, and Royal Palace.

EXCURSIONS: To Monza, to various Certose, to the Lakes, besides many others.

CABS: One horse, by course, 1 lira; by time, first half hour, 1 lira.
Post office: Via Rastrelli, 20.
Telegraph office: Piazza dei Mercanti.

Dentist: C. T. Terry, Doc. Den. **Sur., from New York,** 5, Piazza Cavour.

Bookseller: **Fratelli Dumolard.**

Hotels: The *Hôtel de la Ville* has just been greatly improved by the addition of an enclosed court-yard, lift, and all modern requirements. A most popular house. J. Baer, proprietor.

The *Grand Hôtel de Milan*, near the Opera House; large, first-class; lift. **Post, Telegraph, and Railway Booking Office** in the Hotel. J. Spatz, proprietor.

The *Hôtel Germania*, same proprietor.

Millport.—Scotland, Buteshire, on the greater Cumbray Island.
Sea baths; sandy beach; air mild; scenery barren.
Doctor: Dr. Kerr.

Milo, Island of.—In the Grecian Archipelago.
Thermal waters, of great repute in antiquity, but now neglected.

Mina Nova.—Portugal, Province of Estremadura.
Ferruginous sulphate waters.

Mindelheim.—Germany, Bavaria.
Bicarbonate of lime waters.
A bathing establishment, but little known.

Minehead.—England, Somersetshire, near Taunton.
Sea baths; rocky and sandy beach; climate mild.
Season: July—September.
Doctor: Dr. Gaye.

Mingelsheim.—Germany, Baden.
Cold sulphurous soda water.
Special indications Rheumatism and skin diseases.

Mirabello.—Italy, Lombardy.
Sulphate of lime springs, 30° F.
Special indications: Scrofula.

Mirandela.—Portugal, Province Tras os Montes.
Cold ferruginous waters.

Misdroy.—Germany, **Island of Wollin.**
Sea baths on the Baltic.
Firm fine sandy beach. The scenery is hilly and charming. Establishments good.

Mitterbad.—Austria, Tyrol, district of Botzen.
Ferruginous waters, containing sulphuric acid.
Season: June—August.
One of the most frequented bathing places in the Tyrol.
The scenery is very mountainous and picturesque.
Special indications: Anæmia, fluor albus, difficult menstruation, chronic diarrhœa.

Moching.—Germany, Bavaria, near Munich.
Bicarbonate of lime waters.
The establishment is ample, and provides comfortable accommodation.

Modum.—Norway, district of Tellemarken, near Christiania.
Effervescent ferruginous waters, and summer station.
The situation is very picturesque and mountainous. The environs contain large pine forests, and the air is highly charged with ozone.

Moffatt.—Scotland, Dumfries.
Hot sulphur and saline waters; air bracing.
Hydropathic establishment.
Season Summer
Special indications: Skin diseases, female ailments, dyspepsia.
Doctor: Dr. Thomson Forbes.

Mogador.—Africa, Morocco.
Winter station. Warm climate.

Moggiona.—Italy, Tuscany, District of Poppi.
Sulphurous waters, 65° F.
Special indications: Gravel and skin diseases, abdominal disorders.
The establishment is inadequate.

Moha.—Hungary, Transylvania, comitat of Stuhlweissenburg.
Bicarbonate of lime ferruginous waters.
Act as a mild purgative.

Moingt.—France, Department of Loire.
Cold bicarbonate of soda, table waters.

Molar (El, or Fuente del Toro).—Spain, Province of Madrid, on the road to Burgos, near Jarama, 2,600 above sea level.
Saline sulphurous waters, 45° F.
Special indications: Skin diseases, laryngitis, bronchitis, asthma, amenorrhœa and chlorosis.
The establishment is tolerably well fitted. Accommodation abundant, but not of the best. There is nothing attractive in this station.
Doctor: Dr. Lopez.

Molina de Aragon.—Spain, Province of Guadalajara.
Sulphate of lime waters.

Molinar de Carranza.—Spain, Province of Biscaya, near Tolosa.
Saline springs, 90° F.
Special indications: Scrofula, liver and splenic complaints.
A much frequented establishment, thoroughly well fitted up, but the accommodation is not scrupulously clean.
Doctor: Dr. Zapater.

Molitg.—France, Department of Pyrénées Orientales, near Prades.
Sulphate of soda waters, 50° to 65° F.
Two small establishments.
Special indications: Rheumatism, bronchitis, and skin diseases.

Molla Il.—Italy, Piedmont.
Ferruginous bicarbonate waters, 45° F.
Special indications: Chlorosis.

Monaco and Monte Carlo.—Principality of Monaco, between Nice and Mentone. For routes see table.

Winter station. Sea baths.

Although Monte Carlo is one of the most sheltered stations on the Riviera, it is more frequented by pleasure seekers than by real invalids. Strangers coming to the Riviera only for amusement, or to pass the winter, prefer staying here.

Mean winter temperature, 50° F. The barometric pressure, mean humidity of the air, and average rainfall, are almost the same as at Nice. The temperature in the quarter called "La Condamine," is generally from 3° to 5° F. higher; it is also the most sheltered from winds, but it has the drawback of inferior drainage. The weather statistics are 57 fine, 26 clouded, and 17 rainy days per cent.; the above statistics are, however, open to question. Atmospheric disturbances are somewhat frequent, and in 26 per cent. of days there are strong winds; fogs are rare, but the dews are heavy and frequent. On the whole the climate is an exceptionally dry and exciting one. The mistral is as little felt here as at Villefranche, especially in the Condamine district.

The air being very dry, and high winds of frequent occurrence, the entire absence of dust is the more noteworthy and pleasant. The roads, gardens, etc., are excellently kept, and residence here is certainly as agreeable as in any town on the Riviera.

The Casino, with its almost unrivalled orchestra (concerts twice daily) is the chief attraction. The reading-room contains newspapers of every country, and in the handsome theatre, operatic and other representations are given during the season, with the help of the most renowned artists in Europe. The pigeon-shooting arrangements are very complete, and the meetings well attended.

The old town on the promontory, or Monaco proper, is interesting, and the Prince's palace is well worth visiting. The gardens are very fine. The sea bathing establishments are thoroughly good.

Bankers: Vve. Adolphe Lacroix, Méja & Co.; also at Nice, 1, Place du Jardin Public.

House and Estate Agent: A. Roustan, Villa Fontvieille, Monaco.

Dentist: Mr. H. Ash, Condamine.

Provision and Wine Merchants: Under the Grand Hotel.

Doctor: Dr. Pickering, at Monte Carlo.

Hotels: The *Grand Hotel*, at Monte Carlo, magnificent first-class house, with exceptionally good restaurant, English bar, considerably enlarged and embellished. X. Jungbluth, proprietor.

The *Victoria Hotel*, at Monte Carlo, healthily situated, first-class family house. Very quiet and comfortable. Rey Frères, proprietors.

The *Hôtel des Anglais* is a first-class family house, situated in the best part of the gardens, refurnished, and under new management. L. Troiel Fribert, director.

The *Hôtel Beau Rivage*, half-way between Monte Carlo and the Condamine, is under the able management of Mr. F. Schmitt, of the *Hôtel d'Angleterre*, at Ems.

Moncada-y-reitach.—Spain, Province of Barcelona.
Ferruginous sulphatic waters, 45° F.
Special indications: Chlorosis and dyspepsia.

Monchique.—Portugal, Province of Algarvie.
Thermal sulphurous waters.
Very much frequented station.

Monda.—Spain, Province of Malaga.
Cold saline springs.
Waters act as a diuretic.

Mondariz.—Spain, Province of Pontevedra, near Puentarénas, on the River Ter.
Effervescent ferruginous waters.
Special indications: Chlorosis, anæmia and kindred complaints.
The situation is most lovely, and very romantic. The establishment is of slight importance, and the accommodation not particularly good.
Doctor: Dr. Crucero.

Mondon.—Spain, Province of Orense.
Cold ferruginous bicarbonate waters.

Mondorf.—Germany, Duchy of Luxemburg.
Bromo-saline waters, 60° F.
Special indications: Scrofula, chlorosis, and dyspepsia.
A very good hydropathic establishment.
Doctors: Drs. Marshal and Klein.
Hotel: Kurhaus.

Monegrillo.—Spain, Province of Saragoza.
Saline waters.
Purgative action.

Monestier de Briançon.—France, Department of Hautes Alpes.
Sulphate of lime waters, 50° to 105° F.
Special indications: Rheumatism, old gun-shot wounds, and skin diseases.

Monestier de Clermont.—France, Department of Isère.
Mixed bicarbonate table waters.

Monfalcone.—Austria, Illyria, on the Bay of Trieste.
Iodo-bromurated saline waters, 100° F.
Special indications: Rheumatism and paralysis.
Sea bathing establishment. The springs rise and fall with the tide.

Monsao.—Portugal, Province of Minho.
Saline waters, from 80 to 100° F.

Monsummano.—Italy, Tuscany, near Montecatini, valley of Nievole.
A grotto, where natural sulphurous vapour baths are taken. They are especially efficient in rheumatism.
Three lakes in the grotto contain sulphurous waters from 80° to 100° F.
Special indications: Gout, herpes, rheumatism, anchylosis, sciatica, syphilis, catarrhal affections, and in all cases where strong action of the skin is indicated.

Montafia.—Italy, Piedmont.
Sulphate of lime waters, 30° F.
Employed by the peasants against skin diseases.

Montagnone.—Italy, Venetia, near Abano.
Earthy waters, 130° F.; mud baths.
Special indications: Diseases of the bones, old sores, scrofula, gout, rheumatism.
The bathing establishment is well arranged, and affords comfortable accommodation.
The baths were used by the Romans, as is shown by numerous ruins.

Montalceto.—Italy, Tuscany, Val d'Arbia.
Saline-magnesian ferruginous waters, 65° to 85° F.; mud baths.
A very good establishment, and well frequented station.

Mont' Amiata.—Italy, Tuscany.
Highly effervescent ferruginous waters, 35° to 65° F.
The arrangements are primitive, and consequently used only by the surrounding inhabitants.

Montanejos.—Spain, Province of Castellon de la Plana, near Vivel.
Ferruginous and alkaline effervescent waters, 60° F.
Special indications: Plethora abdominalis, hyperæmia of liver and spleen, vesical catarrhs.
Very little frequented, though the establishment is modern.
Doctor: Dr. del Rio.

Montbarri.—Switzerland, Canton Fribourg.
Cold sulphate of lime waters.
Special indications: Skin diseases and rheumatism.

Montbrison.—France, Department of Loire.
Cold alkaline bicarbonate waters.
Special indications: Chronic disorders of the digestive organs, hæmorrhoids.

Montbrun.—France, Department of Drôme.
Sulphate of lime waters, 30° F.
Mud baths.

Montchauson.—France, Department of Cantal.
Cold bicarbonate and ferruginous waters.

Mont Dore.—France, Puy de Dôme, 1½-hour by carriage from Laqueuille station, route from Clermont-Ferrand to Tulle; 3,150 feet above sea level; 8,000 inhabitants. For routes see table.
Bicarbonate, arsenical, and effervescent ferruginous waters, from 107° to 115° F. The establishment is thoroughly well fitted up with all modern requirements. Climatic station.
Season: 1st June to 1st October.
Special indications: Chronic bronchitis, pulmonary phthisis, asthma, chronic pleuritis, laryngitis, pharyngitis, coryza, chronic ophthalmia, rheumatic and nervous affections, disorders of the skin.
Bookseller: J. Armet.
Doctors: Drs. Léon Chabory, Chabory-Bertrand, Emond (speaks English), Mascarel and Cazalis.
Hotels: The *Hôtel de Paris* and *du Parc* are first-class, well appointed houses, healthily situated, and replete with all modern comforts. Léon Chabory, proprietor, is the only one of this name. Note the address, "Léon Chabory."

The *Grand Hotel* (see plan), a first-class and very comfortable family hotel, near park and bathing establishment; open all the year. Madame Taché-Serizay, proprietress.

The *Chalet*, Montjoli, beautifully situated, overlooking the town, specially comfortable for families. The residence of the Prince of Wales in 1884. See plan. M. Bourdassol, proprietor.

Monte Carlo.—*See* Monaco.

Monte-catini.—Italy, Tuscany. For routes see table.
Saline springs, 72° to 82° F.
Special indications: Rheumatic affections, leucorrhœa, blenorrhœa, menstrual disorders, skin diseases, sciatica.
The purgative waters of the Tettuccio spring are exported, and used in the treatment of hypertrophy of the liver.
The establishments are very good. Vegetation and scenery charming, and in consequence Monte-catini is much frequented.
Doctor: Dr. Fedeli.

Montefiascone.—Italy, Roma.
Thermal sulphurous springs.
Special indications: Skin diseases.

Montegrotto.—Italy, Venetia, near Padua.
Hot saline muds, 176° F.
Stimulating action.
The establishment is good, and the accommodation comfortable.

Montegut Ségla.—France, Department of Haute Garonne.
Ferruginous bicarbonate water, 30° F.
Special indications: Dyspepsia and gastralgia.

Montemayor y Bejar.—Spain, Province of Cáceres, near Salamanca; 2,300 feet above sea level.
Sulphurous waters, 90° F.
The establishment is very good and affords comfortable accommodation. The air is remarkably pure, and the scenery picturesque.
Notwithstanding the difficulty of reaching the town, it is very much frequented.
Doctor: Dr. Sastre y Dominguez.

Monte-Ortone.—Italy, Venetia, near Padua.
Alkali-saline sulphurous waters, 100° F.
There is a bathing establishment and hotel combined. Little frequented.

Monte-Rotondo.—Italy, Tuscany, on the Cornia.
Earthy ferruginous water, 75° F.
No establishment, and primitive arrangements. The site is picturesque.

Montione di Piombino.—Italy, near Arezzo.
Bicarbonate of lime waters, 85° F.
Special indications: Rheumatism and skin diseases.

Montlignon.—France, Department of Seine and Oise, near Montmorency.
Cold ferruginous bicarbonate waters.

Montlouis.—France, Department of Pyrénées Orientales.
Cold ferruginous waters.
Charming situation.

Montmirail-Vacquairas.—France, Department of Vaucluse, near Carpentras and Avignon.
Three springs with a purgative action; one sulphate of lime, and one ferruginous.
The bathing establishment is good and affords fair accommodation.
Doctor: Dr. Millet.

Montner.—France, Department of Pyrénées Orientales.
Ferruginous bicarbonate waters.

Montol.—France, Department of Puy de Dôme.
Cold bicarbonate of lime waters.

Montpellier.—France, Department of Hérault.
Winter station.
Doctors: Dr. Coste and Dunal.

Montpensier.—France, Department of Puy de Dôme, near Riome.
On the route to Gannat two excavations are found, from which carbonic acid gas escapes, as in the "Grotta del Cane," near Naples.

Montreux.—Switzerland, Canton of Vaud, near Vevey and Lausanne.
Climatic and intermediate station, whey cure, grape cure.
Situated on Lake Leman, it has a very charming position; the excursions in its neighbourhood are very interesting and numerous.
Post and Telegraph: At station.
English Church.
Bookseller: Benda, in Vernex.
Chemist: Schmidt.
Doctors: Dr. Carrar, Steiger, Bertholet, and Monnier.
Hotel: The *Hotel Pension Breuer*, first-class house, very comfortable, lift; large winter garden; douches, hot and cold baths, lake bath near the Kursaal and English Church; omnibus at station. G. Breuer, proprietor.

Montserrat.—West Indies, one of the Leeward Islands.
Several thermal springs.

Monzaio les Mines.—Africa, Algiers, near Medah and Blidah; 1,600 feet above sea-level.
Sulphate of sodium waters, 35° to 50° F. Table waters; digestive action.

Morba.—Italy, Tuscany.
Bicarbonic ferruginous and sulphate of lime waters, 45° to 105° F.
Special indications: Rheumatism and herpes.

Morecambe and **Paulton**-le-Sands.—England, Lancashire.
Sea baths; sandy and extensive beach; two piers.
Climate mild but bracing.
Season: June—September.
Doctor: Dr. Jackson.

Morgins.—Switzerland, near Monthey and Evian; 4,344 feet above sea-level.
Saline ferruginous **waters**; climatic **moun**tain air station; whey cure.
Season: June—September.
Special indications Anaemia, **chlorosis**, female com**plaints**, asthma.
The establishment is good and accommodation abundant.

Mortagone.—Italy, Tuscany.
Saline springs, 65° F.

Mortefontaine.—France, Department of Oise, near Senlis.
Cold mineral waters analogous to those of Enghien. Not **used.**

Moselli.—Greece, near Astio.
Indifferent waters, 45° F.
Very charming site. **Air pure and** mild.

Mscheno.—Austria, Bohemia.
Cold sulphurous ferruginous waters.
Special indications: Atonic diseases.

Mula.—Spain, near Murcia.
Ferruginous **bicarbonate** waters, 95° F.
Reconstituent action.
Special indications: **Herpes** and rheumatism.

Mumbles.—Wales, Glamorgan.
Sea baths.

Mumby.—England, Lincolnshire, near Willoughby.
Sea baths, sandy beach.

Münchshofen.—Germany, Bavaria.
Earthy ferruginous waters, 60° F.
Special indications: Liver and spleen obstructions, paralysis.

Munich.—Germany, capital of **the kingdom** of Bavaria. From Paris, by East of France, by Strasburg and Kehl, 36 hours. 151 fr. 1st class.
Munich is, in proportion **to its size,** one of the finest and most interesting **towns in Europe,** especially as regards its collection **of works of art.** The public buildings, gardens, squares, and monuments, **are numerous** and handsome.
OBJECTS OF INTEREST.—Royal Palace is divided into the Old Residence, Königsbau and Festsaalbau; it is filled with art **treasures** in every room; Park and Gardens, the Pinacothek, **the Glyp**tothek, **the Propylaen** and Palace of Fine Arts, old **Picture Gallery, Wimmer & Co.'s Fine Arts** Collections,

Schwanthaler's Museum, National and **Anatomical Museum**, Art Exhibition, Royal Academy of **Science**, Royal Foundry, Art Union, Schack Picture Gallery, **Hof**-brauhaus, Public Library, Government Palace, Town Hall, Gate of Victory, **Cathedral**, Churches of Michael and St. Bonifacius and St. Lewis.

EXCURSIONS: English garden, Brunthal, Great Prison, Public Cemetery, Nymphenburg, Schleissheim, Lake of Stahrenberg, Possenhofen, Berchtesgaden and Königssee, Hohenschwangau, Chateau de Berg and Rottmannshöhe, Oberammergau.

CABS: One horse, ¼ hour 50 pfg.; ½ hour, 1 mark; two horses, one to four persons, ¼ hour, 1 mark; ½ hour, 2 marks.

Post and Telegraph office: Max Josephsplatz.

Dentist: Dr. Bernz.

Doctors: Drs. Von Feder and Von Gietl.

Hotels: The *Hôtel de Bavière*. Situated on the Promenade Platz, the best part of Munich; entirely re-furnished and greatly improved. An excellent house in every respect. Oscar Seif, proprietor.

The *Hôtel d'Angleterre*. A quiet family hotel; moderate prices for pension during the winter months; centrally situated, near the Post and Opera-house. H. Sitzler, proprietor.

Münster am Stein.—Germany, Rhenish Prussia, near Kreuznach.

Waters similar to those of Kreuznach; very efficacious in children's diseases.

A very quiet spot, but full of picturesque attractions.

Doctors: Drs. Welsh and Glaesgen.

Hotel: Kurhaus Hotel.

Münsterberg.—Germany, Silesia, circuit of Breslau.

Cold bicarbonate of iron and hydrosulphuric waters.

A small establishment.

Murcia.—Spain, capital of the Province of Murcia, on the Mediterranean. Winter station of first importance.

Murcia may be called the dryest and least rainy place on the whole east coast of Spain. Mean winter temperature, 55° F. Relative dampness of the air, 66·7 %, with only 18 rainy days during winter season. Spaniards call its sky "El reino serenissimo." Snow is altogether unknown, and fogs and mists are never seen. The town having 90,000 inhabitants, offers all possible resources.

Muritz.—Germany, Mecklenburg.

Sea baths, sandy beach.

Mürren.—Switzerland, Bernese Oberland, *via* Interlaken and Lauterbrunnen, 5,181 feet above the sea level.

Climatic mountain air station, with whey and milk cure; baths and douches.

Season: June—September.

Special indications: Anæmia, chlorosis, failing digestion, phthisis, asthma.

Mürren, as regards the picturesqueness of its scenery, has often been compared to the Righi, the altitude of which is about the same height above the sea level. It is one of the finest centres for excursions in Switzerland, many of its surrounding valleys and mountains being little known to tourists. Its position is very charming, and its climate fresh and bracing. Well sheltered against northern and north-western winds.

Space prevents the enumeration of all the excursions; the local guides must chiefly be depended on.

Post and telegraph office.

English church.

An English doctor is attached to the Grand Hôtel, Mürren.

Hotel: The *Grand Hotel and Kurhaus*, Mürren, is one of the best managed houses in Switzerland. 185 rooms with 250 beds. It is one of the few houses where an effort is made to make guests feel at home. J. Sterchi-Wettach, proprietor.

Müskau.—Germany, Silesia, circuit Liegnitz on the Neisse.

Sulphate of iron waters. Astringent action.

Special indications: Hæmorrhages, hyperæsthesia, female complaints.

Very well appointed.

Doctor: Dr. Prochnow.

Hotels: Berlin, Green Tree.

Mustapha-Supérieur.—Algiers, near the Capital.

Winter station, coming much into prominence lately. Especially suitable for pulmonary complaints.

Doctor: Dr. Savill.

Hotel: Hotel d'Orient and Kirsch.

Nahaud.—United States of America, Massachusetts.

Sea baths, much frequented.

Nairn.—Scotland, near Inverness.

Very much frequented sea baths; "the Brighton of the North."

Sandy beach; dry and bracing air.

Doctor: Dr. Cameron.

Nammen.—Germany, Westphalia.

Cold sulphate of lime springs.

Special indications: Herpes, rheumatism, gout.

Nanclares de Oca.—Spain, Province of Alava, near Añana and Vitoria. Acidulous waters, 50° F.

Special indications: Affections of the stomach, flatulence, neuralgia, and disorders of the urinary organs.

The establishment and accommodation are good, notwithstanding that the place is little frequented.

Doctor: Dr. Almendariz.

Nantes.—France, former capital of Brittany.

Capital of Department of Loire Inférieure, and a very attractive old town. The ancient residence of the Ducs de Bretagne.

OBJECTS OF INTEREST.—The Castle, the Cathedral, Church of St. Croix with belfry, St. Nicholas, St. Clement, and Notre Dame, Town Hall, Prefecture, Exchange, Theatre, Library, Palais de Justice, Museum, Hospital, Lyceum.

EXCURSIONS: Guerande, Croisic, Sables d'Olonne and St. Nazaire.

CABS: Two horses, 1½ fr. the drive; one horse, 1 fr.

Telegraph: Rue St. Jules.

Hotels: *De Bretagne*, *de Paris*.

Naples.—Italy, Bay of Naples. For routes see table.

Winter station, mineral and sea baths.

The climate of Naples is dry and stimulating; the nearer one approaches Vesuvius the more tonic the air becomes.

As a hydromineral station, Naples has a great variety of springs, chief among which are those of Chiatamone, Bagnoli, Ischia, Castellamare, Torre Annunciata, Agnano, Pozzuoli, and Santa-Lucia. The variety of mineral waters and their therapeutic properties have rendered Naples and its environs the most frequented bathing station in Italy.

International Hospital: Villa Bentinck, Corso Vitt. Emmanuele.

Bookseller: F. Furchheim, 59, Piazza dei Martiri.

Guides and Couriers: Chas. Mancinelli and Emanuele Benati.

Doctors: Drs. Vittorelli, Cantani, Schrön, Malbrauc, Tweedie, Stodart, Tomasi and Wyatt.

Hotel: *Hôtel Royal des Etrangers*. A. G. Caprani, proprietor.

Nassau.—Germany, Nassau, on the Lahn, 1½ hour from Ems.

Hydropathic establishment. Climatic station in summer. Electric baths, gymnastics.

Doctor: Dr. Runge.

Nauheim.—Germany, Hessen-Nassau. For routes see table.

Strongly mineralised saline waters; 17 grs. of mineral constituents, of which 14 grs. are chloride of sodium. The water is effervescing, 50° to 95° F.

Two springs are used for drinking—the Kurbrunn and the Salzbrunn. Two others for douches and baths.

Special indications: Enlargement of the liver, derangement of the spleen, scrofula.
Doctors: Dr. Bode, Dr. Beneke.

Naumburg.—Germany, Silesia.
Cold ferruginous waters not used.

Navajas.—Spain, Province of Castellon de la Plana.
Ferruginous bicarbonate waters, 45° F.

Navalpino.—Spain, Province of Ciudad Real, near Piedrabuena.
Bicarbonate of iron waters, 70° F. Little used.
Special indications: Neurosis, hysteria, stomachic and intestinal derangements, menstrual difficulties.
The establishment is much neglected.
Doctors: Dr. Nieto.

Nave dell' Inferno.—Italy, Tuscany, on the Arno.
Earthy alkaline waters, 59° F.
Special indications: Gravel, calculus, vesical catarrhs, intestinal derangements, menorrhagia, hysteria.
The establishment is inadequate and the accommodation poor.

Naxos.—Greece, island, one of the largest of the Cyclades.
Saline waters.

Nebouzat.—France, Department of Puy de Dôme, near Clermont-Ferrand.
Cold acidulous waters.

Neffiach.—France, Department of Pyrénées Orientales, near Perpignan.
Saline waters, 50° F.
Only visited by persons in the neighbourhood.

Nelefina.—Hungary, comitat of Beregh-Ugocz.
Mixed bicarbonate waters.
Pretty good arrangements.

Nenndorf.—Germany, Hessen. For routes see table.
Cold saline sulphurous lime waters.
The bathing establishment is well appointed; inhalations, douches, &c. Mineral and mud baths. Whey cure.
Special indications: Catarrh of the respiratory organs, herpes, chronic sores, disorders of the urinary organs.
The sojourn is agreeable, and amusements are freely provided.
Doctors: Drs. Ewe, Neusell, and Varenhorst.
Hotels: Kassel, Hanover.

Neris.—France, Department of Allier, near Montluçon.
Thermal mixed alkaline waters, 110° to 125° F.
Special indications: Nervous affections, rheumatic neuralgia, uterine affections.

Nerone (Stufa di).—Italy, near Pozzuoli.
Natural mineral steam baths. The heat of the gases emitted is stated to be 170° F.
Special indications: Chronic skin diseases, gout, rheumatism, anchylosis.

Very largely used by the Romans, especially under Nero, by whom they were fitted up with the greatest luxury. They are still visited and used, notwithstanding the entire absence of any suitable accommodation.

Nervi.—Italy, one of the suburbs of Genoa, on the Riviera di Levante. Winter station.

The mean winter temperature is not so high as that of Mentone or San Remo, but the climate is less windy, exciting, and tonic, and more humid. The vegetation is luxurious, and beautiful gardens abound. There is little room for excursions, and Nervi is therefore most suitable for patients who desire repose.
Doctor: Dr. Schetelig

Neskutschnoie.—Russia, near Moskau.
Earthy ferruginous waters, 45° F.
The establishments are good.

Neudorf.—Austria, Bohemia, also called Constantinsbad.
Sulphurous and ferruginous waters, 25° F. Mud baths. Very pure air.
Special indications: Anæmia, disorders of menstruation, rhachitis, gout, and rheumatism.

The establishments are very good and the accommodation is abundant and moderate in price. The situation is picturesque, and the country well wooded.
Doctor: Dr. Dlauhy.
Hotel: Kurhaus.

Neuenahr.—Germany, Rhenish Prussia, on the left bank of the Rhine and on both banks of the river Ahr; 760 feet above sea level. For route see table.

Thermal mixed bicarbonate, mildly alkaline waters, 72·5° and 104° F.
Special indications: Catarrh of the stomach and bowels, affections of the kidneys and liver, Bright's disease, stone, diabetes mellitus, nephritic calculus, scrofula.
Post and telegraph office in the Kurhotel.
English church service.
Doctors: Drs. R. Schmitz, A. Teschemacher and Unschuld, all speaking English.
Hotels: The *Kurhotel*, *Victoria*,

Neuenheim.—Germany, Nassau.
Cold bicarbonate ferruginous waters.

Neuhaus.—Austria, Styria.
Bicarbonate of lime waters, 85° F
Special indications: Nervous diseases.
Good establishment
Doctor: Dr. Paltauf.

Neuhaus.—Germany, Bavaria.
Cold **saline** springs.
Special **indications:** Scrofula and abdominal obstructions.

Neu Lublau.—Austria, Hungary, comitat of Zips, near Leutschau.
Effervescent ferruginous waters.
Special indications: Migraine, hypochondriasis, hysteria, affections of the nerve centres, menorrhagia, blenorrhœa.
The establishments and **accommodation are** satisfactory.

Neumarkt.—Germany, Bavaria, near Nuremberg; 1,445 feet above sea level.
Ferruginous sulphurous waters.
Special **indications:** Rheumatic and **skin** affections, **muscular** contractions, hæmorrhoids, **gravel and vesical** calculi, old sores and mercurial dyscrasia.

Neu Ragoczi.—Germany, **Prussia**, Province of **Saxony**, near Halle.
Alkaline ferruginous waters. Inhalation.
Special indications: Plethora, disorders of liver and spleen, vesical disorders, female complaints, hysteria, chlorosis, asthma.
The establishments are very complete and fitted with all **modern requirements.** The country is picturesque, and the **neighbourhood of** Halle affords ample amusements. Accommodation good and moderate in price.

Neuschmécks or Uj Tatrafüred.— Austria-Hungary, Galicia, Comitat of Zips, near Poprád; 3,100 feet above sea level.
Pure **alkaline** waters, climatic station.
Special **indications:** Affections of the digestive organs, obstructions **of liver and spleen.**
The establishment **is very** good, and **the** accommodation comfortable. Scenery picturesque and **vegetation** luxurious.
Doctor: Dr. H. von Szontagh.

Neuschwalheim.—Germany, Hessen, 1½ hour from Salzhausen.
Earthy muriatic ferruginous waters. Very little frequented.
The establishment is inferior.

Neusiedel.—Austria-Hungary, Lake of, near the town of Neusiedel.
The waters of this lake have a mineral basis of bicarbonate of soda, 45° to 50° F.
Military hospital.
Special indications: Scrofulous affections.

Neusohl.—Hungary.
Cold sulphurous waters.

Neustadt an der Saale.—Germany, Würtemberg, near Waiblingen.
Cold saline waters.
A small and inadequate bathing establishment.

Neustadt-Eberswalde.—Germany, Prussia, Province of Brandenburg.
Ferruginous bicarbonate waters.

Neustadt - on - the - Haardt.—Germany, Bavaria, Palatinate, near Mannheim.
Grape and whey cure station.
Renowned throughout Germany, and much frequented in the months of September and October.

Neuville lez la Charité.—France, Department of Haute Saône.
Cold sulphate of lime waters.

Nevis.—One of the English West Indian Islands. Leeward group.
Four thermal springs.

New Brighton.—England, Cheshire, at the mouth of the Mersey.
Sea baths; fine sandy beach.
Season: June – September.
Doctor: Dr. Bell.

Newcastle.—Ireland, County Down, near Dublin.
Much-frequented sea baths, and sulphurous ferruginous waters; sandy beach; climate mild. The scenery is picturesque.
Doctor: Dr. Clarendon.

New Quay.—England, Cornwall.
Sea baths; fine sandy beach. Climate bracing.
Season: June—September.
One of the most fashionable of Cornish watering-places.
Doctor: Dr. Jewell.

New Quay.—Wales, Cardiganshire.
Sea baths, sandy beach, with occasional pebbles.
Doctor: Dr. Evans.

Newtondale.—England, Lincolnshire.
Saline springs.

Neyrac.—France, Department of Ardèche, branch from Livron to Privas. 3 hours by carriage from Privas to Neyrac.

Bicarbonate of lime ferruginous waters, 95° F.

Very complete establishment.

Special indications: Skin diseases.

Doctor: Dr. Deputowsky.

Nice.—France, Department of Alpes Maritimes. 70,000 inhabitants. For routes see table.

Winter station of first importance; sea baths.

Season: Winter season, November to June. Summer or sea-bathing season, May to October. Principal season, January to March.

Number of visitors about 30,000 annually.

Special indications: Chronic diseases of the chest, lungs, and respiratory organs; affections of the larynx, liver complaints, disorders of the spinal chord, diabetes, gout, rheumatism, paralysis, general weakness, are the principal diseases for which a stay in Nice is recommended. But the climate is very exciting, and patients suffering from any form of nervousness derive no benefit from it.

The dryness of its atmosphere is the chief characteristic of Nice. Rain falls generally only in spring and autumn, and but rarely in winter. The total amount is 702 mm.; 70 days of rain are reported in the season, 80 days of sunshine, and 60 with cloudy sky. **Fogs are rare, and snow** falls occasionally. Mean atmospheric **pressure is 761 mm.**

The mean temperature for the winter months is 55° F. in November, 50° F. in December, 44° F. in January, 50·5° F. in February, 52·2° F. in March, and 59° F. in April.

The **mean** temperature in winter is 50° F.

No winter station offers so **many** resources as Nice. **All** who suffer from that vague disease called "ennui," or "spleen," will find in this sunny spot on the Riviera the remedy they seek in its pleasantest form. Fête succeeds fête, **and the** Municipal Committee, as also the Mayor, use every endeavour to **amuse** visitors.

The carnival lasts ten days, and **is said** to be superior to that of Rome. The days of the "batailles des **fleurs**" are a **special feature**, which no **other town** can offer. The road **all** along the **Bay des** Anges, from the Port down to the Pont Magnan, **is** thronged with **rows of** carriages, all **decked** with flowers.

Clubs: Cercle Masséna, first-floor of the Casino on Place Masséna; Cercle des Beaux Arts, Avenue de la Gare and corner of Place Masséna; Cercle Republicain, Palais Marie Cristine, Rue de France; Cercle Philarmonique, Cercle de l'Athenée.

The season of 1884-85 will witness **the** completion of the splendid new opera house in the Rue François de Paule. The artists will be the most celebrated **of the day.** The Casino is a very extensive building, containing theatre, concert, ball, reading, smoking, writing, card, billiard, and restaurant rooms. The principal feature of **this** establishment is the large covered **winter** garden. On costume ball evenings the decorations and illumination of the hall, with the bustle of thousands of **guests, is a** sight to be seen only here. There is, moreover, the Comédie **theatre,** a skating and ball rink, a circus, and other buildings of general **amusement.**

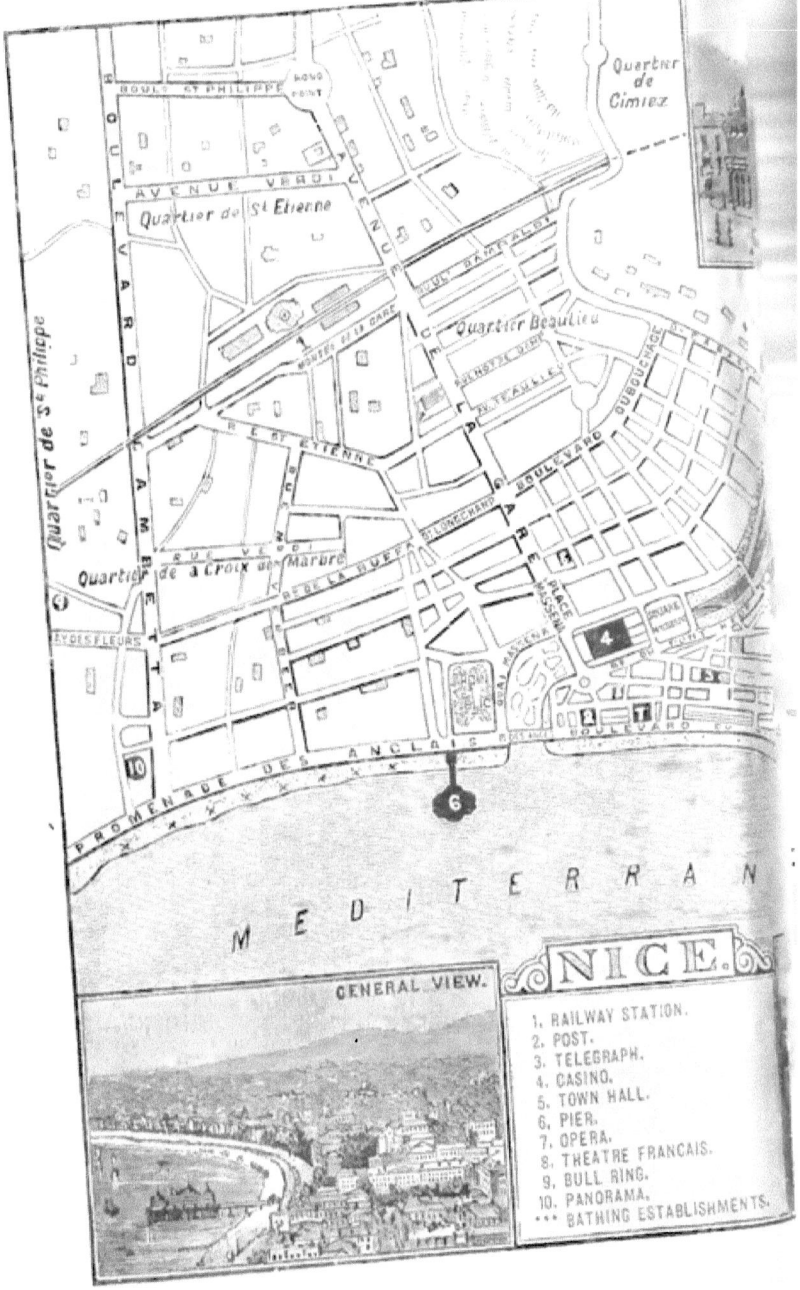

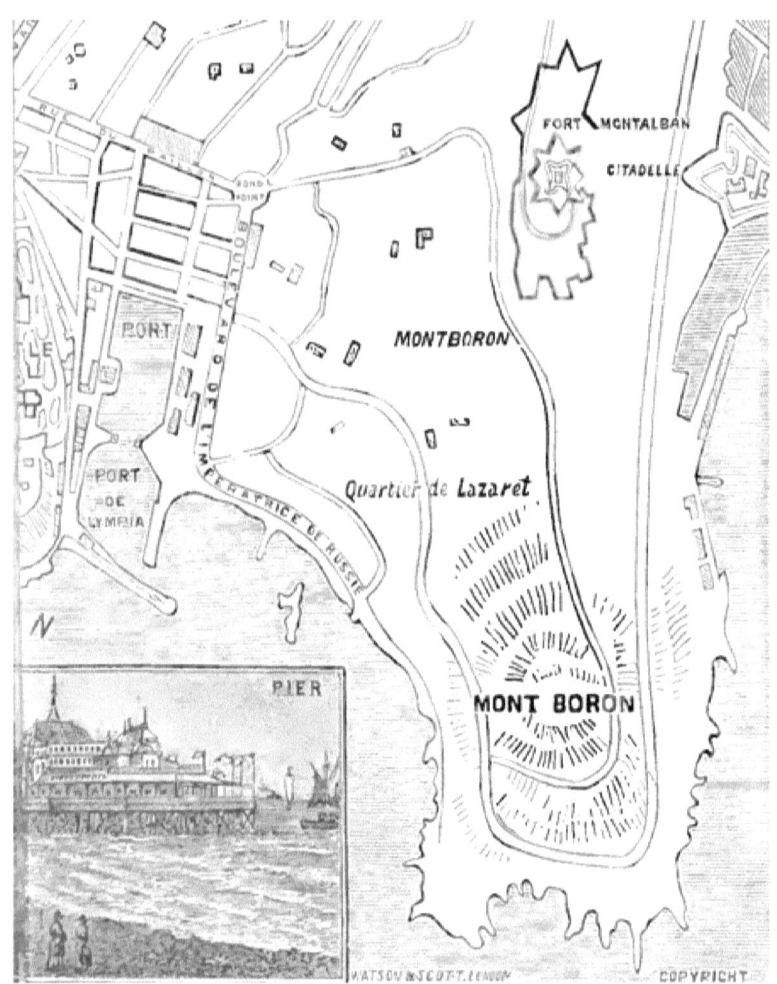

There is a Comédie-Française theatre, ciréus, panorama, skating gallery
EXCURSIONS: The excursions are numerous and very varied. The nearer ones are to the Castle Hill, the Villa Arson Hotel, and Bischoffsheim's Observatory, from all three of which an unsurpassed view of the town and bay is obtained. More distant excursions are to Villefranche (station of the English, French and American men-of-war in the Mediterranean), maritime port and fortresses; Beaulieu, along the Corniche road to Monte Carlo; Mentone by carriage (2 horses, 40 fr. in the season; towards the close, 25 and 30 fr.) to Cannes, Antibes, and the Iles Margarite by boat, rail and carriage, and into the Mountains.

The races take place in the middle of January, the carnival in February, and the regattas in the middle of March. There is an international rifle-shooting match after the regattas, on the Castle Hill during April.

The Nice climate, so very beneficial in many cases of disease, when proper precautions are taken, may be very hurtful if they are neglected. Delicate persons should not go out before ten in the morning nor after half-past three in the afternoon, and should always be provided with overcoats, mantles, and wrappers.

An omnibus runs between Nice and Monte Carlo; fare outside, 2 fr. 50 c. there and back; inside, 2 fr.; three times a day; change of horses at Beaulieu.

Cabs, one horse the course, 1 fr.; by hour, 2 fr. 50 c.; two horses as by agreement. Visitors should arrange beforehand in order to avoid any difficulties at the end of the drive, as the coachmen are often inclined to impose upon strangers.

Turkish Bath and Hammam: Rue de la Buffa, Dr. Bonnal, Algerian shampooers.

Post Office: Rue St. Francois de Paulo.

Telegraph: Rue de la Prefecture.

Photographer: G. Echtler, 3, Rue Adelaide.

Bankers: Credit Lyonnais, 15, Avenue de la Gare.

Booksellers: Galignani, Quai Massena, No. 3.

Mineral Water Merchant: Ed. Thaon, Avenue de la Gare.

House and Estate Agent: The Anglo-American Express Agency 1, Place du Jardin Public.

Wine and Provision Merchant: F. Brand, 13, Rue de France.

American Dentist: A. Preterre, Place Massena.

English Chemists: Nicholls & Passeron, 3, Quai Massena.

Doctors: Drs. West, Drummond, Sturge and Mrs. Sturge, Lippert, Sehnee, Pröll, Granvillier, Barety and Prompt.

Hotels: The *Hotel des Anglais.* This first-class English family hotel is situated in the best position in Nice, at the corner of the Promenade des Anglais and Public Gardens, facing the sea. English management; very popular.

The *Hôtel de la Grande Bretagne.* First-class family house, facing the public gardens and sea. J. Lavit, manager.

The *Cosmopolitan Hotel,* late *Chauvain,* opposite the new Casino; a splendid hotel, with large conversation hall. J. Lavit, manager.

The *Hôtel de Rome,* in an elevated position in its own extensive gardens; 5 minutes' walk from the Promenade des Anglais, quiet, and can be well recommended for families. Mrs. Palmieri, proprietress.

Hôtel de l'Elisee, 59, Promenade des Anglais, situated in a large garden. For a prolonged stay arrangements may be made.

Cimiez : ¼ hour from Nice. The *Grand Hôtel de Cimiez* in the healthiest position, a first class family house with large gardens and grounds, very pure air. Address proprietor.

Niederbronn.—Germany, Lower Alsace, near Weissenburg; at the foot of the Vosges mountains, separating Alsace from Lorraine; 650 feet above sea level. For routes see table.
    Strong saline waters, 45° F.
    Purgative and resolvent action.
    Special indications : Very effective in the treatment of chronic diarrhœa, hepatic constipation, bilious affections, scrofula, obesity.
    Doctors : Drs. Klein and Boell.
    Hotels : Vauxhall, Golden Lion.

Niederhall.—Germany, Baden.
    Saline waters.
    Inferior establishment.

Nieder-Langenau.—Germany, Silesia, near Glatz; 1,100 feet above sea level.
    Ferruginous carbonate of lime waters; mineral muds.
    Milk cure.
    Special indications : Rheumatism, articular pains.
    The bathing establishment is fairly well appointed.
    Doctor : Dr. Seidelmann.
    Hotel : Kurhaus.

Niedernau.—Germany, Würtemberg, near Tübingen, on the banks of the Neckar; 1,050 feet above sea-level.
    Ferruginous and pure bitter waters, very gaseous.
    Special indications : Neuropatic affections, gout, gravel, dyspepsia.
    The establishment is very efficient, and the accommodation dear but good.

Niederselters.—Germany, Nassau, on the Ems, 3 hours from Limburg.
    Alkaline-muriatic waters.
    Only used as table water, and as such largely exported.

Niederweil.—Switzerland; Appenzell; 1,450 feet above sea level.
    Inert waters.
    Used in a primitive fashion.

Nieratz.—Germany, Würtemberg.
    Cold mixed carbonate waters.
    Sedative in action.
    Special indications : Neuroses.

Nierstein.—Germany, Hesse-Darmstadt, near Oppenheim.
Sulphurous waters.
The arrangements are imperfect.

Nissyros.—Greece, island in the Archipelago; 1,800 feet above sea level.
Sulphurous thermal waters, only used by the natives.

Nitrolis.—Italy, island of Ischia, near Moropano.
Thermal ferruginous waters, 85° F.
The baths merely consist of a swimming basin.

Nocera.—Italy, near Perugia.
Inert waters; very fine air.
Season: June—September.
Special indications: Nervous complaints, hysteria.
The establishment is good, and the accommodation comfortable.

Noceto.—Italy, Tuscany.
Alkaline-magnesian waters, 65° F.
Special indications: Affections of the intestines, liver and spleen, and of the urinary organs in general.
Primitive arrangements.

Johanend.—France, Department of Puy de Dôme, near Clermont-Ferrand.
Cold saline waters.
Inadequate establishment.

Nördlingen.—Germany, Bavaria, near Wallenstein.
Effervescent ferruginous waters.
A small bathing establishment.

Nook.—Austria, Tyrol, District of Botzen; 5,000 feet above sea-level.
Inert waters.
The arrangements are imperfect.
Formerly enjoyed a good reputation, but is now almost deserted.

Norderney.—Germany, Island of, on the coast of Hanover.
Much frequented sea bath.
Season: July—September.
The arrangements are very good.
Doctors: Drs. Fromm and Gazert.
Hotels: Hasse, Heitmüller.

Northeim.—Germany, Hanover, between Göttingen and Hanover.
Cold sulphate of lime waters.
Mud baths; establishment good.
Special indications: Rheumatism and herpes.

Nottington.—England, near Weymouth.
Slightly sulphurous waters; little used.

Nouvelle (La).—France, Department of Aude.
Sea baths.

Novelda.—Spain, Province of Alicante.
Sulphate of lime waters, 50° F.
Special indications: Herpes.

Nowosselja.—Russia, Government of Twer, on the Volga.
Effervescent ferruginous waters.
Primitive arrangements, and little or no accommodation.

Nuestra Señora de Abellá.—Spain, Province of Castellon de la Plana, near San Mateo.
Alkaline waters, 50° F.
Special indications: Affections of the digestive and urinary organs, gravel, calculus, hyperæmia of liver and spleen and nervous complaints.
The waters are excellent, but the establishment is very inadequate, and consequently frequented only by the neighbouring inhabitants.
Doctor: Dr. Castell.

Nuestra Señora de las Mercedes.—Spain, Province of Gerona, near Figueras.
Sulphurous and alkaline waters, 70° to 80° F.
Special indications: Stomachic and urinary disorders, uric diathesis, calculus, gravel, hyperæmia of liver, leucorrhœa and amenorrhœa.
The establishment is very efficient, and is situated in a large and well-wooded park. The accommodation is good and moderate in price.
Doctor: Dr. Duran.

Nydelbad.—Switzerland, near Zürich; 2,000 feet above sea level.
Hydro-sulphuric waters. Only frequented by people from the neighbourhood.

Oban.—Scotland, Argyleshire, 2,500 inhabitants.
Climatic resort in summer. Sea baths, good bathing. The air is mild and bracing.
Season: August—September.
The scenery around is very charming, and an ample field exists for excursions. Good boating and fishing.
Doctor: Dr. MacKelvie.
Hotel: The *Great Western*.

Ober-Brambach.—Germany, Kingdom of Saxony, near **Franzensbad.**
Effervescent ferruginous waters.

Special indications: Debility of the digestive organs, affections of the uterus, vesical catarrh, gravel, and **nervous disorders.**

Very good establishment with satisfactory accommodation. The scenery is very fine.

Oberhergern.—Germany, Hesse-Darmstadt, near Münzenberg.
Ferruginous saline waters, effervescent.
Very little used. Poor arrangements.

Oberladis.—Austria, Tyrol, in the Upper **Inn Valley**; 3,780 above sea level.

Climatic station in summer; also an earthy saline spring; may be taken pure or mixed with whey.

This station is frequented chiefly by persons passing the winter in Meran, and coming here for the summer months. The air is pure and delightfully cool, and the situation charming.

There is no regular establishment, but very comfortable accommodation can be had.

Oberlahnstein.—Germany, Nassau, near Coblentz, on the mouth of the Lahn.
Cold bicarbonate of sodium waters, 2 grs. of mineralisation.
A very charming position, with fine view of the Rhine and its hills.

Obermais.—Austria, Tyrol, one of the suburbs of Meran.
Climatic station. See "Meran."

Obermendig.—Germany, Rhenish Prussia, a village near Andernach.
Highly effervescent ferruginous waters.
Only used for drinking. There is no establishment. Accommodation good.

Obernhaus.—Austria, Tyrol, District of Meran, in the beautiful valley of the Adije.
Climatic summer station and ferruginous earthy waters.
No establishment and very quiet.
Scenery very picturesque.

Ober-Rauschenbach.—Austria, Hungary, comitat of Zips, near Pudlein.
Alkali-saline slightly ferruginous waters, similar to Karlsbad, 70° F. Climatic summer station.

Special indications: Rheumatic and gouty complaints, chronic skin diseases and sores, lymphatism, scorfula, goitre, neuroses. The spring is situated behind the Kurhaus and forms a lake. However cold the atmosphere may be, its waters never freeze. Objects immersed in the waters become in a short time incrusted over. The Kurhaus affords abundant accommodation. Moderate rates.

The establishment is built in the style of an old feudal castle. It contains every modern appliance, and from its terrace a superb view is obtained. Concerts are given.

The air is pure and mild. In the vicinity are some grottos, whence certain gases emanate, poisonous both to men and animals.

The place is much frequented both for the beauty of its site and the efficacy of its waters.

Ober-Selters.—Germany, Nassau, on the Ems.
Muriatic alkaline table water.
The production reaches 20,000 quarts daily.

Obertiefenbach.—Germany, Bavaria.
Alkaline-sulphurous waters; climatic station; whey cure. No establishment.

Oberwinter.—Germany, Rhenish Prussia, near the Seven Mountains, and opposite Bonn.
Climatic station in summer. Very mild air. Comfortable accommodation.

Ochsenhausen.—Germany, Würtemberg, 1½ hour from Biberach, 1,799 feet above sea level.
Highly effervescent ferruginous waters. No establishments, but good accommodation.

Odessa.—Russia, on the Black Sea.
Sea baths; mud baths.
Doctor: Dr. Prussian.
The establishments are very good.

Oelves.—Austria-Hungary, comitat of Kolos in Transylvania.
Bitter waters, 55° F.
The water is very good, and might easily, with energy, command a large sale, but the springs are almost abandoned.

Oesel.—Russia, Livonia, Island in the Baltic Sea.
Sea and mineral mud baths.

Oeynhausen.—Germany, Westphalia. For routes see table.
Thermal saline springs, 90° F., charged with carbonic acid.
Special indications: Scrofula, hysteria, gout.
Doctors: Dr. Säuerwald, Dr. Lehmann, Dr. Müller, Dr. Rinteln.
Hotels: Pavillon, Victoria, Rose.

Ofen or Buda.—Hungary, on the right bank of the Danube. For routes see table.
Ofen possesses numerous thermal bathing establishments: The Kaiserbad, the Lukasbad, Königsbad, Raitzenbad, Bruckbad, &c. The water supplying these baths is charged with carbonate of lime, sulphate of soda and magnesia. Its action is diaphoretic, diuretic, and purgative. At 7 k. from Ofen is the establishment of Hunijadi Yanos bitter waters, denominated "The Royal Hungarian Water."

Offenau.—Germany, Württemberg, 3 hours from Heilbronn.
Cold saline and sulphurous waters.
Special indications: Scrofula.
There is a bathing establishment, which affords at the same time comfortable accommodation.

Offenstein.—Germany, Hesse-Darmstadt, district of Pfeddersheim.
Sulphurous waters, 50° F., contains much carbonic acid.
The place is little visited, a fact due, perhaps, to the indolence of the inhabitants.

Oioun-Sckhakna or Frais Vallon.—Algiers, in the immediate neighbourhood of the capital.
Table water, bicarbonate of soda.

Okarben.—Germany, Hessen, district of Grosskarben, on the Nidda.
Muriatic saline waters, little used.

Okmé.—Africa, Nubia.
Thermal sulphur springs.

Olahfalú.—Austria, Hungary, Transylvania.
Evervescent ferruginous waters.
Little used, even by the surrounding peasantry.

Olah-Szt-György.—Austria-Hungary, Transylvania, comitat of Bistritz-Naszod, 1 hour from Rodna.
Alkali saline waters.
A small drinkhall. The waters are exported.

Oldesloe.—Germany, Schleswig-Holstein; one hour from Hamburg; 5,000 inhabitants.
Saline and sulphurous waters.
The establishments are good, and the accommodation most comfortable.
The royal family of Denmark having ceased to frequent the place, its former popularity has received a severe blow. Of late years visitors are again on the increase. A peculiar feature about Oldesloe is the large swimming baths in a salt lake, covering an extent of 1½ acres.

Olenyova.—Austria-Hungary, comitat of Bereg.
Alkali saline waters, containing iron. Table water.

Olette.—France, Oriental Pyrénées. For routes see table.
Several alkaline sulphurous springs, from 65° to 175° F.; mild climate.
Special indications: Rheumatic neuralgia, and neuropathy in general.
The arrangements are not good, and the number of visitors is limited. The accommodation is comfortable.

Olivera.—Spain, Province of Cadiz.
Cold sulphurous waters.

Olliergues.—France, Department of **Puy de Dôme**, near **Ambert**, on the Dore.
Acidulated ferruginous and gaseous waters.

Ollmütz.—Austria, Moravia.
Cold sulphurous waters.

llon.—Switzerland, Canton Valais, between **Aigle and Bex**; 1,000 feet above sea level.
Climatic station.
Very mild air and sheltered position.
A quiet spot, picturesquely situated.

Olmenhausen.—Germany, Baden, between Reutlingen and Tübingen.
Cold sulphurous waters.
No establishment, and little visited.

loneschti.—Roumania.
Sulphurous waters.
No establishments, and all the arrangements are primitive. The place is much frequented; the excellence of the waters would, however, justify a larger concourse of visitors, were it not for the absence of all comfort.

ni.—Russia, Caucasus, near Kutaïs.
Alkaline, slightly saline waters.
Special indications: Liver and splenic disorders, urinary affections.
No establishment; the waters are used for drinking only.

Ontaneda y Alceda.—Spain, Province of Santander, valley of Toranco.
Thermal sulphate of lime springs, 80° F.; very rich in nitrogen.
A well-appointed establishment, receiving a large number of herpetic patients. The site is picturesque.
The village is small, and contains some 60 houses. The accommodation is good.
Doctor: Dr. Salazar.

Oporto.—Portugal, second town in the kingdom; 90,000 inhabitants.
Sea baths, sandy beach; also a winter station. There are good bathing appliances.
The country around is beautiful.

Oppenau.—Germany, Baden, at the foot of the Kniebis.
Ferruginous waters.

Orb.—Germany, Bavaria, district of Aschaffenburg; 450 feet above sea level.
   Saline waters, 35° F.
   Good arrangements, with comfortable accommodation.
   Special indications: Scrofula, skin diseases and chronic nervous affections.

Orebro.—Sweden.
   Indifferent waters; mud baths.
   The site is charming, and although there is no proper establishment, Orebro is much frequented.

Orel.—Russia, Government of Poltawa.
   Bitter and alkali-saline waters.
   Exported, but not beyond Russia. One of the strongest alkali-saline waters known.

Orenburg.—Russia, north of the Caspian and Aral Sea.
   Kumis or fermented mare's milk cure.

Orense.—Spain, on the left bank of the Miño.
   Three bicarbonate of soda springs, 155° to 160° F., but not used.

Orezza.—France, in the interior of Corsica, Department of Piedicroce.
   Gaseous ferruginous bicarbonate table waters.
   Mild climate.
   Season: July—August.
   Special indications: Anæmia, chlorosis.
   Doctor: Dr. Donné.

Origny.—France, Department of Loire, near Roanne.
   Cold bicarbonate of iron waters.

Oriol.—France, Isère, Clelle Station, on the line from Grenoble to Gap.
   Highly effervescing ferruginous bicarbonate waters. The high degree of effervescence of these waters explains the fact of their being so well tolerated by dyspeptic and anæmic persons. They are likewise efficacious in cases of gravel.
   Oriol has no bathing establishment, the waters being used exclusively as table waters, and for exportation.

Oroslau.—Austria, Croatia.
   Indifferent waters, 130° F.
   Special indications: Nervous disorders, hysteria.
   The site is charming, but the establishment leaves room for improvement.

Ormaiztegui.—Spain, Province of Guipúzcoa, near Zumárraga.
Sulphurous alkaline waters, 50° F.
Special indications: Skin diseases, herpes, scrofula, granular affections of pharynx and larynx.
The establishment is good, with moderate accommodation.
Doctor: Dr. Asenjo y Cáceres.

Ostende.—Belgium, West Flanders, on the North Sea.
Sea baths of first importance. Season from 1st June to 1st October. Sandy beach and bracing climate.
The bathing machines are good and the sands excellent.
English church, Rue Longue. Post and telegraph office.
Doctors: Drs. Saulmann, Janssens and Jumné.
Hotels: Hôtel de la Plage.

Osterfingen.—Switzerland; 1,323 feet above sea level.
Saline ferruginous waters.
The arrangements are of a primitive kind.

Osterspay.—Germany, Nassau.
Chalybeate waters, not used beyond their own immediate neighbourhood.

Osthofen.—Germany, Hesse-Darmstadt, near Worms.
Sulphurous waters, 40° F., not in use.
The surrounding country abounds in sulphurous springs.

Osztrovsk.—Austria-Hungary, near Klopas.
Alkaline earthy waters.
Little used, and no establishment.

Otschin.—Roumania, near Brāsa.
Ferruginous sulphurous waters.
No establishment, and little visited.

Ottenstein.—Germany, Saxony, near Schwarzenberg; 10 m., carriage to Ottenstein; 1,550 feet above sea level.
Climatic station in summer; air very pure and mild. Koumiss cure. Hydropathic establishment. Electric and pine-cone sap baths; inhalation.
The environs are covered with pine forests. The walks and promenades are beautiful and well kept. A large park surrounds the Kurhaus. The accommodation is abundant and comfortable.

Outrancourd.—France, Department of Vosges, near Contrexéville.
Cold sulphate of lime and magnesian springs, 2 grammes of mineralisation; slightly purgative.

Paderborn.—See Inselbad.

Paignton.—England, Devonshire, Torbay.
　Sea baths; firm sandy beach; bracing **climate**.
　Season: May—October.
　A great variety of amusements is offered **during** the season.
　Doctor: Dr. Goodridge.
　Hotel: The **Crown**.

Paimpal.—France, Department of Côtes du Nord.
　Sea baths.

Paipa.—South America, New Granada, State of Tunja, **and near the** latter town.
　Thermal sulphate of **so**dium waters, 135° to 180° F.
　An exceptionally high mineralisation of 470 grammes, **of** which **329 are** sulphate of sodium.

Palazonia.—Italy, Sicily, Val di Nolo, near Cattagnone.
　Highly effervescent ferruginous **waters**. No establishment, and used only by the inhabitants.

Palazzolo.—Italy, Sicily, 6 **miles from** Syracuse, Val di Nolo.
　Cold saline **waters**.
　The splendid **ruins** standing **close to this spring indicate that it** was much **in** favour in ancient times. To-day it is almost completely abandoned.

Palermo.—Italy, Sicily, **capital of the island.**
　Winter station, though very damp. Sea baths.
　Season: November—April.
　Bathing season: May—October.
　Mean winter temperature, 55° F. The **changes in evening and morning** temperature are grad**ual. Good drinking** water. Mean **temperature** throughout **the year, 75° F.; 75 to** 80 rainy days during winter. Atmosphere **free from** sudden changes.
　OBJECTS OF INTEREST.—The Marini, **Public** Gardens, Botanic**al** Gardens, Cathedral, Churches of **St.** Guiseppe **and** Martorana, **St. Domenico, the Royal Palace,** Capella Palatina, Museum, Collection **of Antiquities, Palace of** Ziza, Catacombs, Pala**zzo** Vercelle**, Circus and Opera-house.**
　EXCURSIONS: Monreale, La Favorita, Termini, Girgenti, Lercana.
　CABS: One horse, 50 c.; by the hour, 1 fr. 80 cts. Two horses, **80 c.; one hour, 2 fr. 20 cts.**
　Chemist: Caputo, Via Vitt. Emanuele, 107.
　English **church:** Via Cavour.
　Post office: Piazza Bologni.
　Telegraph: Via Maccqueda, 226.
　Doctor: Dr. Berlin, Via San Sebastiano, **30.**
　House and Estate Agency and Wine *Merchant:* S. Zingales Zanelli, via Giovani Moli, 27.
　Hotels: Hotel Trinacria, **de France.**

Pality.—Austria-Hungary, comitat of Bács, near Szegedin.
Baths in the Lake of Pality, very celebrated for its waters, which are especially efficacious in scrofulous cases.
The establishments are very efficient and the accommodation good, and moderate in price. The scenery is charming, and a great many visitors resort to the place:

Pallanza.—Italy, Lago Maggiore.
Climatic station.
Season: Spring and autumn.
Special indications: Chest and lung complaints.
Doctor: Dr. Scharrenbroich.

Pallestrina.—Italy, Romagna, in the Sabine hills.
Climatic station in summer. Very picturesquely situated, and an agreeable place of residence.

Panassau.—France, Department of Dordogne.
A water enjoying great reputation in the surrounding country, but not yet analysed.

Pandraux.—France, Department of Haute Loire, near Puy.
Effervescent ferruginous waters. No establishment, and little used.

Pannanich Wells.—Scotland, Aberdeenshire.
Saline ferruginous waters.
Two very well-appointed bathing establishments
Special indications: Stone, gravel, dyspepsia, scrofula.

Pantalaria.—Island in the Mediterranean, between Sicily and Tunis.
Sulphurous steam grotto. Very hot natural steam baths
No establishment, although the baths are frequently used

Pantano.—Italy, Tuscany, near Cetona.
Ferruginous waters, 65° F. Highly gaseous. Used only by the surrounding inhabitants.

Panticosa.—Spain, on the French frontier, near Cauterets, in the Pyrénées; 8,500 feet above the sea level.
Four sulphate of soda springs—El Higado, 80° F., el Herpes, 73° F., la Laguna, 73° F., and el Estomago, 88° F. These waters have various actions; el Higado and el Herpes are reputed sedatives to the nervous system; La Laguna is purgative; while the Estomago spring is stimulating to the digestion. Azotic inhalations. The establishment might be improved as regards comfort.

Paracuellos de Gilloca.—Spain, Province of Zaragoza, near Calatayud.
Sulphate of lime; cold waters.

Special indications: Chronic skin diseases, **scrofula**, glandular swellings, ulcers, fistulae and ophthalmia, intestinal **obstructions** of a chronic character, leucorrhoea, &c.

The situation is charming. The establishment modern, with good accommodation. A residence here is not, however, very agreeable.

Doctor: Dr. Viejo.

Párad.—Hungary, comitat of Heves, near Erlau.
Sulphurous and effervescent ferruginous waters, one of the springs containing 0·76 per cent. of carbonate of iron. No other spring is so rich in iron. These waters are more highly gaseous than those of Spa, in Belgium. There is also a spring containing a large quantity of alum.

Special indications: Anaemia, chlorosis, **debility**, urinary disorders, liver and splenic plethora, affections of the respiratory **organs.**

A very well-appointed establishment. Párad is one of the most frequented **bathing-places in Hungary.**

Paradies.—Switzerland, **Thurgau,** between Schaffhausen and **Diesen**hofen; 1,180 feet above sea level.

Artificial steam and mineral water baths. Chiefly frequented on account of its beautiful situation.

Paramé.—France, **Department of** Ille and Vilaine, from Paris, *via* St. Malo.
Sea baths, **sandy beach.**
There is a casino.

Páramo de Ruiz.—South America; **New Granada, in the** central Cordillera; 11,500 feet above sea **level. .A** very cold and elevated mountain plain.

A thermal spring, containing, like the **solfatare** waters, free **sulphurous acid and hydrochloric acid.** Only used in a primitive way by the people of the surrounding districts.

Parchim.—German, **Mecklenburg Schwerin.**
Chalybeate waters.
A very good establishment; the accommodation is comfortable, **ample, and moderate in price.**

Paris.—France, Seine. Capital, above 2,000,000 **inhabitants; on both** banks **of the** Seine river.
In Batignolles, Belleville, Ternes, **and near** the Pont d'Austerlitz are cold sulphate of lime **springs.** They have been incompletely analysed **and are** not **used.**

Paris is a strongly fortified town, extending as far as Mont

Valerien and St. Denis. It is divided into 20 parishes or arrondissements, each having a church and several chapels. It is a very old town, but owing to the efforts of Napoleon III. its appearance has been greatly improved by the formation of boulevards, squares and wide airy streets. The municipal debt is consequently very heavy. Paris is a resort of all the idlers, pleasure-seekers and travellers of both hemispheres. The English and American colonies in the city are very considerable. In a word, Paris may be described as the "gayest" city of the world.

The chief thoroughfares are the Boulevards des Capucines, des Italiens, Poissonière and Bonne Nouvelle, Rue de Rivoli, Rue de la Paix, Rue Royale and Champs Elysées, which are crowded until two or three o'clock in the morning. In the afternoon all the cafés along the boulevards are filled with idlers, and when the theatres are closed the same scene is renewed.

Theatres: New Opera, Theatre Italien, Français, Comédie Française, Vaudeville, Palais Royal, Gymnase, Eden, Odéon, Variétés, and several others.

Places of Amusements: Folies Bergères, Rue Bergère, Skating Rink, Hippodrome, Parque d'Hiver, and in summer the Cafés-chantants in the Champs Elysées.

Churches: Notre Dame, Madeleine, Notre Dame de Lorette, St. Etienne du Mont, St. Germain l'Auxerrois, St. Eustache, St. Germain des Prés, St. Sulpice, Panthéon.

Buildings: The Louvre; the Luxembourg, Palais Royal; Law Courts; Post Office, La Sainte Chapelle; Palais de l'École des Beaux Arts, Hôtel des Invalides, Town Hall, Palais de l'Elysée, Palais de l'Industrié, du Trocadero (Exhibition, 1878), the Exchange; Grain and Wine Markets; the National Library; Museum of the Jardin des Plantes, Conservatoire des Arts and Métiers.

Parks—Gardens: Bois de Boulogne, Bois de Vincennes, Park Monceau, Jardins des Tuileries, and Champs Elysées.

Post Office: The most central, Place de la Madeleine.

Telegraph: In most of the large Hôtels.

English Church: Rue Royale, 23.

Scotch Church: 162, Rue de Rivoli.

Wesleyan Chapel: Rue Roquépine, 4.

House and Estate Agents: Sprent, Sprent and Phipps, 252, Rue de Rivoli.

American Agency: The American Exchange in Europe, reading rooms, etc., 35, Boulevard des Capucines.

Bankers: Monroe & Co., Rue Scribe.

Dentist: A. Preterre, 29, Boulev. des Italiens.

English Bookseller and Library: Fotheringham, Rue des Capucines, 8.

English Chemist: Rodgers, Rue de la Paix.

Doctors: Dr. J. Chapman, 212, Rue de Rivoli; Dr. Boggs, 362, Rue St. Honoré; Dr. Vidal, 49, Rue du Luxembourg.

Hotels: The *Splendid Hôtel.* A first-class house, facing the Opera and boulevards, corner of Rue de la Paix, and Avenue de l'Opera; most central and very popular. Ebensperg, proprietor.

The *Windsor Hôtel*, 226, Rue de Rivoli. A small first-class family hotel, exceedingly well appointed, comfortable and homelike. Sprengel, proprietor.

The *Belle Vue Hotel*, Avenue de l'Opera. Excellent restaurant and table d'hôte; lift, telephone; thoroughly heated in winter; a modern house. Louis Hauser, proprietor.

Páros.—Greece, island in the Archipelago.
Saline spring of the Temple of Anargyron.
Much frequented in antiquity, but now neglected.

Partenkirchen.—Germany, Bavaria, in the Bavarian Alps; 2,500 feet above sea-level.
Climatic station in summer. Goats' whey.
The place is situated in a most charming valley, and has a mild climate. The accommodation is good, and reasonable in price.

Passage.—Ireland, county Cork, on the west side of Cork harbour.
Sea baths; sandy beach; good climate.
Season: June—September.
Doctor: Dr. Stone.
Hotel: The *Gerston.*

Passugg.—Switzerland, Grisons, near Coire.
Alkali-saline waters; one spring containing iron. All the springs are highly effervescent, and are some of the strongest of their kind.

Special indications: Chronic catarrh of the mucous membrane of the stomach and intestines, habitual constipation, acute anchylosis, gout, affections of the kidneys and urinary organs, morbus Brightii, diabetes, obesity.

The establishment is adequate, and comfortable accommodation can be had in Coire.

Passy.—France, Department of Seine, near Paris.
Sulphate of lime and ferruginous waters, not effervescing, and difficult of digestion.

Paterna de la Rivera.—Spain, Province of Cadiz, near Medina Sidonia.
Ferruginous, saline, sulphate, magnesian and silicated waters, 15 grammes of mineralisation, of which five are chloride of sodium and five sulphate of magnesia.

Special indications: Herpes, anæmia, chlorosis.

Appliances, very primitive; while scarcely anything has been done for the comfort of visitors.

Paterna y Gigonza.—Spain, Province of Cadiz, near Medina-Sidonia. Saline sulphurous waters, 60° to 70° F.

Special indications: Chronic skin diseases, scrofula, leucorrhœa, inveterate ulcers, caries.

The establishment of Paterna is very small and incomplete. That of Gigonza is larger, but equally imperfect. The situation is attractive enough, but accommodation poor.

Doctor: Dr. Vasquez.

Paterno.—Italy, Sicily, Val di Demona, on the eas coast.

Ferruginous and saline waters.

No establishment, and used by the natives only.

Patmos.—Turkey, island in the Archipelago.

Cold sulphurous waters.

No establishment; the waters are obtained in a primitive way, by digging holes in the ground, in which the baths are then taken.

Pau.—France, Department of Basses Pyrénées; 650 feet above sea level. For routes, see table.

Winter station of first importance.

Season: From 15th November to end of May.

Special indications: Chest complaints, consumption, with hæmoptysis tendency to inflammatory action and asthma, bronchitis, rheumatism, neuralgia, and various forms of nervous affections.

Pau, the ancient capital of Bearn, is situated on the north bank of the river Gave, at the foot of the Pyréneés, of which it affords a very fine view. It has been looked on for many years as a most useful winter station for invalids, owing to its mild and sedative climate, and the absence of strong winds. The above-mentioned diseases more especially are likely to be benefited by a prolonged stay in the town. The various febrile disorders of children here assume a particularly mild type. The soil of Pau being of a light sandy character, or moraine (glacier formation), allows of the free percolation of surface water, so that there is very little free moisture in the air, although evaporation is rapid.

During the last few years the drainage of Pau has been reconstructed, and it is now perfect. The town may to-day be safely called the best drained in France, and will probably henceforward enjoy complete immunity from epidemics.

The mean temperature of Pau in the winter months is 47° F.; mean barometric pressure is 743·25; humidity of the air is 68·955 monthly, saturation point being 100. Average rainfall per month, 3·538 inches. Taylor gives 122, and Otley, who included, however, in his statistics nightly showers, 140 rainy days in the year.

Amusements very varied, and excursions innumerable.

English and Scotch churches.
Post-office: Rue de la Nouvelle Halle.
Bookseller: Ariza, 17, Rue de la Prefecture.
Exchange Bank: John Musgrave Clay.
Dentist: Edwards, Rue Serviez.
English Chemist: Jarvis, 20, Rue Serviez.
Doctors: Drs. J. Bagnell, W. H. Bagnell, Hunt, Clay, Bruce and Oliphant.
Hotels: The *Grand Hotel Gassion*, *Hotel de France*.

Pesolina.—Austria-Hungary, Comitat of Zemplén, near Szinna.
Ferruginous effervescent waters.
No establishment and no accommodation.

Peebles.—Scotland, Peeblesshire, near Inverleithen.
Hydropathic establishment, saline waters.
Doctor: Dr. MacGregor.

Pegli.—Italy, on the Riviera, near Genoa.
A winter station with mild climate.
Doctor: Dr. Spurway.
Hotel: The *Grand Hotel*

Peiden.—Switzerland, Canton of the Grisons; 2,500 feet above sea level.
Cold alkaline, earthy and saline waters, with 4 grammes of mineralisation.
Special indications: Neuroses, herpes, liver complaints, difficult digestion, chronic vesical catarrh, laryngitis.

Peissenberg.—Germany, Bavaria, district of Weilheim; 1,800 feet above sea level.
Sulphurous waters.
There is an establishment affording fair accommodation.

Pejo.—Italy.
Very pure ferruginous waters.
Owing to insufficient means of communication, this water, one of the most easily digested iron waters in Italy, has not received proper attention.

Pelagio.—Italy, Tuscany, near Monte Rotondo.
Thermal mixed springs, 90° F.
Special indications: Rheumatism, gout.

Pelago.—Italy, near Florence.
Mixed carbonate waters.
Special indications: Diseases of the bladder, gravel.

Peleikiton.—Greece, Morea, near Gythion.
Saline waters.
In use in ancient times, but little visited at present.

Pembrey.—Wales, Carmarthenshire.
Sea baths, sandy beach; climate mild.
Season: May—September.
Doctor: Dr. Jones.

Penarth.—Wales, Glamorganshire, near Cardiff.
Sea baths, shingly beach; climate mild.
Doctor: Dr. Nell.

Pendine.—Wales, Carmarthenshire, near St. Clear.
Sea baths, **sandy** beach, very good bathing; climate mild.

Penmaenmawr.—Wales, Carnarvonshire.
Sea baths; sandy **beach; climate** mild.
Season: July—September.
Doctor: Dr. Hughes.

Penna (La).—Italy, the old Pinna Vestina, on the Adriatic.
Muriatic ferruginous waters, 60° F.
Special indications: Anæmia, chlorosis, debility, neurosis, gravel.

Penna (La).—Italy, near Genoa.
Saline sulphurous waters, forming a small lake, 65° F.
Used in primitive fashion.

Penzance.—England, **Cornwall, Land's** End.
Climatic air station and sea baths, sandy beach.
Climate mild and invigorating.
Special indications: Chronic affections of the pharynx, larynx, and bronchi.
Very **much** frequented station.
Doctors: Drs. Couch and Grenfell.
Hotel: *The Queen.*

Pergine.—Italy.
Alkaline ferruginous waters, 60° F.
No establishment, **and only** used by the surrounding peasantry.

Perpignan.—France, Department of Pyrénées Orientales; 28,000 inhabitants; near the Spanish frontier.
This **is** the Roman *Flavium Eprusum.* Its quaint and **curious Spanish and** Moorish looking houses give it an oriental **appearance.**
OBJECTS OF INTEREST: Cathedral of the 14th century; the Citadel close by, Museum, Theatre, Prefecture, Law Courts; a **stopping-place for travellers to Le** Boulou, Amélies-les-Bains and other East Pyrenean watering-places.
Hotels: The *Grand Hotel de Perpignan.* A good, comfortable house, overlooking **the river**; good cooking and cellar. J. Jonca, proprietor.

Perrière (La).—France, Savoy, near Moutiers; 1,400 feet above sea level.
Saline waters, 100° F.
The bathing establishment is complete, and offers comfortable **accommodation.**

Perruches.—France, Department of Cantal, near Aurillac.
Carbonate of soda springs, rich in minerals. Have not been accurately analysed.

Pertinó.—Italy, Romagna. near Civitella.
Saline-iodurated waters, 60° F.
No establishment. The accommodation is extremely deficient.

Perugia.—Italy, **capital of Umbria, mid**way between Florence **and Rome**; 18,000 inhabi**tants; 1,715 feet above the sea level.**
A transitory station of great historical interest.
Perugia is the seat of a prefect**ure** and headquarters **of an army** division. It is an old Etruscan town, full of interest. There are **about** 100 churches. **Pope Leo** XIII. was bishop here for upwards of 20 years.
Objects of interest: The episcopal palace, with new façade; the seminary—**both** enlarged and renovated by Pope Leo XIII., at his own expense; Gothic Cathedral of San Lorenzo; Pinacoteca and University, with very valuable collections; the Arch of Augustus, the Gallenga, Domini, and Conestabile **palace.**
The view from the gardens, where the palace of the prefecture stands, on the stately remains of the Paolina fortress, is one of the finest in Italy. The climate is beautiful, **an**d attracts numbers of families during the summer **months,** from Rome and Florence.
Some very **interes**ting **excursions to be made** in the environs.
Hotels: *Grand Hotel Perugia*. H. Brufani, proprietor.

Pesaro.—Italy, 3½ hours by rail from Bologna.
Sea baths; sandy beach.
No establishments; great freedom prevails.

Pescara.—Italy, **8½ hours** rail from **Bologna.**
Sea baths; fine beach.
No establishment.

Pesth.—Capital of Hungary.
Several ferruginous bicarbonate springs.

Petersthal.—Germany, Baden, in the Black Forest.
Bicarbo**nat**e of lime silicated waters (0·08 and 0·09 grammes of silicate). Four cold springs, 5 gr. of mineralisation.
Special indications: Dyspepsia and gastro-intestinal affecti**ons.**
A **very** well managed **esta**blishment.
Doctor: Dr. Jaegerschmidt.

Petites Dalles (Les**).**—France, Department of Seine Inférieure, from Paris, *via* Fécamp.
Sea-baths, sandy beach.

Petragliá.—Italy, Sicily, Val di Nolo.
A mineral water spring, the **surface of which** is covered with **petroleum.** The water is used as a vermifuge by the surrounding peasantry.

Petriolo.—Italy, Tuscany.
Thermal, sulphurous and saline waters, 105° F.
Special indications: Rheumatism and herpes

Pfaeffers.—Switzerland, Canton of St. Gall. From Paris, *viâ* Basle and Zurich; 2,200 feet above sea level.
Indifferent waters, 90° F. Three springs.
These waters are reputed to be stimulating to the nervous system, and to exercise a beneficial influence on the digestive organs.
Special indications: Nervous disorders, hysteria, chlorosis.
The establishment is very good and the situation picturesque. Scenery wild.

Pierrefonds.—France, Department of Oise, near Compiègne.
One cold sulphate of lime spring, analogous to Enghien, and one ferruginous and arsenical.
Special indications: Pulmonary catarrh.
The establishment is thorough, and contains, besides all modern appliances, abundant and excellent accommodation.
Doctor: Dr. Janvier.

Pietra.—Italy, Tuscany, in the Val de Chiana.
Carbonate of lime waters, 37° F. Two grammes of mineralisation.
Special indications: Dyspepsia and hepatbic obstructions.

Pietrapola.—France, Corsica, 30 miles from Corte, 50 miles from Bastia.
Sulphate of soda waters, 80° to 135° F.
Establishment has tanks and douches, and is magnificent.
Special indications: Nervous affections and syphilitic lesions.
The springs were much used under the Romans.

Pigna (La).—Italy, near Bordighera.
Sulphate of soda waters, 70° F. Not much frequented.

Pignol.—Switzerland, Canton of the Grisons, 15 miles from Coire.
Ferruginous magnesian waters.

Piguien.—Switzerland, Grisons; 3,240 feet above sea level.
Earthy, sulphurous, ferruginous waters, 60° F. Milk and whey cure.
The establishment is good and affords comfortable accommodation.

Pillo.—Italy, Tuscany, near Gambasti.
Very effervescent ferruginous saline waters, 35° F., with purgative action.
Special indications: Gravel and vesical calculi, general debility.

Pilsen.—Austria, Bohemia.
Cold sulphate of iron waters.

Pisa.—Italy, Tuscany on the Arno, near Leghorn.
 Winter station with a sedative climate. 57 days with a clear sky during winter season; damp and steady air. Generally thought a melancholy station. Other statistics are wanting.
 Hotels: *Hôtel Victoria* and *Dell'Arno*.

Pisciarelli.—Italy, near Naples, at the foot of Monte Secco.
 Sulphate of iron waters, 180° F., and natural steam baths.
 The establishments are inadequate.

Pistyan or Postyén.—Hungary, on the route from Vienna to Pesth, near Tyrnau.
 Earthy, alkaline-ferruginous waters, 135° to 150° F. Mineral muds.
 Special indications: Cutaneous and rheumatic affections.
 Doctor: Dr. Wagner.

Pitigliano.—Italy.
 Earthy saline waters, 105° F.
 There is only a natural basin, without any accommodation.

Pitkeathly.—Scotland, Perthshire; also known under the name of Bridge of Earn. 1½ mile from Perth.
 Saline waters.
 Situated in a most picturesque country, and reported to be the most frequented Spa of Scotland.

Pitlachy.—Scotland.
 The Athole hydropathic establishment.
 Very fine Highland scenery.
 Doctor: Dr. McRoy, M.D.

Pixigueiro.—Spain, Province of Orense.
 Thermal sulphur springs, 80° F.

Pizzofalcone.—Italy, on the road to the Castello dell'Ovo.
 Alkaline-ferruginous waters.
 Only used by the surrounding people.

Pizzo (Il).—Italy, Calabria, on the Bay of Euphemia.
 Sea baths.

Pjätigorsk.—Russia, Caucasus. *See* Beschtua or Maschuka baths.

Plaine, (La).—France, Department of Loire.
 Cold ferruginous waters. Very little used.

Planchamps.—France, Savoy, near Rumelly.
 Ferruginous alkaline springs.

Plan de Phazy.—France, Hautes Alpes.
 Thermal ferruginous saline waters, 90° F.
 The establishment is inadequate.

Plan (Le).—France, Department of Haute Garonne, near Mured.
 Cold ferruginous waters.

Platimgan.—Island of Java, East Indies.
Saline waters, 105° F.

Plaue—Germany, Principality of Schwarzburg-Rudolstadt,
Saline waters, 65° F.
The establishments **and arrangements are not of** the best.

Plombiéres.—France, Department of Vosges, near Aillevillers; 1,33 feet above sea level; 1,800 inhabitants. For routes see table.

Various waters, with a mineral basis of soda and sulphur, 175° F.; together with ferruginous cold waters, 30° F. Milk and whey cure; complete hydropathic establishment; climatic **station**; natural gas baths.

Season: May 15th—September 30th.

Number of Visitors: About 8,000 in the season.

Special indications: General debility and inaction of the skin, rhachitis, anæmia, chlorosis, scrofula, chronic rheumatism and gout, chronic nervous diseases, intestinal congestions, hysteria, partial and local paralysis, contractions, anchylosis and articular complaints, hemiplegia and paraplegia, skin diseases, dyspepsia, chronic diarrhœa, intermittent fevers, hepatic colic, uterine catarrh, leucorrhœa, sterility, dissmenorrhœa, chorea.

The waters of Plombières were well known to the ancient Romans, and were used by them, as the ruined baths testify. It is a neat little town, situate in a valley very picturesque and beautiful. There are six bathing establishments: The Bains Napoléon, Bain Imperial, Bain Temperé, Bains des Capucins, Bains Romains and Bains des Dames, all very well fitted up. The first is the most complete; the Bains Romains the most elegant. There is also a military hospital.

The approach to Plombières is very striking. Coming from Epinal, the steam as it rises seems to issue from some vast crater. Gradually the roofs, then the houses and other objects become visible. The climate is hot in the day, with cool mornings and evenings.

The surrounding country is highly cultivated to a certain elevation, above which the forests spread. The excursions are numerous, and there are some interesting ruins in the neighbourhood. Residence on the whole is pleasant enough.

There is a Casino open from 1st June till 15th of September. Reading-room under the Colonnades. Concerts and balls form the chief amusements, and the interior arrangements leave nothing to be desired. The promenades are fine.

Post and Telegraph Office.

Doctors: Drs. Lietard (Inspector), Bottentint and Leclerc.

Hotels: The Grand, Hôtel Stanislas.

Plymouth.—England, Devonshire.
  Sea baths. The beach is rough. The climate is mild, but damp and relaxing, rain being prevalent.
  Doctors: Drs. Harper and Eccles.
  Hotel: The *Duke of Cornwall*.

Po-Csevicze.—Austria-Hungary, comitat of Gömör, near Balogfalu.
  Effervescent earthy ferruginous waters
  The establishment is inferior. Little or no accommodation.

Poggeto-Theniers.—France, Department of Alpes Maritimes, near Nice.
  Sulphate of iron waters. Owing to the great amount of sulphate of iron which they contain, these waters must be taken with discretion.

Poggetti.—Italy, Tuscany, Val d'Ombrone.
  Ferruginous earthy saline waters, 85° F.
  Very primitive arrangements. Recommended in cases of gravel and vesical calculus.

Poggibonzi.—Italy, Tuscany; station on the railway from Florence to Siena.
  Cold saline waters; purgative action.

Poggio Curatale.—Italy, near Fiora.
  Effervescent ferruginous waters.
  Primitive appliances.

Poggio Pinci.—Italy, Tuscany, Val d'Ombrone.
  Ferruginous waters. Little used.

Poggio Rosso.—Italy, near Arrezzo.
  Saline waters. Not used.

Pojan.—Austria, Transylvania, comitat of Haromszek.
  Effervescent ferruginous waters.
  The establishment is primitive.

Polena.—Austria, Hungary, comitat Bereg.
  Alkali, saline, ferruginous waters.
  Largely exported. No arrangements.

Polzin.—Germany, Prussia, circuit of Cöslin on the Wugger.
  Effervescent ferruginous waters.
  Season: May—September. Much frequented.
  The establishments are all very good, and the accommodation comfortable, abundant and moderate in price.
  There are electric, pine-cone and mud baths.

Pomaret.—France, Department of Gard.
  Sulphurous waters, 65° to 70° F.
  Used by the surrounding inhabitants in cases of skin disease, gravel, and constipation.

Pongyelok.—Austria-Hungary, near the Puszta-Mastineez.
Effervescent earthy waters.
Special indications: Hypertrophy of liver and spleen, hypochondriasis and urinary disorders.

Pontaillac.—France, Department of Charente Inférieure, near Royan.
Sea baths; sandy beach.

Pontamafrey.—France, Savoy, near St. Jean de Maurienne.
Cold saline waters, containing iodide, iron and arsenic.

Pont à Mousson.—France, Department of Meurthe.
Four ferruginous springs, used only by the peasantry in cases of chlorosis.

ontano.—Italy, island of Ischia.
Earthy ferruginous waters, 90° F.
No establishment.

Pont de Bared.—France, Department of Drôme, near Dieu-le-Fit.
Bicarbonate of lime and magnesian springs.
Special indications: Dyspepsia and gravel.

Pontresina.—Switzerland, Canton of Grisons, Upper Engadine; 5,915 feet above the sea level, one hour's walk from St. Moritz. For routes see table.
Climatic mountain station.
Special indications: Nervous disorders, scrofula, anæmia, chlorosis, asthma, rhachitis and phthisis, in its earliest stages, convalescence.

Pontresina, with 400 inhabitants, is one of the oldest communes of the Upper Engadine Valley. Its situation is beautiful, and the scenery is magnificent.

English Church.
Post and Telegraph Office.
Banker and Exchange Office: C. Gredig-Enderlin.
Doctor: Dr. J. M. Ludwig, who speaks English.
Hotel: The *Hôtel Enderlin*, a first class house in an open and healthy position, with a magnificent view towards the mountains and glaciers. Especially adapted for private families. A splendid suite of public rooms with library; baths. Drainage and ventilation on most improved principles; full southern aspect. For further particulars apply to C. Gredig-Enderlin, formerly at the *Crown Hotel*.

Pontrieux.—France, Department of Côtes du Nord, near St. Brienne.
Sea baths; sandy beach.

Ponts, Les.—Switzerland, Canton of Neuchâtel.
Cold sulphate of lime waters.
A fairly frequented station.

Popoli.—Italy, Abruzzo, between Pescara and Aquila.
  Sulphurous ferruginous waters, not thoroughly analysed; used in drinking and bathing.
  Special indications: Rheumatism, skin diseases, uterine and intestinal affections, chronic vesical catarrhs.
  No bathing establishment. The springs being close to the town, visitors reside here and walk or drive to the baths.

Porla.—Sweden, District of Oerebro, Westmanland.
  Cold ferruginous waters.
  These waters are highly yellow; Berzelius, when analysing them, discovered two new acids, viz., cremic and apocremic.
  Special indications: Chlorosis, scrofula.
  A well frequented station, with good establishment.

Pornic.—France, Department of Loire Inférieure, in front of the Island of Noirmontiers.
  Sea baths, fine sandy beach; a good establishment, having a casino with concert, reading, and assembly-rooms.

Porretta (La).—Central Italy, between Pistoia and Bologna.
  Thermal sulphurous saline waters and mud baths.
  A well kept and much frequented establishment.

Port Bail.—France, Department of La Manche, *via* Valognes.
  Sea baths, sandy beach.

Porte.—France, Corsica.
  Ferruginous bicarbonate waters, 35° F.

Port-en-Bessin.—France, Department of Calvados, *via* Bayeux.
  Sea baths, very fine beach.

Portland.—England, south coast.
  Sea baths.
  Very salubrious climate.

Portobello.—Scotland, a suburb of Edinburgh.
  Sea baths, firm sandy beach; ferruginous waters at Joppa.
  Climate bracing.
  Doctor: Dr. Balfour.

Portoria.—Spain, Province of Galicia.
  Sulphurous waters, 100° F.
  No establishment, and little used.

Port Rush.—Ireland, County of Antrim.
  Sea baths, sandy beach, pure air.
  Doctor: Dr. Gibson.

Port Said.—Egypt, on the Mediterranean Sea, at the entrance of the Suez Canal.
  Sea baths.
  Hotel: Des Pays bas.

Port Steward.—Ireland, County of Londonderry.
Sea bathing, sandy beach, air bracing.
Season: June—September.
Doctor: Dr. **Campbell**.

Port Thareau.—France, Department of Nièvre.
Cold ferruginous bicarbonate waters.

Portugos.—Spain, Province of Granada.
Cold **ferruginous** bicarbonate waters. Much used by the surrounding peasantry.

Posehiavo.—Switzerland, Canton of the Grisons; 2,300 feet above sea level.
Cold sulphurous waters.
A well-appointed establishment.

Pötschnig.—Austria-Hungary, **one hour** from Oedenburg.
Alkali-saline, ferruginous **waters,** 55° F.
Special indications: **Affections of the spinal cord,** nervous **debility,** hysteria, spleen **and liver** complaints.
The establishment is good, and has comfortable accom-**modation;** moderate prices. The site is charming.

Potsdam.—Germany, Prussia, near Berlin.
Cold ferruginous bicarbonate waters.

Pougues.—France, Nièvre, **near** Nevers. For routes see table.
Mixed bicarbonate ferruginous gaseous waters.
Special indications: Affections of the digestive organs.
Very well managed establishment. Complete balneal appli-ances.
Doctor: Dr. **Logerais**.

Pouillon.—France, Department of Landes.
Saline waters, 50° F.
Special indications: Rheumatism, scrofula, ague.
Doctor: Dr. **Landry**.

Pouliguen.—France, Department of **Loire Inférieure,** near Savenay.
Sea baths, sandy beach.

Pouroille.—France, Department of Seine Inférieure.
Sea baths.

Pozo-Amargo.—Spain, Province of Sevilla.
Sulphurous springs, 50° F.

Pozzuoli.—Italy, near Naples, on the road to Bajä.
Sulphurous saline **waters,** 80° **F.**
Special indications: Gout, rheumatism, scrofula, skin diseases.
Establishments **open all the year;** in the vicinity are the " Stufe di Nerone." (See these.)

Prades-Vernet.—France, Department of Ardéche, near Vals.
Bicarbonate of iron water, effervescing.

Prague or Prag.—Capital of Bohemia. From Paris, viâ Strasburg, Kehl Wurzburg, 38 hours. 161 fr. 1st class.
Prague stands on both banks of the Moldau, and is the second city in the German provinces of Austria. It is a town of great antiquity and possesses quite an Oriental look.
OBJECTS OF INTEREST.—The Hradschin, or Royal Palace on the hill, Cathedral of St. Vitus, Palace of the Grand Duke of Tuscany, the Carolinum is the oldest university in Germany, Clementinum or university for theology, Tihen Church, Town Hall, Museum, Library, Military Hospital, House of Correction, Lunatic Asylum, Wisserad, Czernin Palace, Wallenstein Palace.
CABS: Two horses, 1 fr. 25 c.; half a day in the town, 10 fr.
Post-office and Telegraph: At the station.
Hotels: Blue Star, Golden Angel, Stadt Wien, de l'Angleterre, Kaiser von Oesterreich.

Preblau.—Austria, Carynthia, near Wolfsberg.
Alkaline waters. Climatic air station. The waters are exported.
Season: May—October.
Special indications: Chronic stomachic catarrhs, pyelitis, gravel, and vesical catarrh.
The situation of Preblau is very romantic and charming. The surrounding mountains and hills are clad with pine and fir trees. The establishments are elegant and thoroughly complete. The accommodation is according to modern requirements, and not excessive in price.
The "Gillitshhof" is a kurhaus and hotel combined.

Prechac.—France, Department of Landes, near Dax.
Cold saline waters, 130° to 140° F. Mud baths.
Small establishment. The surrounding country being much exposed to inundations, the town is little visited.

Prelo.—Spain, Province of Oviedo, near Boal.
Alkaline cold waters.
Special indications: Herpes, scrofula, syphilis, intestinal obstructions.
The establishment is not entirely up to the requirements of modern balneology. The accommodation is good and very cheap.
Doctor: Dr. Langurjo.

Prenzlau.—Germany, Prussia, Province of Brandenburg, Uckermark.
Effervescent ferruginous waters.
The establishment is good, and much frequented by the people of Berlin.

Pré St. Didier.—Italy, Sardinia, Val d'Entrèves; 3,110 feet above sea level.
Saline earthy waters, 95° F.
A good establishment. The water is used only in baths.

Preste (La).—France, Department of **Pyrénées Orientales,** on the route from Perpignan to Amélie les Bains.
Inert waters, from 90° to 105° F.
Special indications: Diseases **of the urinary** organs, gravel **and catarrh of** the bladder.
Very efficacious, but insufficiently known.
The establishment is good, the site picturesque, and the vegetation luxurious.

Pretiolo.—Italy, Tuscany, Valley **of the Cecina.**
Sulphurous waters, 110° F.
Formerly in great repute; it is now little visited. The establishment is old.

Prodersdorf.—Hungary, comitat of Oedenburg.
Sulphate of lime waters, 60° F.
Special indications: Rheumatism, herpes.

Prompsad.—France, Department of Puy de Dôme, near Riom.
Lithium and phosphate of iron **waters, 50° F.**

Propiac.—France, Department of **Vaucluse, near Merindol.**
Sulphate of lime waters.

Provins.—France, Department **of Seine and Marne,** 2½ hours from Paris.
Cold ferruginous waters.
There is a good establishment.

Prugues.—France, Department of Aveyron, near **Comares.**
Cold bicarbonate of **iron** waters.

Prutzerbad.—Austria, Tyrol, Upper Inn Valley.
Alkali-saline waters. Effervescent.
Long used, but never properly worked.

Psekups-Springs.—Russia, Caucasia, near Eisk and Jekatherinodar.
Sulphurous waters, 65° to 140° F.
The **country** is very charming, and the place much frequented. The establishment is good and well-appointed.

Puda (La).—Spain, **Province of Barcelona.**
Sulphate of lime **and** saline **waters, 70° F.**
Special indications: Herpes, scrofula, bronchial catarrh.
Very well-appointed establishments, and good accommodation.
Doctor: Dr Gongona y Joanico.

Puente Viesgo.—Spain, Province of Santander, on the road to Burgos.
Saline waters, 85° F.
Special indications: Rheumatism.
A fine situation, with a well-frequented bathing establishment.
Doctor: Dr. de Rugama.

Puertolano.—Spain, Province of Ciudad-Real.
Ferruginous bicarbonate waters, 50° F.
Special indications: Dyspepsia and diseases of the uterus.
Doctors: Dr. Mestre y Marzal.

Püllna.—Austria, Bohemia, near Brüx.
Bitter waters. Exported since last century. The exportation reaches annually more than 4,000,000 bottles. Prize medals: Universal Exhibition of Philadelphia, 1876; Paris, 1878; Sydney, 1879; Melbourne, 1880; Eger, 1881; Trieste, 1882; Vienna, 1883; Amsterdam, 1883; Calcutta, 1883-84, 1st Class diplôma and silver medal.

One of the best and mildest preservatives and remedies against diseases of the digestive organs, constipation, congestions, liver and bladder complaints, nervous disorders, obesity, diseases of the eye, headache, giddiness, dyspnœa, gastritis.

To be had from all chemists and mineral water merchants in the world, and at Püllna from the Communal Bitter-water Director, M. Anton Ulbrich, son of the Founder.

There are no bathing or drinking establishments at Püllna, the whole range of buildings being adapted only for bottling the water. Treatment is not carried out at Püllna itself.

Puttbus.—*See* Friedrich-Wilhelmsbad.

Puzzichello.—France, Corsica, Arrondissement of Corte.
Two springs, one sulphurous, and the other ferruginous.
Small establishment, where mud baths are also administered.
Special indications: Skin diseases and hæmorrhoidal affections.
The scenery is charming, and a sojourn agreeable but quiet.

Puzzola di Pienza.—Italy, near Pienza.
Sulphate of iron waters. Effervescent.
No establishment.

Pwllheli.—Wales, Carnarvonshire.
Sea baths, fine sandy beach. Climate mild.
Season: June—September.
Doctor: Dr. Hughes.

Pyrawarth.—Austria, Lower Austria, two hours from Vienna.
Chalybeate waters. Climatic station.
Season: May—October.
The establishment is very elegant, and is situated in a beautiful park. The Kurhaus is close by, and has ball, theatre, concert and reading-rooms.
The **scenery is** very **charming, and the end of a** visit is **generally looked to** with regret.

Pyrmont.—Germany, Principality of Waldeck-Pyrmont, on the Emmer, near Hanover, 2,000 inhabitants. For routes see table.
Chaly**beate waters of** importance and considerable reputation; **effervescent** saline **waters, brine baths,** inhalations.
Season: **15th May—1st October.**
Number of Visitors: Above 14,000 annually.
Special indications: Female complaints, especially anæmia, chlorosis, scrofula, stomachic and intestinal catarrhs, obesity, affections of spleen and liver
Pyrmont is one of the oldest and most renowned bathing places in Germany. It is beautifully situated in the Emmer valley, a charming portion of the native country of H.R.H. the Duchess of Albany.
English Church.
*Post and Telegraph **Office.***
Dentist: H. Brandt.
Doctors: Drs. Schücking, Köhler, Lynker, Lahs and Weitz.
Hotels: The *Grand Hotel des Bains*; good attention and a beautiful view of the mountains. Völkers Bros.

Queenstown.—Ireland, county Cork.
Sea baths, good sandy beach. Climate mild and equable.
Season: June—September.
Doctor: **Dr. Allen.**

Querzola.—Italy, Parma, near Reggio, at the foot of the Apennines.
Alkali-saline waters, mud baths.
Primitive arrangements.

Quez.—France, Department of Pyrénées Orientales, near Latour.
Sulphurous waters, 65° F.
No establishment, and little used.

Quinéville.—France, Department of La Manche, *viâ* Valognes.
Sea baths, sandy beach.

Quinto.—Spain, Province **of** Saragozza, **on** the right bank of the Ebro.
Saline earthy bitter waters, 60° F.
Special indications: Affections of the digestive organs, **syp**hilis, rheumatism, **skin** diseases.
A well-attended station.

The situation is lovely. The establishment modern and elegant. There is ample accommodation for every class of visitor.

Doctor: Dr. Iborra.

Rabbi.—Austria, Tyrol, Valley of Rabbi, near Trient; 2,800 feet above sea level.

Cold ferruginous waters. Whey cure.

The establishments are elegant and comfortable, and the accommodation abundant.

Rabka.—Austria-Hungary, Galicia, near Makow.

Iodo-bromurated saline waters, very effervescent.

The establishments are not entirely in accordance with modern requirements. The place, however, is much frequented.

Raddusa.—Italy.

Sulphurous and saline waters.

Little used, and scarcely known beyond the immediate neighbourhood.

Radeberg.—Germany, Kingdom of Saxony, one hour from Dresden.

Earthy saline ferruginous waters and mud baths.

Season: June—October.

The establishments are fitted with all modern requirements. The accommodation is comfortable and moderate in price. The situation is picturesque, and the walks and drives are numerous and well kept.

Radein.—Austria, Styria, 3 hours from Gleichenberg.

Saline, iodo-bromurated waters, largely exported.

The water-drinking establishments are modern, and the accommodation good.

Ragatz.—Switzerland, adjoining Pfeffers.

Supplied by the indifferent waters of Pfeffers.

More frequented on account of the scenery than for the efficacy of the waters.

Rajecz-Teplitz.—Hungary, comitat of Trentschin, near Sillein, 1,300 feet above sea level.

Mixed waters, 80° to 95° F., containing alum and iron.

A well-frequented station. The site is charming.

Doctor: Dr. Telek.

Rakós.—Austria-Hungary, Transylvania, district of Csik.

Effervescent ferruginous waters, used only by the peasantry.

Ramleh.—Egypt, on the Mediterranean, North-east of Alexandria.

A summer station for patients desirous of remaining in Egypt during the hot season.

Hotel: Beau Séjour.

Ramlöza.—Sweden, near Helsingborg, 4 miles from Kopenhagen.
Ferruginous spring and sea baths.

Ramsay.—England, Isle of Man, on the North-east coast.
Sea baths; sandy level beach.
Climate healthy and bracing.
Season: June—September.
Doctor: Dr. Ashdown.
Hotels: The *Albert*.

Ramsgate.—England, Kent.
Sea baths, sandy beach. Climate warm and bracing.
Season: June—September.
Doctor: Dr. Curling.

Randamel.—Iceland, District of Vestfirdinga Fiordunger.
Ferruginous waters, 45° to 60° F.
The most frequented of any springs in Iceland.

Ranigsdorf.—Austria, Moravia, District of Ollmütz, near Trübau.
Effervescent ferruginous waters.
The establishment is good.

Ransbad.—Switzerland, Canton St. Gall, near Sevelen.
Earthy waters. Little used.

Rapolano.—Italy, Tuscany, on the road to Siena.
Thermal saline sulphurous springs. 95° F.
Season: July—August.
Special indications: Herpes, gravel and vesical calculus.
A well-frequented station.

Rappalo.—Italy, Riviera di Levante, 1¾ hours from Genoa; 11,000 inhabitants.
Winter station.
There is as yet little hotel accommodation; owing, however, to its sheltered position and beautiful vegetation, Rappalo has a future.

Rappenau.—Germany, Baden, on the road from Heidelberg to Heilbronn.
Saline waters.

Rastenberg.—Germany, Thuringia.
Chalybeate waters.
Special indications: Scorbutus, dropsy, disorders of the blood, gout, rheumatism, hysteria, hypochondriasis.
Also saline, sulphur, and pine-cone sap baths.
Doctor: Dr. Kiel.
Hotel: Bath, *Kurhaus*.

Ratzes.—Austria, Tyrol, 5 hours from Botzen.
Ferruginous and sulphurous waters.
The climate being somewhat severe, **the** place is little **frequented.**

Ravone-in-Casaglia.—Italy, Province of Bologna, near Modena. For routes see table.
Saline strongly **iodurated waters**; used in bathing and **drinking.**
Notwithstanding the fact that there is no bathing establishment, **the place is much frequented for** the beauty of **its** environs.
The waters are said to be superior to those of Montecatini.

Recklinghausen.—Germany, Prussian Province of Westphalia, District of Münster.
Saline waters.
The establishment **is good,** and situated **in the** midst **of** pine and beech woods. **Very** quiet.

Recoaro.—Italy, Lombardy, Province of Vicenza; 6,000 inhabitants; 1,470 feet above sea level. Railway to Vicenza, thence by steam tramway.
Alkaline, acidulated ferruginous waters, 45° F., and mud baths.
Special indications: **Chronic and** nervous debility, **female** diseases, obesity, anæmia, **chlorosis,** gravel and vesical **calculus,** congestion of the liver, **biliary** calculi, hæmorrhoidal complaints, intestinal and stomachic catarrhs.
Post and Telegraph Office close **to** the bathing establishment.
Doctors: Drs. Chininelli and Schivardi.
Hotels: The *Grand Hôtel des Bains* A. Visentini; *Hôtel Giorjetti* and *Hôtel de l'Europe.*

Redcar.—England, Yorkshire, on the North Sea, near Gainsborough.
Sea baths; **flat** sandy beach.
Season: May—September.
Doctor: Dr. Bennett.

Rede de Corvaçeira.—Portugal, Province of Traz os Montes.
Sulphurous waters, 105° F.
Primitive establishments.

Redruth.—England, Cornwall.
Saline and sulphate of lime waters; very rich in lithium.
Deserve to be better known.

Regneville.—France, Department of La Manche.
Sea baths; sandy and shingly beach. Oyster beds.

Rehburg.—Germany, Hanover.
  Cold carbonic and sulphate of lime waters. Mineral and mud baths.
    Doctors: Drs. Michaelis and Kaatzer.
    Hotel: Kurhaus.

Rehme.—Germany, Westphalia, on the railway from Cologne to Berlin.
  Saline waters, 75° F.
  Inhalation hall; very fine Kurhaus.

Reiboldsgrün.—Germany, Saxony, near Auerbach; 2,100 feet above sea level.
  Climatic mountain station and ferruginous waters.
  The establishment and accommodation are good.

Reichenau.—Austria, 2½ hours from Vienna by South Austrian Railway.
  Climatic station in summer; whey cure; hydropathic establishment.
  The establishment is thoroughly well fitted up; situated in a fine park, and contains concert, assembly, and reading-rooms. It likewise affords comfortable accommodation.

Reichenhall.—Germany, Upper Bavaria, near Salzburg; 1,100 feet above sea level. For routes see table.
  Saline waters. Very important air cure station.
  Special indications: Lymphatic disease, scrofula.
  Whey cure and pine-cone baths; inhalations of brine-saturated air. The spring has from 22 to 24 per cent. of saline contents. Beautifully situated near Berchtesgaden and the Königs Sea; most picturesque country. But Reichenhall is chiefly valuable as a transitory station to patients returning from or going to Italy.
    Doctors: Drs. Pachmayr, Rapp, von Liebig (in Munich in winter) and Burdach.
    Hotel: Burkert.

Reinerz.—Germany, Silesia; 1,750 feet above sea level; near Nachod.
  Alkali-saline ferruginous waters, 20° to 45° F.
  The climate materially assists the waters in cases of general debility.
  Everything is well-appointed.
    Doctor: Dr. Secchi.

Remollion.—France, Department of Hautes Alpes, near Embrun.
  Carbonate of lime and sulphate of magnesia springs; 7 grs. of mineralisation.

Renaisson.—France, Department of Loire, near Roanne, near St. Alban and St. Galmier.
  Mixed bicarbonate waters, only for table use.

Renlaigne.—France, Department of Puy de Dôme.
Cold ferruginous waters, more effervescent than the Orezza waters.

Rennes-les-Bains.—France, Department of Aude, near Limoux; 1,000 feet above sea level.
Sulphate of lime and magnesia ferruginous waters. Five springs, from 30° to 130° F.
Special indications: Atony, paralysis, rheumatism, anchylosis, scrofula, lymphatic affections.
Well appointed.
Very mild climate. The place is charmingly situated, and is a much frequented station.

Retorbido.—Italy, Piedmont.
Saline sulphurous waters, 50° F.
Special indications: Hypertrophy of liver and spleen, scrofula, lithiasis, skin diseases.
The establishment is inadequate to the requirements of the place.

Reutlingen.—Germany, Württemberg, near Stuttgart; 1,170 feet above sea level.
Cold sulphurous bicarbonate waters; whey cure.
Insufficient arrangements.

Reyrieux.—France, Department of Ain, near Trevaux.
Strong sulphurous and ferruginous waters.

Rheinfelden.—Switzerland, near Basle; 890 feet above sea level.
Saline waters and brine baths.
The establishments are good, and fair accommodation can be had.

Rhyl.—Wales, Flintshire, near Chester.
Sea baths; hard sandy beach; climate mild.
Concerts on pier and promenade.
Season: June—September.
Doctor: Dr. Griddlestone.
Hotel: The Belvoir.

Riando.—Italy, Piedmont, on the Volturno.
Saline waters.
The establishment is of the most primitive kind.

Riedbad.—Switzerland, Canton of Appenzell; 2,610 feet above sea-level half-hour from Appenzell.
Alkali-saline sulphurous waters.
Were used so long ago as the 16th century; the establishment is good.

Rietenau.—Germany, Württemberg, near Marbach.
Effervescent saline earthy waters.
The establishment and accommodation are good.

Rieumajou.—France, Department of Hérault, near Salvatat.
Several cold bicarbonate of lime springs.
Special indications: Affections of the urinary organs, dyspepsia.

Riga.—Russia, on the Baltic.
Sulphurous waters, 35° F.
The establishment is fairly good.

Rigi Scheideck.—Switzerland, on the Lake of Lucerne; 5,073 feet above sea level.
Ferruginous waters.
Alpine air cure; goats' and cows' whey. More frequented for its beautiful scenery.

Riguardo.—Italy, valley of the Era.
Ferruginous muriatic waters, 55° F.
The establishment is inferior, but the place is well-frequented.

Rima-Brezó.—Austria-Hungary, comitat of Kaschau.
Effervescing ferruginous waters.
Used in primitive fashion.

Rimini.—Italy, on the Adriatic, rail from Bologna to Ancona.
Sea bathing; sandy beach.
Season: June—October.
Splendid and well-appointed bathing establishments. Very healthy climate. It is *the* sea-side resort of Italian society.

Rio.—Italy, Island of Elba.
Sulphate of iron waters, 65° F.
There is no establishment, and the bathing appliances are very rough.

Riolo.—Italy, Province of Ravenna, near Castel Bolognese, ½-hour from Bologna, ½-hour carriage from Castel Bolognese; 760 feet above sea level.
Chalybeate, sulphurous and saline waters; several springs hydropathic establishment; inhalation.
Season: June 15th—September 15th.
Special indications: Chlorosis, leucorrhœa, vesical disorders when not connected with organic diseases, for the chalybeate springs; affections of larynx and pharynx, chronic affections of the skin, gout, rheumatism for the sulphurous; stomachic, intestinal and female complaints for the saline waters.
The bathing establishment is one of the best in Italy, and is fitted with all modern appliances. It is situated in a large

park, surrounded by villas and gardens. The resinous atmosphere of the pine forests on the surrounding hills is very suitable to patients suffering from chest and lung affections, or nervous disorders.

Hotels and villas, as also the establishment, afford abundant and comfortable accommodation at moderate prices. Highest inclusive charge, 12 fcs. per day.

The village is *clean*, and picturesque as are all Italian country places. Promenades and excursions are abundant.

Post and Telegraph office.

Yearly number of visitors, 5,000.

Concerts, illuminations, fêtes champêtres, &c., are the chief amusements.

Doctor: Dr. Mezzini.

Rio-**Mayor.**—Portugal, Estremadura, near Santarem.
Cold saline springs.

Rio-**Meo.**—Italy, District of Vernio.
Alkaline waters.
Primitive arrangements.

Rio-**Real.**—Portugal, Estremadura.
Sulphurous waters, 60° F.

Rio-**Sordo.**—Italy.
Saline waters.
Inadequate arrangements.

Rio-**Tinto.**—Spain, Province of Huelva.
Sulphate of iron waters, 14 grs. of mineralisation.
The proximity of copper and iron mines explains the mineralisation of these waters.

Rio-**Vinagre.**—South America, New Granada, Boyaca; 9,800 feet above sea level, at the foot of the volcano.
Thermal waters, 175° F., containing sulphuric acid, 1·11 gr., hydrochloric acid, 0·91.
These waters are analogous to those of the Parama de Ruiz.

Rippoldsau.—Germany, Baden, in the Black Forest; about 2,000 feet above the sea level. For routes see table.
Saline chalybeate waters, very effervescent, containing sulphate of magnesia in large quantities; whey cure, milk cure and hydropathy. Climatic station of first importance, owing to the surrounding pine woods. The water is exported.

Season: May 15th—September 30th. Number of visitors annually from 2,000 to 3,000.

Special indications: Indigestion, dyspepsia, disorders of the stomach, plethora abdominalis, hæmorrhoids, constipation, hepatic disorders, nervous complaints, hypochondria, hysteria, asthma, anæmia and chlorosis, general debility, irregular menstruation, glandular swellings, intestinal obstructions, gout and rheumatic affections of the mucous membrane.

The situation of Rippoldsau is very charming, at the foot of the Kniebis, the highest mountain in the Black Forest. The scenery is wild and romantic, precipices, ravines and waterfalls, alternate with meadows and pleasant homesteads, on the borders of an ubiquitous pine forest. Rippoldsau is one of those spots which, though easy enough of access, convey to the visitor the sensation of being at an infinite distance from the turmoil of cities and civilised life. Though nature has so amply provided for this beautiful spot, art has also aided. The gardens, avenues of trees, parks and walks over the hills have been laid out with great care and skill. All things considered, Rippoldsau is the spot in which to pass a few weeks in summer, for rest and retirement apart from all considerations of health.

Rippoldsau is a small village with 820 inhabitants, who still dress in the picturesque Black Forest costume, and the whole place gives an idea of rural simplicity scarcely to be found, perhaps, elsewhere. The avenue of limes in front of the baths is the rendezvous of visitors.

Some portions of the bathing establishment are modern, while others preserve their original and archaic characters.

The excursions are very varied, and extend through the whole Black Forest. Amusements are likewise provided for; a Bohemian band plays three times daily. The buildings are all connected by covered passages, and communicate by means of electric bells. They contain large dining-rooms, ladies' room, music, coffee, reading, smoking and billiard-rooms and a large lawn tennis ground. The whole is lighted by petroleum gas. Balls and dances are given from time to time.

Post and telegraph office in the hotel.

Doctor: Dr. Feyerlin (Médicinalrath), who speaks English.

Hotel: The establishment affords first-class accommodation, and is supplied from its own farm with dairy produce. The kitchen and cellar are both very good, and the attendance perfect. There are close upon 300 rooms in the various buildings. Otto Goeringer, proprietor.

Rita.—Italy, island of Ischia.

Ferruginous alkali-saline waters, 155° F.

Excellent water, but nothing has been done to attract visitors. No establishment.

Riva.—Austria, Tyrol, on the Lake of Garda.

Climatic station.

Season: Spring and autumn.

The climate is windy and changeable; the scenery around is picturesque. This station is little frequented.

Riva-los-Baños.—Spain, province of Logroño, near Torrecilla de Cameros.
Alkaline waters, 70° F., contains a little iron.
Special indications: Dyspepsia, gastralgia, liver and spleen affections, chlorosis, anæmia, vesical catarrh, calculus.
The establishment is modern, and situated in a large park. The accommodation is cheap and good.
Doctor: Dr. Menendez Tejo.

Rivera.—Spain, Province of Jaen.
Sulphate of lime waters, 45° F.
Special indications: Skin diseases.

Roccabigliera.—France, Department of Sea Alps, near Nice, on t left bank of the Vesubia.
Sulphate of lime waters, 80° F.
The arrangements are very primitive, and little is done to attract visitors.

Rocca San Felice.—Italy.
Alkali-saline waters, 95° F.
The establishment is very inadequate.
The waters are used internally and externally.

Rodenberg.—Germany, Hessen.
Saline waters, brine baths, 60° F.
The establishment is tolerable, and the accommodation moderate.

Rodna.—Austria-Hungary, Transylvania, on the frontier of Moldavia.
Highly effervescing alkaline ferruginous waters, 7 gr. of mineralisation. Exported.
Special indications: Catarrhal affections, anæmia, chlorosis.
Good arrangements, and comfortable accommodation at moderate rates.

Roggendorf.—Austria, Danubian Provinces.
Sulphate of sodium and magnesia waters; purgative action.

Rohitsch.—Austria, Styria, on the Croatian Frontier in the Noric Alps.
Cold effervescent saline ferruginous waters. Exported.
Special indications: Intestinal atony and mucous catarrhs.
Very good arrangements. The site is picturesque.
Doctors: Drs. Glax and Hoisel.
Hotel: Kurhaus.

Rohnau.—Germany, Silesia, district of Landshut.
Ferruginous sulphurous waters.
The establishment is good, and affords comfortable accommodation.

Roigheim.—Germany, Wurtemburg, near Möckmühl.
Cold sulphate of lime waters.
A small establishment; very quiet.

Roisdorf.—Germany, Rhenish Prussia, near Bonn.
Cold effervescing alkali-saline and chalybeate waters.
Only for table use and exportation.

Rolle.—Switzerland, Canton of Vaud, on the borders of the Lake of Geneva; 1,000 feet above sea level.
Alkali-saline ferruginous waters.
Special indications: Catarrh of the uterus.

Romagna.—Italy, Tuscany.
Thermal mixed carbonic waters, 105° F.

Rombole.—Italy, Tuscany, near Rappolano.
Sulphate of lime waters, 90° F.
Primitive establishment.

Rome.—Italy, Capital, 250,000 inhabitants. For routes see table.
Winter station of first importance.
Season: October to May. Principal months March to May.
Special indications: Debility and scrofula in children, bronchial chronic catarrh, emphysema. The climate is sedative to the nerves and respiratory organs.
Mean winter temperature 45° to 50° F., mean barometric pressure 756·8; relative humidity of the air, 74% in winter; in summer, 57%. The climate is therefore damper than that of the Riviera, and dryer than that of Venice and Pisa. High winds are prevalent during the season. Weather statistics: 92 days with cloudless sky, 81 dull, and 72 rainy, out of the 245 days of the season.
Patients arriving in Rome should first consult their doctor before engaging an apartment for the winter; this is essential, on account of the fevers which are endemic in some quarters of the city.
The drinking-water is excellent. The promenades on the Monte-Pincio are well kept, and invaluable to patients requiring out-door exercise.
Rome is the most celebrated of European cities, alike famous in ancient and modern story. Since 1871 it has been the capital of the Italian kingdom. It is situated on seven hills, on both banks of the Tiber, and 16 miles from its mouth.
OBJECTS OF INTEREST.—St. Peter's, Vatican, Lateran, Quirinal, Albani Villa, Borghese Villa and Palace, Barberini Palace, Capitolium Museum, Doria, Orsini and Colonna Palaces, Farnese Villa and Palace, Forum Romanum, Museo Kirchereano, St.

Luca, **Villas** Ludovici and Massimo, Medici Villa, Palatine, **Pamfili** Doria Villa, Rospigliosi and Spada Palace, Wolkanski Villa, Colosseum, Capitol, various museums, Tarpeian rock, **Column of** Trajan, Pantheon, Temple of Neptune, of Æsculapius, Theatre of Marcellus, Baths of Diocletian, Tomb of Hadrian, or **Castle** of St. **Angelo**; Basilicas St. Paul, St. Lorenzo, St. **Agnes**, St. Cecilia, **St.** Clement, Churches of **St.** Agostino, St. Angelo, **St. Maria, Capucini**, St. Lorenzo, St. Maria in Loreto, St. Martino, **St. Onofrio.**

EXCURSIONS: The Via Appia, Albano, Frascati, Via Latina, Claudian Aqueducts, Temple and Tomb of Bacchus, Palestrina, Mont Tivoli, Hadrian's Villa, Veii, Ostia, Segni, and many others.

CABS: Per course, 80 cts., one or two persons, two horse carriage, **1 fr. 50 c.**

Post **and Telegraph** office: Piazzo St. **Silvestro.**

Chemists: **Sinimberghi**, Evans & Co., **64, 65** and 66, Via Condotti.

Booksellers: **Piale's**, Piazza **di Spagna**; Herm. Loescher, **307,** Via del **Corso.**

Bankers: Vansittart & Co., Piazza di Spagna.

House and Estate Agents: **Contini & Donzelli, 6, Via** Condotti.

Doctors: Drs. Aitken (surgeon), Gason, **Drummond**, Grigor **and Young.**

Hotels: The *Quirinal Hôtel*, extensive **first-class house**; very popular; lift; post; telegraph; best and healthiest position in Rome. B. Guggenbühl & Co., proprietors.

The *Anglo-American Hôtel*, 128, Via Frattina. **Central** position, near Post **and Telegraph Office**; moderate **prices.** Visciotti & **Merli, proprietors.**

Römerbad.—Austria, Styria, near Cilly; 760 feet **above sea-level.**

Inert waters, 95° F. Climatic station in summer.

Equable atmosphere. The establishment **is very** good, and the situation picturesque. The gardens, parks and roads are well **kept**, and Römerbad is a favourite resort.

Doctor: Dr Mayrhofer.

Römerbad.—Switzerland, Aargau, near Zofingen; 1,450 feet above sea level.

Earthy-alkaline waters.

Establishment and accommodation **are** good.

Romeyer.—France, Department of Drôme, near Die.

Cold sulphurous waters.

Die is more frequented for its resin baths **than its** spring.

Roncegno.—Italy, Trientino.

Ferruginous, highly arsenical waters.

Special indications: Skin diseases.

Roncevaux.—France, Department of **Saône and Loire.**
Ferruginous waters.
Used only by the surrounding peasants in cases of dyspepsia.

Ronneburg.—Germany, **Saxe**-Altenburg, near Gera.
Iodurated ferruginous waters; pine-cone sap, lime, sulphurous and steam baths. **Whey cure.**
The establishment is fitted with all modern improvements, and the scenery is romantic. Accommodation good and reasonable.

Ronneby.—**Swed**en, Province of Bleckingen, near Carlskrona.
Sulphurous and sulphate of iron waters.
Special indications: Debility, anæmia, chlorosis.

Ronya.—Austria-Hungary, comitat of Neograd.
Effervescent ferruginous waters.
There is an establishment, but the accommodation is indifferent.

Rorschach.—Switzerland, Canton of St. Gall, on the South shore of the Lake of Constance, 4,500 inhabitants; 1,240 feet above sea level.
Climatic station in summer; baths in the lake; Turkish baths. Very pure and equable atmosphere.
Rorschach is a small old town, the foundation of which dates from the 8th century. It is a very favourite summer station and a centre for excursions. The hills and mountains at its back are clad with pine forests, and handsome villas with their gardens surround it to a large extent. There are many quaint old houses, and a museum with a reading-room. Post and telegraph offices near the port. One of the most favoured excursions is by railway to Heiden.
Other excursions may be made to Annaschloss, Möthelischloss, Stone Table, Weinburg, Romanshorn, and Arbon.
Hotels: The *Hôtel du Cerf.*

Roscoff.—France, Department of Finistère.
Sea baths; sandy beach.
A quiet station.

Roselle.—Italy, Tuscany, on the Ombrone, near Groseto and Siena.
Saline ferruginous waters, 105° F.
Special indications: Arthritic and rheumatic complaints, neuralgia, muscular contractions, rheumatic paralysis.
The establishment is a new and elegant building. The waters were used by the Romans. Accommodation good.

Rosenheim.—Germany, Bavaria, between Munich and Salzburg; 1,370 feet above sea level.
Effervescent ferruginous waters, and climatic station.
The situation is very romantic. The establishment good, affording comfortable accommodation at moderate rates.

Rosheim.—**Germany, Als**atia, 3 hours from Strasbourg.
   Mixed carbonate of lithium waters.

Röslibad.—Switzerland, near Zürich.
   Earthy alkaline **waters, containing alum.**
   The establishment is thoroughly good, and supplies all varieties of artificial mineral baths.

Rosnau.—**Hungary,** comitat of Gomör, **on** the left bank of the Sago.
   Sulphate of iron waters; **whey cure.**

Rosstrevor.—**Ireland,** county Down, near Warrenspoint.
   Sea baths, sandy beach, mild climate.
   Doctor: Dr. Vesey.

Rosswein.—Germany, Saxony.
   Earthy saline and ferruginous waters.
   Arrangements rough.

Rostona.—Italy, **near** Chianni.
   Sulphurous waters.
   Special indications: Skin diseases, gravel, and vesical calculus.
   The establishment **is** inferior.

Röthelbad.—**Germany,** Württemberg, near Stuttgart **and Ulm.**
   Inert waters.
   The establishment is good.

Röthenbach.—Germany, Württemberg, near Najold; **1,250 fe**et above sea level.
   Earthy waters, 70° F.
   The establishment is not of the best.

Rothenburg.—Germany, Bavaria, on the Tauber; 1,050 feet above sea level.
   Inert waters.
   The establishment is good.

Rothenfelde.—Germany, **Westphalia,** near Osnaburg.
   Saline **waters** and **brine** baths, 65° F. Contains a large quantity of iodine and bromine.
   The establishment is adequate, but little has **been done to at**tract visitors. The air contains much **ozone, due to the** surrounding pine forest.

Rothenfels.—Germany, Baden, **close to Baden-Baden.**
   Saline waters, 50° F.
   A very picturesque situation. The establishment is excellent.

Rothesay.—Scotland, Island of Bute, Clyde.
Cold saline and sulphurous springs. Mild climate.
Sea bathing on the Clyde. The scenery is very fine.
Doctor : Dr. Hunter.

Rothwell.—Germany, Württemberg, in the Black Forest; 2,100 feet above sea level.
Saline treatment of scrofula; whey cure.
The site is beautiful. Establishment good.

Rotterdam.—Holland, on the Maas. From Paris, *viâ* Brussels and Antwerp.
The second commercial town in Holland. Formerly the seat of the Dutch East India House. Most of the principal thoroughfares are canals bridged over by draw-bridges.
OBJECTS OF INTEREST.—The Cathedral, Old church, South church, Town Hall, Exchange, Sailors' home. Courts of Justice, Museum of the Botanical Society, Botanical Gardens, Zoological gardens, House of Correction, Boyman's museum, Royal yacht club, Library.
EXCURSIONS: The Hague and Sheveningen, Moerdyk, Gouda, Schiedam, the mother city of Schiedam gin.
CABS: One or two persons, 60 c. the drive; 1 fl. 20 c. per hour.
Post and Telegraph office: Wynstraat, corner of Wynehaven.
Doctor : Dr. Maury.
Hotels, *Victoria*, *New Bath*, *Leygraaf*.

Roueas Blanc.—France, Hérault.
Saline waters and transitory station.

Rouen.—France, Department of Seine Inférieure, 40 m. from Dieppe, 56 m. from Hâvre, 82 m. from Paris.
A very ancient commercial town, mostly industrial; many factories. The town is rich in antiquities.
OBJECTS OF INTEREST.—The Cathedral, Church of St. Ouen, St. Godard, and St. Patrice, St. Maclau and St. Vincent, Courts of Law, Town Hall and Picture Gallery, Museum, Library, Custom-house, Exchange, Market-place, Hospital, Lunatic asylum, Theatres, two old Castles, Botanical Gardens.
English church service at All Saints.
Post-office : Rue Jeanne d'Arc.
Telegraph : Exchange.
Hotels : The *Grand Hôtel de Paris*. Celebrated for its wines and excellent cooking. Guenard-Bataillard, proprietor.

Rouzat.—France, Department of Puy de Dôme, near Riome.
Ferruginous waters.

Royan.—France, Department of Charente Inférieure.
Sea baths.

Royat.—France, Department of Puy de Dôme, near Clermodut-Ferran; 1,400 feet above sea level.

Four springs; mixed alkaline, gaseous, ferruginous, **and** slightly arsenical and lithic waters, 45° to 95° **F.**

Special indications: Lymphatic affections, anæmia, chlorosis, **catarrhal** affections, **arthritic** gout and skin diseases dependent on a gouty diathesis.

Excellent arrangements. The scenery is mountainous.

Royat offers all the comforts and amusements of **an** important station to its visitors.

Bookseller: Puel Vve., in front of the Establishment.
Doctors: **Drs. Brandt**, Frèdet, Petit and Laugaudin.
Hotels: The *Grand Hôtel*, splendidly situated near the casino. L. Servant, proprietor.

The *Splendid* and *Continental Hotels*, well situated in the park, near the bathing establishment. Chabassière, proprietor.

Roye or **St. Mars-les-Roie**s.—France, near Noyon.
Effervescent ferruginous waters.
Very little used.

Roznau.—Austria, Moravia, two hours from Pohl; 1,225 feet **above** sea level.
Whey cure and climatic station.
The establishment is elegant, and thoroughly well fitted-up. Accommodation good and abundant, at moderate **rates.** Sojourn agreeable.
Doctor: **Dr. Koblovsky.**

Rubinat.—Spain, on the French **Frontier.**
Sulphate of sodium and magnesia waters.
Strongly purgative action.

Rudolstadt.—Germany, capital of Principality of Schwarzburg.
Indifferent waters. Very picturesque site.
Doctor: **Dr. Clemens.**
Hotels: Ritter, Lion.

Ruhla.—Germany, Thuringia.
Ferruginous waters, and hydropathic establishment.
Doctor: **Dr. Seyd.**
Hotels: Kurhaus, Bellevue.

Runcorn.—England, Lancashire.
Sea baths.

Ruszpolyana.—Austria, Hungary, Comitat of Marmaros, District Vissō.
Earthy ferruginous waters.

Russwyl.—Switzerland, 3½ hours from Lucerne; 2,040 feet above sea level.
 Saline waters, only used by the surrounding peasantry in cases of skin diseases, difficult digestion, and nervous complaints.

Ryde.—England, Isle of Wight.
 Sea baths; firm sandy beach.
 Hydropathic establishment. Mild climate.
 Doctor : Dr. Davey.
 Hotel : The *Crown.*

Ryhope.—England, Durham.
 Sea baths; firm sandy beach.
 Doctor : Dr. Sage.

Saarguemines.—Germany, Alsatia.
 Cold saline waters, little used.

Sables d'Olonne.—France, Vendée.
 Sea baths.

Sacedon.—Spain, near Guadalaxara, near Huete.
 Sulphate of lime waters, 70° F.; mud baths.
 Special indications : Skin diseases, nervous affections, rheumatism, hemicrania, chronic sores, kidney affections.
 Very well arranged, elegant and comfortable establishment.
 Much frequented station, known and used in the times of the Romans, and subsequently by the Moors.
 Doctor : Dr. Gimeno.

Sachsa.—Germany, in the Harz, near Nordhausen.
 An air-cure station, owing to the surrounding pine forests.
 Doctor : Dr. Schötensack.
 Hotel : Schützenhaus.

Sack.—Russia, Government of Tauri, district of Keslow.
 Brine and mud baths.
 This lake is called, owing to the ill-savoured evaporation arising from it, "the putrid lake." In July and August it dries up almost completely. The mud is reported to be efficacious against gout, rheumatism, hæmorrhoids and skin diseases.

Säckingen.—Germany, Baden, 3 hours from Basle.
 Inert waters, 70° F.
 The establishment is good, but residence is extremely dull.

Sadschütz.—Austria, Bohemia, near Neudorf-Eisendorf.
 Earthy saline ferruginous waters.
 Imperfect arrangements, and little visited.

Saidschütz.—Austria, Bohemia, near Teplitz.
Bitter waters, exported in large quantities.
There is an establishment which, however, is little frequented.

Sail-les-Bains.—France, Department of Loire, near Roanne.
Six springs: **three bicarbonate, one ferruginous, and two sulphur, 55° to 85° F.**
Special indications: **Herpes, intestinal atony, affections of the uterus.**
A very well **appointed** establishment.

Sail-sous-Couzan.—France, Department of Loire, near Montbrison.
Cold alkaline ferruginous waters.
Special indications: Dyspepsia; the waters are reputed **to have the power of** exciting hæmorrhoidal and menstrual discharges.
The establishment is good, and accommodation abundant and comfortable.

Sándorsprings.—Austria-Hungary, comitat of Mármaros, valley of Csiszla, near Borsabánya.
Bicarbonate of lime **waters.**
No establishment **and** no accommodation.

St. Alban.—France, Department of Loire, near Roanne.
Bicarbonate of soda ferruginous highly effervescent waters, 65° F.
This **well-conducted establishment contains a** chamber for **the carbonic acid gas** treatment. The scenery is picturesque.
The water is one of the best table waters known.
Doctor: Dr. Servajan.

St. Amand.—France, Department of Nord, near Valenciennes.
Sulphate of **lime** waters, 50° F.
The **sulphurous muds,** with a temperature of 60° F., **constitute** the principal source of treatment.
The establishment and accommodation **are** luxurious.
Doctor: **Dr.** Isnard.

St. André.—France, Haute Savoie, near Remilly.
Cold hydrosulphuric alkaline waters.

St. Andrews.—Scotland, Fifeshire, near Edinburgh.
Sea baths; the beach is rocky.
Climate **cold** and damp.
Doctor: Dr. Heddle.

St. Anne's-on-Sea.—England, Lancashire, **near** Blackpool.
Sea baths; sandy beach.
Climate mild but bracing.

St. Antoine di Guagno.—*See* Guagno.

St. Bee's.—England, Cumberland.
Sea bath; firm, clean sandy beach.
Climate mild and bright.

St. Blasien.—Germany, Baden, near the Lake of Constance, in the Hell Valley.
Saline waters.
Every variety of baths. Very fine scenery, and pure and mild air; in a pine forest.
Special indications: Anæmia, nervousness, affections of the lungs, convalescence after illness.
Establishment and accommodation good.
Doctor: Dr Haufe.
Hotels: St. Blasien, Crown.

St. Boes.—France, Lower Pyrénées.
Sulphate of lime bituminous waters.

St. Bonnet.—France, Department of Hautes Alpes, Arrondissement of Gap.
Sulphate of lime waters, 80° F.
Special indication: Skin disease.

St. Briac—France, Department of Ille and Vilaine, *via* St. Malo.
Sea baths; sandy beach.

St. Cassien.—Italy.
Thermal sulphurous waters.
Used especially in affections of the eyes.

St. Cergues.—Switzerland, Canton Soleure, 3,150 feet above sea level; in a very sheltered position.
Climatic station.
In the neighbourhood are noble pine forests.

St. Christau.—France, Department of Lower Pyrenees, South of Oleron.
Cuprous waters.
Special indications: Affections of the eyes.
Doctors: Drs. Tillot and Vigneau.

St. Davids.—Wales, Pembrokeshire.
Sea baths; firm sandy beach.
Pure and bracing air.
Doctor: Dr. Foley.

St. Denis-les-Blois.—France, Department of Loire and Cher, near Blois.
Cold effervescing ferruginous table waters.

St. Diery.—France, Department of Puy-de-Dôme, near Issoire.
Ferruginous waters.

St. Domingo.—Island of St. Domingo or Haiti, West Indies.
Several thermal sulphurous springs, 70° to 125° F.

St. Galmier.—France, Department of Loire, near Montbrison.
Alkaline table waters, which have been highly advertised; there are, however, many others superior to St. Galmier. More used on account of its pleasant agreeable taste than for its medical properties.

St. Genis.—Italy, Piedmont, near Turin.
Saline sulphurous iodurated waters.
Special indications: Scrofula, goitre.

St. Georgen.—Germany, Bavaria, near Altötting.
Inert waters; a spring of very small size.

St. Gervais.—France, Savoy, at the foot of Mont Blanc, near Sallanches, on the road to Chamounix; 1,830 feet above sea level.
Saline sulphurous waters, 50° to 120° F.
Special indications: Skin diseases, obstructions of the intestines and tape-worms.
A much frequented station; it offers to the visitor all modern comforts. The scenery is grand.
Doctor: Dr. Bertier.

St. Gildas.—France, Morbihan, near Vannes.
Sea bathing establishment for women kept by nuns. They lived formerly at the Abbey of Abeylard.

St. Honoré.—France, Nièvre, in the centre of France, 8 hours from Paris and from Lyons; 920 feet above sea level. For routes see table.
Alkaline sulphurous and arsenical waters, 70° to 80° F.
Special indications: Skin diseases, affections of the pulmonary mucous membrane, scrofula.
A very well-organized thermal establishment.
Doctor: Dr. Odin.

St. Icaire.—France, Savoy.
Analogous waters to those of Challes.

St. Jean d'Aulph.—France, Haute Savoy, near Thonon.
Cold hydrosulphurous, alkaline, and iodurated waters; 11 grammes of sulphurous principles.

St. Jean de Luz.—France, near the Spanish frontier, in the Bay of Biscay.
Winter station and sea baths.
Doctor: Dr. Goyeneche.

St. Laurent les Bains.—France, Department of Ardèche, near Privas; post from Montélimar.
Saline-alkaline waters, 130° F.
Special indications: **Rheumatism**, paralysis.
Three establishments, well frequented.
Doctor: Dr. Bignon.

St. Lawrence-on-Sea.—England, Kent, Isle of Thanet.
Sea baths, sandy beach. Climate bracing.
Season: June—November.
There is an hydropathic establishment.
Doctor: Dr. Johnstone.
Hotel: The *Granville*.

St. Leonards-on-Sea.—England, Sussex, on the Channel coast.
A very fashionable sea-bathing town. Sandy beach.
Season: Autumn months.
Doctors: Drs. Greenhill and **Warwick**.

St. Loubouer.—France, Landes.
Sulphate of lime waters, 45° F.

St. Lunaire.—France, Department of Ile and Vilaine, *via* St. Malo.
Sea baths.

St. Malo.—France, Department of Ile and Vilaine.
Sea baths; soft sandy beach; bracing climate.
The establishments are very good.
Dentist: A. Preterre.

St. Marie de Mont.—France, Calvados, *via* Isigny.
Sea baths.

St. Martin de Fenouilla.—France, Department of Pyrénées Orientales, near Le Boulou.
Highly effervescent ferruginous waters.
Little frequented. Exported.

St. Martin-Lantosque.—France, Department of Sea-Alps, 7 hours by carriage from Nice.
Slightly **sulphurous** thermal waters.
A favourite resort with the inhabitants of Nice. Not much accommodation.

St. Mary's.—England, Scilly Islands.
Sea baths, fine white sandy beach; mild and invigorating air; rock scenery.
Doctor: Dr. Moyle.
Hotel: *Hugh House*.

St. Maurice.—France, Department of Puy de Dôme, on the right bank of the Allier.

Bicarbonate of iron waters, 40° to 80° F.; 6 grammes of mineral principles.

Special indications: Scrofula, chlorosis.

An establishment very little visited.

St. Moritz-Bad.—Switzerland, Engadine; 5,620 feet above the sea level. For routes, see tables.

Highly effervescent chalybeate waters, 42° F., but St. Moritz is more frequented as a climatic mountain station; whey and milk cure; complete hydropathy.

Season: 15th June—15th September; some houses keep open, however, all the year round.

Number of Visitors: Annually above 30,000.

Special indications: Anæmia, chlorosis, asthma, uterine and vaginal catarrhs, dysmenorrhœa, amenorrhœa and displacements; scrofula rhachitis, and incipient phthisis; all subacute affections.

The ferruginous springs of St. Moritz were known in ancient times. In the 15th and 16th centuries Italian pilgrims used to to visit them.

The mean barometric pressure is 609 m/m; mean temperature during the three season months, 51° F.; relative dampness of air, 73·5 %; mean cloudy days, 56 %; days with rain, &c., 98 in the year; and proportion of rain, 898.

The pleasure-grounds in front of the Curhaus afford a fine view over wooded slopes and ice-covered mountain peaks. The panorama is superb, and the field for excursions very extensive. Boating and fishing afford ample amusement. Ball, concerts, illuminations and similar fêtes are given here during the season. The Curhaus contains assembly, ball, concert, reading, smoking, billiard and ladies' rooms.

Post and Telegraph Office: In the Curhaus.

English Church: Midway between baths and village.

Chemist: Mutschler, in Hotel de Moritz.

Photographer: R. Guler, near the Post.

House and Forwarding Agents: F. & E. Tognoni, Maison Tognoni.

Bankers and Exchange Office: F. & E. Tognoni, and in the Engadiner Kulm Hotel.

American Dentist: Dr. H. L. Schaffner.

Doctors: Drs. Taverney, Biermann, Drummond and Berry.

Hotels: The *Curhaus*, first-class house, with 25 saloons and 240 beds. M. Dalarage, director.

The *Hotel du Lac*, very comfortable, and beautifully situated. Gnst. Arras, director.

The *Hotel Victoria*, with a beautiful view over lake and mountain; covered court. Well managed by Thomas Fanconi, proprietor; also of the *Hotel Bernina*, at Samaden.

The *Hotel Engadiner Kulm*, at the village of St. Moritz, is a first-class house, replete with all modern comforts; ventilation and drainage have received especial attention; electric light; fire-places in all the rooms; during winter the corridors are heated by caloriferes. The proprietor holds the permission to fish in the lakes of Sils and Maloja. The two villas Beau Sejour and Grande Vue will be found exceedingly comfortable for families. English Church, Telegraph, Post, Bank and Exchange in the Hotel. J. Badrutt, proprietor.

St. Myon.—France, Department of Puy de Dôme, near Acquiperse.
Bicarbonate of iron waters; 5 gr. of mineralisation.
Table water. Exported.
Special indications: Dyspepsia.

St. Nectaire.—France, Department of Puy de Dôme, near Issoire.
Mixed alkaline ferruginous bicarbonate waters, 45° to 100° F. Supplying three establishments.
Special indications: Gastralgia, liver complaints, rheumatism and leucorrhœa.

St. Ours.—France, Department of Puy de Dôme.
Bicarbonate ferruginous waters.

St. Pair.—France, Department of La Manche, *viâ* Granville.
Sea baths.

St. Pardoux.—France, Department of Allier, near Bourbon l'Archambault.
Bicarbonate ferruginous waters. Exported.
Table waters. There is no establishment, and the arrangements are solely for exportation.

St. Parize.—France, Department of Nièvre, near Nevers.
Sulphate of lime and magnesian waters. Highly gaseous
Special indication: Ague.

St. Peter.—Austria, Carinthia, district of Klagenfurt.
Ferruginous effervescent waters.
The arrangements are inadequate. The site is picturesque; with capital a great deal could be made of this place.

St. Pierre d'Argenson.—France, Department of Upper Alps.
Cold ferruginous bicarbonate waters.

St. Prieste des Champs.—France, Department of Puy de Dôme.
Ferruginous (very slightly mineralised) waters.

St. Prieste-la-Roche.—France, Department of Loire.
Ferruginous bicarbonate waters.

St. Quentin.—France, Department of Aisne.
Cold ferruginous bicarbonate waters.

St. Radegund.—Austria, Styria, near Graz.
Hydropathic establishment; whey cure.
Doctor: **Dr. Macher.**

St. Raphael.—France, Department of Var, between Hyères and Cannes, on the sea. Population 3,500. For routes see table.
Winter station.
Special indications: Debility, rhachitis, scrofula, lymphatic affections, emphysema, anæmia, and chlorosis.
There are no statistics as to barometric pressure, rainfall, or humidity of the air. Medical men hesitate to send invalids suffering from chest complaints, as the town is open to the full force of the "mistral" and its injurious influences.
English church: Avenue du Grand Hôtel. A. F. Dyce, Chaplain.
Doctors: Drs. **Boutemps** and **Goyat.**
Hotels: **The** *Grand Hôtel des Bains,* **Hôtel de la** *Plage.*

St. Remy la Varenne.—France, Department of Marne and Loire.
Cold ferruginous bicarbonate waters.

St. Roman le Puy.—France, Department of Loire.
Cold bicarbonate waters.

St. Sautin.—France, Department of Orne.
Ferruginous waters.

St. Sauveur.—France, Department of Hautes Pyrénées, near Argelès; 2,400 feet above sea-level. For routes see table.
Sulphurous saline-alkaline waters, 50° to 85° F., containing a little arsenic.
Sedative action; three springs.
Special indications: Nervous and lymphatic affections, **and** female diseases.
Doctors: Drs. **Blondin** and **Lafont.**

St. Simon.—France, Savoy, near **Aix.**
Cold bicarbonate of lime waters.
Special indications: Gravel.
Doctor: Dr. **Bonjau.**

St. Thomas.—France, Pyrénées Orientales.
Sulphate of soda water, 115° to 140° F.
Special indications: Articular rheumatism, scrofula and skin diseases.

St. Ubrich.—Germany, Alsatia.
Cold ferruginous bicarbonate waters.

St. Vaast-la-Houque.—France, Department of La Manche, via Valognes.
Sea baths.

St. Valéry-en-Caux.—France, Department of Seine Inférieure, near Fécamp.
Sea baths; firm sandy beach.
Season: July—October.

St. Valery-sur-Somme.—France, Department of Somme, near Boulogne.
Sea baths; sandy beach. A good establishment.
Season: May—September.

St. Vallier.—France, Vosges.
Same water as Contrexéville.

St. Vincent.—Italy, Province of Aosta, near Chatillon.
Alkali-saline waters, 30° F.
Special indications: Dyspepsia and gravel.
The establishments are neglected, and little is done to attract visitors.

St. Yorre.—France, Department of Allier, near Vichy.
Bicarbonate of soda waters.

Ste. Adresse.—France, Department of Seine Inférieure, near Le Havre.
Sea baths.

Ste. Catherine.—North America, Canada, near Niagara Falls.
Mineral water establishment.

Ste. Helène des Milières.—France, Savoy.
Ferruginous arsenical waters.

Ste. Lucia.—One of the Danish West Indian Islands.
Hyper-thermal springs.

Ste. Madeleine de Flourens.—France, Department of Haute Garonne, near Toulouse.
Cold ferruginous waters.
The establishments are good and the scenery charming.

Ste. Marie.—France, Department of Hautes Pyrénées, near Luchon.
Four sulphate of lime springs, 40° F.
The new establishment is elegant, and affords comfortable accommodation. The station is rising in favour.

Ste. Marie.—France, Department of Cantal, near Chaudesaigues.
Cold ferruginous waters.

Salah Bey.—Algiers, near Constantine.
Thermal springs, 65° F.

Salces.—France, Department of Pyrénées Orientales, near Perpignan.
Effervescing saline waters, 50° F.

Salceti.—Italy, Tuscany, valley of the Arno, District of San Luce.
Alkali-saline waters, **55° F**. The supply is very limited.

Salcoaths.—Scotland, Firth of Clyde.
Sea baths.
A good bathing establishment.

Salcombe.—England, Devonshire, also called the Montpellier **of the** North.
Climatic air station and sea baths.
Recommended to persons suffering **from** affections of the respiratory organs and pulmonary complaints.
Doctor: Dr. Langworthy.

Salerno.—Italy, near Naples.
Winter station; waters effervescing and ferruginous, 60° F. Mild and moist air, with more sunny days than Naples.

Salice.—Italy, District of Ravanazzo, near Retorbido.
Alkali-saline waters.
The establishment and accommodation are indifferent.

Salies.—France, Department of Haute Garonne, near St. Gaudens.
Sulphate of lime and saline waters. The latter possess 34 gr. of mineral **principles,** of which 30 gr. are chloride of sodium.

Salies de Béarn.—France, Department of Basses-Pyrénées, near Orthez.
Saline bromo-iodurated waters; 250 gr. mineral principles, of which 225 gr. are chloride of sodium.
Special indications: Scrofula and lymphatic affections.

Salinetas de Novelda.—Spain, Province of Alicante, near Novelda.
Sulphurous earthy waters, 55° F.
Special indications: Skin diseases, scrofula, rheumatism, mercurialism, amenorrhœa.
The establishment is not up to the level of modern requirements: it is situated in a large park. The accommodation is cheap and not of the best.
Doctors: Drs. Lozano y Rubio.

Salins.—France, Department of Jura, near Dôle, 1,200 feet above sea level.
Saline **bromine** and brine waters, having 320 gr. of mineralisation.
Special indications: Scrofula, lymphatic affections and atony.
A much frequented establishment. Very well arranged, affording comfortable and abundant accommodation, at reasonable prices.
Doctor: Dr. Guyenot.

Salins Moutiers.—France, Savoy; rail up to Alberville, thence by carriage to Salins, 3 hours. 1,500 feet above sea level.

Ferruginous saline effervescing waters, 90° F.; 15 grammes of mineral principles, of which 11 are chloride of sodium. Tonic and resolvent action.

Special indications: Scrofula, rhachitis, and affections of the bones.

The establishment has been recently erected, and contains all modern appliances, with comfortable accommodation.

Doctor: Dr. Desprez.

Salles.—France, Department of Haute Garonne, near Luchon.

Ferruginous waters, 35° F., used as an adjuvant to those of Luchon.

Salmas.—Asia, Persia, north of the Lake Ourmiah.
Thermal sulphurous waters.

Saló.—Italy, on the south shore of the Lake of Garda.

Climatic station. Too cold to winter in. The accommodation is inferior.

Salonichi.—Turkey, Macedonia.

Several saline sulphurous thermal springs, close to the town. Mud baths.

Two establishments, one of which only is in use. Its arrangements are elegant, as in all Turkish baths.

Saltburn-by-the-Sea.—England, Yorkshire.

Sea baths and chalybeate waters; fine firm sandy beach. Climate mild and bracing. The country is wooded.

Doctor: Dr. Rhodes.

Saltfleet-Haven.—England, Lincolnshire, near Skidbroke.
Sea baths; sandy beach.

Salz.—France, Department of Aude.

A small salt-water brook near the thermal spring of Rennes, in conjunction with which it is sometimes used as an adjuvant.

Salzbrunn.—Germany, Silesia; 1,250 feet above sea level.

Cold alkaline waters. Climatic and whey cure station.

Special indications: Nervous affections and pulmonary catarrh.

The establishments are excellent and the place much frequented.

Doctors: Drs. Biefel and Valentiner.
Hotels: Kursaal, Kurhaus.

Salvadora.—Spain, Province of Jaen, near Jumilena.

Sulphurous earthy waters, 50° F.

Special indications: Vesical catarrhs, calculus, gravel, hyperæmia of liver.

The establishment is old and neglected. There is a project to erect a new one with modern improvements.

Doctor: Dr. Morales.

Salzburg.—Austria-Hungary, near Hermannstadt, in Transylvania.

Iodurated brine baths.

Season: July—August.

The establishment is good, and affords comfortable accommodation.

The parks and gardens, owing to the saline nature of the soil, have been artificially made. Vegetation in the vicinity of the town is poor.

Salzburg.—Austria, Salzkammergut, on the Salzach, 25,000 inhabitants; 1,310 feet above the sea level. For routes see table.

Climatic station in summer.

This town is mostly resorted to as a centre for excursions on the northern slopes of the Tyrolese Alps. It is a very old town, and full of curious and interesting antiquities. It was the birthplace of Mozart. It has been considerably improved of late by the formation of new streets, squares, and parks. The **Cathedral**, the Library of the Imperial College (formerly the **University**), **Museum**, Theatre and Riding School, are all remarkable buildings.

The Curhaus is a fine edifice, and contains baths of every description, ball, concert, and refreshment-rooms, and has some fresco paintings by Otto.

Humboldt said of Salzburg: "Its scenery is some of the finest in the world."

The excursions are very varied and numerous; Aigen, Gaisberg, Berchtesgaden, Reichenhall, Hallein, Ischl, Gastein, Königsee, Leopoldskron, Grodig, Gölling, Taxenbach, Mattsee, and Mondsee are amongst the most interesting.

English church. Excellent schools.

Post and Telegraph Office: Residenz Platz.

Bookseller: Henry Dieter, Markt and Residenz, Platz Corner.

Exchange and Bank: Charles Spängler, 7, Mozart Square.

Chemist: Sedlitzki, Markt Platz.

Doctor: Dr. Sedlitzki.

Hotels: The *Hôtel de l'Europe* is situated in a fine park, and is, so to say, a sanatorium. It is near the station, and is excellently kept in every respect. Large stables; the best starting place for excursions. C. Jung, proprietor.

Salzhausen.—Germany, Hessen Darmstadt.
Cold saline waters, 11 grammes of mineral principles.
Special indications: Scrofula.

Salzschlirf.—Germany, Hessen, near Fulda.
Iodo-bromine saline waters, 15 grammes of mineralisation.
Special indications: Scrofula.

Salzuffeln.—Germany, Lippe-Detmold, near Herford.
Brine baths.
The establishments are not of the best, and the accommodation is only moderate.

Salzungen.—Germany, Saxen-Meiningen, near Eisenach.
Cold saline springs, 30 grammes of mineral principles.
Well appointed establishment, with good accommodation.
Special indications: Scrofula.
Very picturesque site. Vegetation luxurious. Parks and gardens are well kept.
Doctors: Drs. Ley and Wagner.
Hotels: *Kurhaus*, *Crown*, and *Starke*.

Samaden.—Switzerland, Upper Engadine, and the capital of the district; 850 inhabitants; 5,513 feet above the level of the sea. For routes, see table.
Climatic mountain air and winter station.
Special indications: Nervous debility and failing nutrition, anæmia and chlorosis, rhachitis, incipient phthisis, scrofula, and similar affections following on malaria, typhus, &c.
Samaden forms the head of the Upper Engadine, and is situated at the foot of the Piz-Padella at the junction of the St. Moritz and Pontresina Valleys. It is a very lively place during summer. In winter the calmness of the atmosphere is noteworthy.
English Church: Near the Post Square.
Post and Telegraph Office: On the Square.
Chemist: Sal. Bernhard.
Bank and Exchange Office: **J. Töndury**, in the Hôtel des Alpes.
Bookseller: Ludw. Mayer.
Doctors: Drs. Leudi and Brügger.
Hotels: The *Bernina Hôtel*, very comfortable and popular first-class house; very fine views; arrangements made for summer and winter residence. Thomas Fanconi, proprietor; also of the *Hôtel Victoria*, at St. Moritz-Bath.

San Adrian y la Losilla.—Spain, Province of Léon, near Vecilla.
Bicarbonate of magnesia and iron waters, 90° F.; 4 grammes of mineralisation, of which 2 are carbonate of magnesia.

Special indications: Nervous complaints, disorders of the digestive organs, anæmia and chlorosis.

The establishment offers little attraction.

Doctor: **Dr. Rulifanchas.**

San Bartolomé de la Cuadra.—Spain, Province of Barcelona, near San Feliú.

Effervescent ferruginous **waters.**

The establishment is beneath **notice.**

Doctor: Dr. Arnesto.

San Bernhardino.—**Switzerland**, Canton of Grisons; 5,200 feet above sea-level.

Ferruginous waters.

Special indications: Dyspepsia, lymphatism and catarrhal affections.

Largely frequented establishment, which is well fitted up. The situation is one of the most picturesque in Switzerland.

San Casciano.—Italy, Province of **Siena**, valley of the Paglia.

Ferruginous waters, 110° F., and sulphurous waters, 100° to 120° F.

These waters were much used in antiquity, as the ruins in the neighbourhood prove. The new establishments are very complete, and afford abundant and comfortable accommodation. Great efforts are being made to resuscitate the place once more. The scenery is picturesque.

San Daniele.—*See* Abano.

Sandefjord.—Norway.

Sea baths and sulphurous waters.

Climate very mild.

Season: June—August.

The best frequented station in Norway. Establishment and accommodation equally good.

Sandgate.—**England**, Kent, between Folkestone and **Hythe.**

Sea **baths**; shingly beach. Climate mild.

Season: June—September.

Doctor: Dr. Howard.

Sandown.—England, Isle of Wight.

Sea baths fine sandy beach; climate very mild. Bathing arrangements are very good.

Doctor: **Dr. Neal.**

Sandrocks.—England, Isle of Wight, near Chale.
Alkali saline waters, 40° F.
Special indications: Intermittent fevers, chronic dysentery, rheumatism, intestinal affections.

San Fedele.—Italy, Tuscany, valley of the Cecina, near Volterra.
Saline waters. Exported.
The establishment is good.

San Filippo.—Italy, Tuscany, Valley of the Orcia, near Radicofani.
Bicarbonate of iron and sulphate of lime waters; 5 springs, ranging from 45° to 120° F.
The establishments might be improved.

San Genesio.—Italy, Piedmont on the Po.
Sulphurous iodurated waters.
Used only for drinking. No establishment of any note.

San Georgen.—Austria-Hungary, near Pressburg.
Earthy alkaline sulphurous waters.
Special indications: Disorders of liver and spleen, chronic skin diseases, lymphatism.
The establishment is situated in an extensive garden, and is elegant and comfortable.
Society at San Georgen is very select.

Sangerberg.—Austria, Bohemia, near Marienbad.
Alkali-saline ferruginous waters.
The establishment is excellent; accommodation first-class and moderate in price. Park and gardens are well kept. The surrounding country affords very interesting excursions.
Doctor: Dr. H. Penn.

San Germano.—Italy, Province of Naples, near the lake of Agnano.
Steam baths, 90° to 150° F. The waters from which the steam is given off contain sulphides of calcium and iron.
Special indications: Rheumatic pains, joint affections, catarrhs, neuralgias, paralysis.
The baths are not prolonged beyond a quarter of an hour.

San Giuliano.—Italy, Tuscany, near Pisa.
Sulphate of lime waters, 80° F.
Special indications: Nervous affections.
The establishments are good.

San Giuseppe.—Italy, Naples, near Mondragone.
Cold ferruginous waters.
Only used by the surrounding peasantry.

San Gregorio de Brozas.—Spain, Province of Cáceres, near Alcántara.
Sulphurous earthy waters, 60° F.
Special indications: Intestinal and digestive disorders, hyperæmia of liver, vesical catarrhs, anæmia and chlorosis.

The situation is picturesque, and the surrounding hills are well wooded and afford long and sheltered walks. The establishment and accommodation are deficient.

Doctor: Dr. Balbas.

San Juan de Azcoitia.—Spain, Province of Guipuzcoa, near Tolosa.
Sulphate of lime waters, 40° F.
Special indications: Skin diseases, bronchial catarrhs.

The valley is very beautiful, but the establishment and accommodation are inferior.

Doctor: Dr. Diaz.

San Juan de Campos.—Spain, Balearic Islands, island of Mallorca.
Thermal saline and silicated waters, 115° F.; 0·78 of silicate. Winter station.
Special indications: Rheumatism, skin diseases.

The waters are good, and would warrant greater use being made of them; at present the establishment, as well as the accommodation, is neglected.

Doctor: Dr. Sanchez Jabra.

San Leopoldo.—Italy, district of Lari, near Val d'Era.
Muriate of iron waters.
Special indications: Vesical catarrh, hysteria, hypochondriasis, liver complaints.

San-Lorenzo.—Austria, Istria, 1½ hour from Trieste, 40 minutes from Pirano.
Sea baths; very fine sandy beach; whey and grape cure; mud baths; a climatic station in summer.

The climate is equable, and the air pure.

San Martino.—Italy, Valtclino, near Bormio; 4,950 feet above sea level.
Mixed sulphurous waters, 100° F.
Season: June—August.

Two well-arranged establishments, and good accommodation.

San-Marziale.—Italy, Tuscany, district of Siena, Val d'Elsa.
Earthy waters, 80° F.

Though well known and much visited in antiquity the baths are now neglected. In spite of considerable efforts to resuscitate them, they are not much frequented.

San-Michele delle Formiche.—Italy, Tuscany, on the Monte Sestolo. Sulphurous waters, 120° F.

Special indications: Inveterate gout, skin diseases, nervous affections, rheumatism.

The establishment consists of a small building, with two cisterns and several distinct baths.

San Pedro do Sul.—Portugal, Province of Beira.
Hyperthermal sulphurous waters.

San Remo.—Italy, on the Riviera di Ponente, near the French frontier; 18,000 inhabitants. For routes see table.

Winter station of first importance; sea baths

Season: For winter, November—May. Sea bathing, September—November.

More than 7,000 visitors annually.

Special indications: Latent scrofula, chronic bronchial, stomachic and intestinal catarrhs, emphysema, pharyngitis, laryngitis, pleuritic exudations, incipient phthisis, rheumatism, Bright's disease, diabetes and general debility.

The mean winter temperature, according to Dr. Hassall's statistics, is 54° F.; the mean barometric pressure, 761·5 mm; relative humidity of the air, 66·7 per cent.; death-rate, 24 per 1,000; rainy days during seven season months, 30; sunny and cloudy, 182. Wind and fogs are rare. Snow quite exceptional, in some winters not more than once or twice.

San Remo proper is an old town, with the characteristics and peculiarities of the middle ages more marked than in any other town along the coast. Some of the streets of the old quarter are unoccupied at present, and a traveller with a little ingenuity may easily fancy himself transported to the ruins of Pompeii. San Remo is well sheltered, as the mean temperature shows. Both its older and newer portions are higher than any town on the Riviera. The town itself is built on a hill, with its base resting on the seashore. This hill rises behind it to an altitude of 4,270 feet (Monte Bignone). Three ranges of mountains form a natural shelter for San Remo. The first is formed by seven olive-clad hills, separated from each other by a number of picturesque valleys, in which the lemon tree is successfully cultivated. Next is the Monte Bignone, already referred to, which is pine-clad and forms an unbroken chain. Behind this second range come the Alpes-Maritimes, with an elevation of from 7,000 to 8,000 feet. The latter are invisible from the town, and can only be seen from the summits of the second range. San Remo is thus open only to south-east, south and south-west winds

In early spring, the country round the bay resembles a vast garden. As regards vegetation San Remo is perhaps the most favoured of any health resort on the Riviera. As early as the month of February wild flowers come into bloom, a fact which sufficiently attests the mildness of its climate. The air is bracing and healthy to a degree which attracts the attention of visitors staying for any length of time. The **climate** may be described as moderately warm, dry, very sunny, and in comparison with other towns on the coast, only slightly stimulating. There is a théatre and a club. Daily concerts are held in the Giardino Publico.

English Church: St. John the **Baptist, All Saints and Presbyterian,** Via Carli.
Post and Telegraph Office: Via **Vittorio Emmanuele.**

Chemist: Squires, Via **Vittorio Emmanuele.**

Bankers and Exchange Office: **Asquasciati Bros.,** Via Vittorio Emmanuele.

Bookseller and Circulating Library: Gandolfo, **Via Vittorio** Emmanuele.

House and Estate Agency: Congrêve, Via Privata.

Doctors: **Drs.** Hassall, Freeman, Schmitz, Kay, Shuttleworth, Brandis, Salzmann, Thomas, Goltz, Onetti and Daubeney.

Hotels: The *West End Hotel,* **ten** minutes' walk from the station, a really splendid house within and without. It is the only one possessing a lift. Rob. Wülfing, proprietor.

The *Hôtel Palmeri,* near the sea, is a first-class **house.**

The *Hôtel Victoria* in **the** East Bay.

San Sebastian.—Spain, Santander, in the Bay of Biscay.
Sea baths; **sandy** beach. Climate bracing, yet mild.
Season: June—September.
One of the fashionable Spanish seaside resorts. The scenery around is picturesque.
Doctor: Dr. Smith.
Hotel: Hôtel Ingles.

San **Vicens.**—Spain, Province of Lerida, near Seo de **Urgel.**
Sulphurous alcaline waters, 50° F.
Special **indications:** Skin diseases, bronchial and vesical catarrhs.
The **establishment and** accommodation are **modern, the** former not quite complete, **and the** latter **moderate in** price. The surroundings are bare.
Doctor: Dr. Laza.

Santa Agueda.—Spain, Province of **Guipuzcoa,** near Villareal; 500 feet above sea level.
Cold hydrosulphuric waters, also a ferruginous spring.
Season: June—September.
Special indications: Skin diseases, and especially female complaints.
A clean, well-conducted and well-frequented bathing establishment.
The accommodation is very good, and prices are moderate. It is undoubtedly the best in every respect in Spain, and can compete with the most famous spas of Germany or France.
The environs are picturesque, and residence pleasant.
Doctor: Dr. Lucientes y Pueyo.

Santa Ana.—Spain, Province of Valencia, near Játiva.

Acid waters, 55° F., containing iron.

Special indications: Skin diseases, catarrhs, anæmia, chlorosis, calculus and scrofula.

The establishment and accommodation are inferior, and consequently Santa Ana is little visited.

Doctor: Dr. Suñer.

Santa Barbara.—North America, California.

Thermal sulphurous waters, 90° F.

Santa Cesarea (Acqua di).—Italy, Province of Lecce, district of Galipoli, near Ortelle, on the line from Brindisi to Otranto; stop at Maglie, thence by omnibus (1½ fcs.), or carriage (5 frs. to 10 fcs.).

Earthy ferruginous sulphurous water, 65° F. Healthy and mild climate. The waters are very efficacious.

Season: May—October.

Frequented by more than 4,000 patients during the season.

Special indications: Rheumatism, skin diseases, chronic affections of the intestines.

The establishment is good. Close to the spring accommodation can be had at from 4 fcs. to 8 fcs. a day. Post and Telegraph at Poggiardo, 1 hour distant.

Santa Filomena de Gomillaz.—Spain, Province of Aláva, near Vitoria.

Sulphurous earthy waters, also a ferruginous spring.

Special indications: Skin diseases, more especially those of syphilitic origin. Affections of the digestive and urinary organs, leucorrhœa, amenorrhœa and dysmenorrhœa.

The site is very picturesque. The establishment is modern and elegant. Large galeries connect it with the hotel. The whole is surrounded by lovely gardens, and a park with extensive and well-sheltered walks. Especially suitable for patients going south for the winter, who may here pass a short time during the early autumn. The accommodation is in keeping with the establishment. Residence here is pleasant enough. Prices are moderate. Much visited.

Doctor: Dr. Lázaro.

Santa Gonda.—Italy, Tuscany, between Pisa and Florence, on the left bank of the Evola.

Earthy saline waters, 55° F. Mud baths.

Special indications: Skin diseases, chronic gout and rheumatism.

Santa Margherita.—Italy, on the Riviera di Levante, 1½ hour from Genoa.

Winter station.

This little town, at the head of the bay of Rapallo, is charmingly situated in a sheltered position. It has a promising future before it.

Santa Restituta.—Italy, island of Ischia, near Laceo.
Saline waters, 115° F.; sand baths.
The establishment is good.

Santenay.—France, Department of Côtes d'Or.
Saline waters; laxative action.
The town has a future.

Santorin.—Greece, an island in the Grecian Archipelago.
Chalybeate waters.
The finest island in the Grecian Archipelago, renowned in ancient times for its luxuriant vegetation.
The establishment is small and incomplete. Some accommodation can be obtained.

San Vigone.—Italy, Tuscany, Valley of Orcia, on the road from Siena to Rome.
Sulphate and carbonate of lime waters, 105° F.
Special indications: Rheumatism, skin diseases.

Saragoza.—Spain, capital of the Province of same name, from Paris, *viâ* Bordeaux and Irun.
Situated on the Ebro. It is one of the most ancient towns of Spain.
The country all around is very productive, and the scenery charming. The town is said to have been founded by one of Noah's nephews; it is certainly one of the ancient "castra" of the Romans. It has become famous on account of its siege by the French in 1808 and 1809.
OBJECTS OF INTEREST.—The Cathedrals of la Seo and of el Pilar, Lonja, Torre Nueva, Aljaferia, Archiepiscopal palace; General Hospital, University, Workhouse. Casa del comercio, Casa del raporto, Casino, Theatre
EXCURSIONS: Paseo de Sta. Eugracia, Torreo, Buena Vista, Monte Oscuro, Casa Blanca, Moncayo.
CABS: No fixed tariff, therefore make an arrangement before engaging.
Post-office: In Calle del Coreo.
Telegraph: At Civil Governor's house.
Doctor: Dr. Daina.
Hotels: De *l'Univers*, de *l'Europe*, Cuatro Naciones, Viscaina.

Saratoga.—North America, State of New York, near Albany.
Saline waters, 20° to 30° F., and cold bicarbonate of iron waters.
Special indications: Dyspepsia, calculous affections, scrofula.
A very fashionable and much-frequented station. Season, August and September.
Hotels: Great Union, Washington, pension £1 a day.

Sarcey.—France, Department of Rhône, near Lyons.
Cold ferruginous bicarbonate waters.

Sardara.—Sardinia.
　　Sulphurous waters, 145° F.
　　Only frequented by inhabitants of the island.
　　Special indications: Rheumatism.

Sarepta.—Russia, Government of Saratow, district of Zarizyn.
　　Saline bitter waters. Exported.
　　There are some establishments, and the accommodation on the whole is comfortable.

Sark.—England, Channel Islands.
　　A holiday resort only, as there are no baths, nor is the air particularly healthy. The place is interesting for its scenery.

Sassnitz.—Germany, Pommerania.
　　Sea bathing, sandy beach, with heavy surf.
　　The situation is fine, and owing to its quiet the place is visited chiefly by real invalids. Accommodation very good.

Sasso di Maremma.—Italy, Tuscany, on the Ombrone, near Grosseto.
　　Ferruginous bitter waters. Exported.
　　The establishments are inferior, and accommodation inadequate.

Saturnia.—Italy.
　　Sulphurous waters, 100° F.
　　The establishment might be improved.
　　Accommodation indifferent.

Saubuse.—France, Department of Landes.
　　Saline waters, 80° F.
　　Special indications: Rheumatism.

Saucats.—France, Gironde.
　　Cold ferruginous bicarbonate waters.

Saudon.—Spain, Province of Guadalaxara.
　　Sulphurous waters.
　　Special indications: Rheumatism and skin diseases.

Saulce (La).—France, Department of Hautes Alpes, near Gap.
　　Saline waters, 35° to 55° F.

Sault.—France, Department of Vaucluse.
　　Cold sulphate of lime waters.

Saundersfoot.—Wales, Pembrokeshire, on **Carmarthen** Bay, near Tenby.
　　Sea baths and chalybeate waters, sandy beach. Climate fine and bracing.
　　Season: June—September.
　　Doctor: Dr. Allen.

Sausillonges.—France, Department of Puy de Dôme.
 Cold bicarbonate of soda waters.
 Special indications: Liver complaints, anæmia.

Saute-Vaux.—France, Department of Cantal.
 Bicarbonate of soda waters.
 Special indications: Liver complaints, anæmia, feeble digestion.

Savergnolles.—Department of Cantal.
 Cold ferruginous bicarbonate waters.
 Special indications: Chlorosis.

Saxe (La).—France, near Courmayeur in Savoy; 3,750 feet above sea level.
 Sulphate of lime waters, 70° F.
 The establishment is good and affords comfortable accommodation.

Saxon.—Switzerland, Canton Valais.
 Bromo-iodurated saline waters, 60° F. Climate mild and healthy.
 Special indications: Syphilis, scrofula, and gout.
 The environs are very picturesque, and numerous excursions may be made.
 Doctors: Drs. Reichenbach and Boyer.

Scarborough.—England, Yorkshire, on the coast of the North Sea. Great Northern line of rail.
 Cold sulphate of magnesia and iron waters.
 Very fashionable and much-frequented sea baths.
 Doctors: Drs. Cooke and Dale.
 Hotels: The *Royal*.

Scey.—France, Department of Haute Saône, near Vesoul.
 Cold mixed bicarbonate waters.

Schandau.—Germany, Saxony, railway from Dresden to Prague.
 Cold ferruginous bicarbonate waters.
 A climatic station. Tonic action.
 The establishment is very good.
 Doctor: Dr. Roscher.
 Hotels: *Bath*, *Sendig*.

Schelesna-Wodsk.—Russia, Caucasus, near Pjätigorsk; 1,800 feet above sea-level.
 Alkaline ferruginous waters, 65° to 110° F.
 Koumiss cure.
 Special indications: Intestinal catarrhs, hypochondriasis, catarrh of the urinary organs, rheumatism.
 The climate is very healthy, the place being surrounded by pine and other woods. The establishment is elegant, but the accommodation not quite up to the level of western requirements.
 A much-frequented station.

Scheveningen.—Holland, near the Hague, on the North Sea.
Very fine sands, and a very well-frequented sea bathing station.
The bathing arrangements are good.

Schieder.—Germany, Principality of Waldeck, near Meinberg.
Saline waters and brine baths.
Primitive arrangements.

Schimberg.—Switzerland, Canton Lucerne, near Ebenstetten.
Alkaline sulphurous and ferruginous waters. Mountain air cure.
A much frequented establishment. Social distinctions count for little at Schimberg.

Schinznach.—Switzerland, Canton of Aargau, between Berne, Basle, and Zurich, station on the North East Swiss Railway; 1,100 feet above sea level. For routes see table.
Sulphurous waters of considerable celebrity, very rich in sulphuretted hydrogen and carbonic acid, also in chloride of sodium and salts of lime, 95° F. The springs, recently lined with cement, yield about 220 gallons a minute.

Season: 15th May to 1st October.

Special indications: Chronic skin diseases, eczema, acne, psoriasis, furunculosis, &c., scrofula, chronic nasal, throat, and laryngeal catarrh, bronchitis, emphysema, asthma, rheumatism, gout, syphilis, mercurialism.

The mean temperature in summer is 67° F.; the mean barometric pressure 0·728. Atmospheric disturbances are exceedingly rare, and the changes never very rapid.

Schinznach is most picturesquely situated on the rapid and beautiful stream of Aar, at the foot of the Wülpelsberg. This hill, crowned with the old and far-famed ruins of the castle of the Habsburgs, slopes down into a fertile and well wooded valley with a southern aspect.

The establishment stands in its own grounds, which are of large extent, mostly laid out in park, with numerous shady walks, protected from dust and noise. An excellent farm supplies visitors with its produce. Schinznach appears to have been especially intended as a paradise for children.

In this park are six dwelling-houses or pavilions (capable of admitting 350 persons), communicating with each other and with the bath house by long covered verandahs and galleries. Cleanliness in every detail and a faultless table are the principal features of these houses.

The director of the establishment, Mr. Hans Amsler, who has lived for some years in England, and knows English requirements and habits, is very obliging, and will gladly give further information on application.

1. Upper Kurhaus.
2. "Hôtel de Nassau."
3. "New Hotel."
4. Lower Kurhaus.
5. "Hôtel de Berlin."
6. Swiss Chalet.
7. "Hôtel Victoria."
8. Dr. Baumann's Villa.
9. Chemist.
10. English Church.
11. Post.

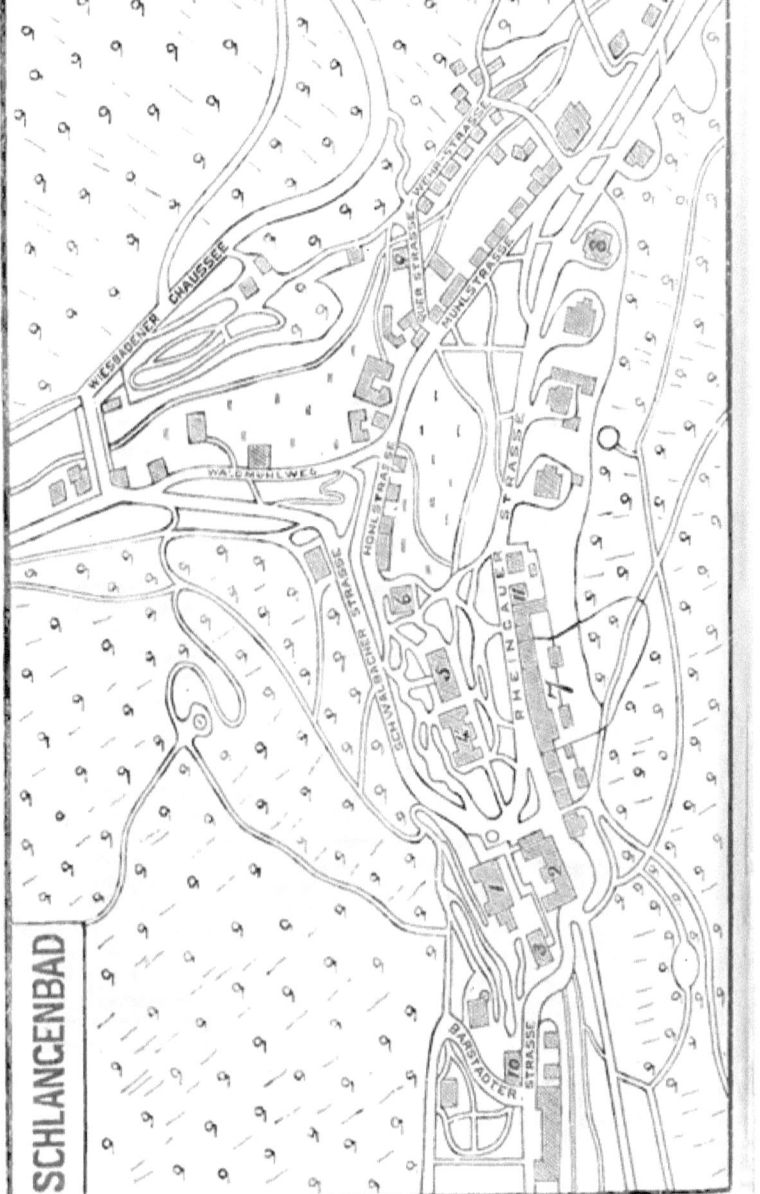

Post and telegraph office in the establishment: eight departures and distributions daily.
Protestant church.
No cure tax as in many German watering-places.
Concerts, balls, and all other amusements and distractions; walks and excursions of great variety and beauty.
Cabs and carriages, horses.

Three doctors at the establishment, all of whom speak English.

The internal use of the iodine waters of Wildegg (3 miles from Schinznach) is very efficacious in connection with the baths of Schinznach.

Patients with skin diseases, especially eczéma, are advised to continue the waters and fomentations at home, after having left the Schinznach. The water bears travelling well.

Schlackenbad.—Prussia, Saxony, near Hettstadt.

Artificial inert thermal baths.

The establishment is good, and contains ball, concert assembly and billiard rooms. The situation is romantic.

Schlangenbad.—Germany, Prussian Province of Hesse-Nassau, near Wiesbaden; 950 feet above sea level. For routes, see table.

Indifferent waters, 85° to 92° F. Nine springs.

Season: 1st May to 1st October.

Special indications: Menstrual difficulties, female diseases, chronic inflammation of the pelvic organs, neuralgias, gout, rheumatism, skin diseases, and bad complexions.

The establishment is well arranged, and Schwalbach waters (half hour's distance) are often successfully combined with milk, whey and aromatic herb essences.

The village (60 houses with 500 inhabitants) is situated in a valley of the Taunus, with three entrances towards the south and the Rhine. The valley is surrounded by hills, clothed with forests, and rising to a height of 1,900 feet. The atmosphere is very pure and fresh, and saturated with ozone. The environs are charming and idyllic, with many well-shaded and beautiful promenades and drives.

Concerts, balls, reading-rooms, and various other amusements. Tax for one person 12 marks, other members of a family 9 marks each.
The Curhaus, Casino, upper, middle, and lower Curhaus, Hotel de Berlin, and Swiss House are all under the management of the Government.
The baths are in three separate establishments.
Post and telegraph in the Kurhaus. English divine service.

Doctors: Drs. Baumann, Grossman and Wolff.

Hotels: The *Hôtel Victoria* will be found exceedingly well managed, and in every way comfortable and homelike. It is well situated, facing the park and baths. W. Winter, proprietor. (See Plan No. 7.)

Schleusingen.—Germany, Thuringia, 1¾ hours from Hildburghausen; 1,210 feet above the sea level.

A climatic air station.

The establishment and accommodation are very good.

Schmalkalden.—Germany, Hesse-Cassel, on the Werra railway line, 1,000 feet above sea level.
   Iodo-bromurated saline waters, 45° F. Brine baths.
   Special indications: Scrofula.
   The establishment is good and the site picturesque.
   Doctors: Drs. Fuckel and Rehm.
   Hotels: Crown, Eagle.

Schmardau.—Russia, Lithuania, district of Apitsch, near Podaizen.
   Sulphurous waters; not much frequented.

Schmecksz, or Tatra-Füred.—Hungary, comitat Zips, rail to Poprád; 3,100 feet above sea level.
   Pure alkaline bicarbonate waters. For table use only.
   Climatic station.
   Special indications: Debility, disorders of the digestive organs, catarrh of the intestines, derangements of liver and spleen, rhachitis, anæmia and scrofula, phthisis, emphysema, bronchial, laryngeal, and pharyngeal catarrhs.
   Bookseller: A. Mayer.
   Doctor: Dr. L. Jarmay.
   Hotel: The establishment.

Schmeckwitz.—Germany, Saxony, near Dresden.
   Earthy saline sulphurous waters, 55° F.
   The establishments are good, but the accommodation is not of the best.

Schmerikon.—Switzerland, Canton, St. Gallen, near Uznach.
   Cold ferruginous bicarbonate waters.
   Climatic station.
   The establishment is well arranged and beautifully situated, with a view over the lake of Zürich.

Schönau.—Austria, Bohemia, near Teplitz.—*See* Teplitz-Schönau.
   Thermal indifferent waters.
   Special indications: Gout and nervous affections. There are extensive military hospitals here.

Schönbeck.—Germany, Prussia, Province of Saxony.
   Saline waters.

Schönbrunn.—Switzerland, near Zug; 2,270 feet above sea-level.
   Hydropathic establishments, climatic and whey cure station. Irish and Turkish baths.
   The site is very fine. Climate mild and bracing. The establishment is excellent, and the accommodation equal to that in the best Swiss hotels.

Schöneck.—Switzerland, Canton Unterwalden, above Beggenried, Lake of Lucerne.

A summer station, with hydropathic establishment, compressed air baths.

Doctor: Dr. Von Corval.

Schöngau.—Switzerland, Canton of Lucerne.

Cold bicarbonate of lime waters.

Schooley Mountains.—United States of North America, State of New Jersey.

Ferruginous and magnesian waters.

A much-frequented station.

Schûls, also **Tarasp-Schûls.**—Switzerland, Canton of Grisons.

Cold bicarbonate of soda and iron waters.

Schum.—Austria, Tyrol, district of Botzen.

Alkaline, ferruginous and sulphurous waters.

Special indications: Anæmia, rheumatism and general debility, skin diseases and chronic sores.

Schüols.—Switzerland, Engadine, 10 hours from St. Moritz, 3,730 feet above the sea level.

Effervescent ferruginous waters. There are some 30 springs, of which only a few are used.

Schwalbach or Langenschwalbach.—Germany, Hessen-Nassau, between Wiesbaden and Ems; 2,200 inhabitants; 955 feet above sea level. For routes see table.

Chalybeate waters; every description of bath; climatic station; mud baths (peat).

Season: May to October.

Number of visitors, between 5,000 and 6,000, as shown by the visitors' list, the majority being ladies.

Special indications: All female complaints are treated here as a specialty; anæmia, chlorosis, nervous complaints, debility of muscles and mucous membranes, especially catarrhal affections of the genital and urinary organs.

Schwalbach is a neat little town of about 500 houses, tastefully built. Its situation is very romantic, and the surrounding country is picturesque and affords an ample field for excursions. The bathing establishment is an elegant building, with every modern appliance. The Kursaal contains assembly, concert, reading and refreshment rooms. Daily concerts and bi-weekly *soirées dansantes* are given.

English Church.

Post and Telegraph.

Banker and Exchange Office: Ernst Grebert.

Wine Merchants: Ferd. Grebert Söhne.

Doctors: Drs. Frickhöffer, Grebert, Boehm and Genth.

Hôtels: The *Hôtel Allcesaul* and *Hôtel de la Promenade*, are well situated in the midst of gardens, first-class in every respect. A covered way leads to the bathing establishment. Ferd. Grebert Sons, proprietors; dependance, Villa Grebert.

The *Villa Grebert:* In the centre of its own gardens, suitable for families.

The *Duke of Nassau Hôtel:* Well situated, a first-class, and very good family house, with home comforts; entirely refurnished and embellished. J. C. Wilhelmy, proprietor.

The *Taunus Hôtel:* A good second-class house, facing the Royal baths; very clean, and moderate charges; open all the year, with pension. J. Boll, proprietor.

Schwalheim.—Germany, Hessen, near Frankfort-on-Maine.
Mixed cold bicarbonate waters. For table use only.
Doctor: Dr. Fleury.

Schwallungen.—Germany, Saxe-Meiningen.
Ferruginous waters.
The establishment is indifferent.

Schwarzenberg.—Switzerland, near Aarau; 2,180 feet above sea level.
Alkali-saline waters.
The establishments are good and the place much frequented.

Schwarzenberg.—Austria-Hungary, comitat Zips, near Igló.
Hydropathic establishment, with all modern requirements.

Schwarzseebad.—Switzerland, six hours from Fribourg; 3,269 feet above the sea level.
Sulphurous waters; climate severe, but healthy.
The establishments are good.

Schwefelbad.—Austria-Hungary, comitat of Zips, near Leibnitz.
Cold sulphurous waters.
Special indications: Rheumatism, gout, chronic skin diseases and old sores.
The site is beautiful and the establishment good.

Schwefelbergerbad.—Switzerland, Canton of Bern; 4,170 feet above the sea level.
Sulphurous waters; little used.

Schweizermühle.—Germany, Saxony, near Königstein; 1,200 feet above sea level.
Hydropathic establishments and climatic station.
Special indications: Affections of the lungs and respiratory organs.
The establishment is elegant and contains comfortable assembly rooms. The accommodation is good and moderate in price.

Schwelm.—Germany, Westphalia.
Cold ferruginous bicarbonate waters.

Sciacca.—Italy, Sicily, on the south-west coast.
Three springs, one sulphurous and ferruginous, and one magnesian, the latter slightly purgative; also natural steam baths.
The establishments are good, and the place is much frequented. The accommodation is scarcely in accordance with English requirements.

Sclafani.—Italy, Sicily, Province of Palermo.
Sulphate of lime waters, 145° F., containing alum.
Season: May—July.
Bathing establishment is old-fashioned, but the accommodation is fair.

Sesavnik.—Austria-Hungary, comitat of Zemplén, near Eperies.
Alkaline ferruginous sulphurous waters.
The establishment is arranged in a primitive manner.

Seaford.—England, Sussex, near Eastbourne.
Sea baths, shingly beach. Climate fine and bracing.
Doctor: Dr. B. Tuck.

Seascales.—England, Cumberland.
Sea baths, sandy flat beach; bracing climate.
Season: June—September.

Seaton.—England, Devonshire, near Sidmouth.
Sea baths, heavy surf; climate mild and bracing.
Doctor: Dr. Pattison.
Hotel: Royal *Clarence*.

Seaton-Carew.—England, Durham, on Hartlepool Bay.
Sea baths, firm sandy beach; climate bracing.
Doctor: Dr. Norman.

Sea-View.—England, Isle of Wight.
Sea baths, extensive and sandy beach; climate dry and bracing.
Hotel: The *Crown*.

Sebastianweiler.—Germany, Württemberg, near Tübingen.
Sulphate of sodium waters, 40° F.; laxative and diuretic action.
Well-appointed establishment. The sojourn is quiet, and offers but few amusements.

Sedlitz.—Austria, Bohemia.
Sulphate of sodium and magnesia waters. Purgative action.

Seebruch.—Germany, Westphalia, district of Herford, near Vlotho.

Cold sulphurous waters.

Not much used, and the arrangements are primitive, though the waters are said to be very efficacious.

Seelisberg.—Switzerland, Canton of Uri, Lake of Lucerne; 2,210 feet above sea level. For routes see table to Lucerne, thence by rail to Fluelen and boat to Treib, or from Lucerne by boat to Treib; thence by carriage, 30 minutes.

Climatic station in summer, with whey cure and baths in the lake of Seeli. Air very mild and pure; very sheltered. Baths in the hotel.

Season: May 15th to October.

The Canton of Uri is celebrated for the legend of William Tell and the Oaths of Alliance. It is the cradle of the Swiss confederacy. In 1852, the present excellent hotels, three in number, and belonging to the same proprietor, were built. They form the chief attraction of Seelisberg, both as a quiet and comfortable place of call and as a centre for numerous excursions. Owing to its altitude of 2,500 feet, this establishment enjoys the same climatic conditions as the other stations in Switzerland. A moist atmosphere, with an equable and comparatively low temperature, is the characteristic of Seelisberg from a hygienic point of view. Vegetation is luxurious, and forests, consisting chiefly of pine and fir, begin at the very door of the establishment. The terrace in front of the hotel is 650 feet long, and 1,200 feet above sea level. It is shaded by trees, and is an excellent place to enjoy the beautiful scenery from, or for walking exercise. There is a "Trinkhalle," where most European mineral waters are supplied. Baths of every description; English church in the hotel; Post and Telegraph Office; during the season, concerts, balls, etc.

A resident physician.

The *Hôtel Sonnenberg* contains about 300 bed-room and drawing rooms; it is very well managed, and the prices are moderate. Good kitchen and excellent cellar. The dining-room is cool and well ventilated, 22 feet in height, and capable of holding 500 persons at a time. It is the largest in Switzerland. M. Truttmann, Proprietor.

Seeon.—Germany, Upper Bavaria, near Endorf; 1,725 feet above sea level.

Earthy waters; whey cure and brine baths.

The establishment is good and the situation very picturesque.

Seewen.—Switzerland, half-hour from Schwyz; 1,410 feet above sea level.

Ferruginous waters, highly effervescent.

The establishments are very good and the accommodation comfortable and abundant.

Segesta.—Italy, Sicily, half-hour from the ruins of this old town.
Sulphurous waters, 165° F.
Used in a primitive fashion.

Segray.—France, Department of Loiret, Arrondissment of Pithiviers.
Cold ferruginous waters.

Segre.—France, Department of Maine and Loire.
Cold ferruginous waters.

Segura de Aragon.—Spain, Province of Teruel, near Montalvan; 2,200 feet above sea level.
Alkaline and saline waters, 55° F., also a ferruginous spring.
Special indications: Rheumatism, nervous affections, hysteria, chorea, female complaints, leucorrhoea, eye affections, scrofula.
The establishment is antiquated, but accommodation is good and moderate in price.
Doctor: Dr. Viñolas.

Segura.—Spain, Province of Castellon de la Plana.
Saline sulphurous waters, 55° F.
Special indications: Skin diseases and lymphatic affections.
Primitive arrangements.

Seltz or Selters.—Germany, Nassau, Circuit of Wiesbaden on the Ems.
Alkaline highly effervescing waters. Used only for exportation.

Savergnolles.—Department of Cantal.
Cold ferruginous bicarbonate waters.
Special indications: Chlorosis.

Semur.—France, Department of Côte d'Or.
Cold saline waters. Five grains of mineralisation.

Sennfeld.—Germany, Bavaria, near Schweinfurth-on-the-Maine.
Sulphurous waters; little used.

Sentein.—France, Ariège, near St. Girons.
Cold bicarbonate of iron waters.

Sepey.—Switzerland, Canton Vaud, 7½ miles from Aigle by carriage.
A mountain air station.
Excursions in the charming valleys of Ormonts and the Mosses; place of call for Chateau d'Oex and Diablerets. Very pure air. Post, telegraph, carriage and relays of horses, with guides.
The *Hôtel and Pension des Alpes* is the only hotel which can be recommended. V. Chenaud, proprietor.

Seraglio.—Italy, near Siena.
Indifferent waters, 60° F.
No establishment and little or no accommodation.

Serapis, Temple of.—Italy, near Pozzuoli and Naples.
Alkali-saline waters, 85° F.
A place now scarcely visited, but very celebrated in antiquity, as its ruins show.

Seravalle.—Italy, Tuscany.
Ferruginous waters, 40° F.

Serboneschte.—Roumania, near Putschos.
Sulphurous waters.
Arrangements primitive.

Serdopol.—Russia, Finland.
Earthy waters.
Only used by the surrounding peasantry.

Sergiewsk.—Russia, Government of Orenburg, near Kasan.
Sulphurous waters.
No accommodation, and visitors have to provide it for themselves as best they can.

Sermaise.—France, Department of Marne, near Vitry le Français.
Mixed sulphate and bicarbonate waters, with a laxative and slightly diuretic action.

Sestri-Ponente.—Italy, 20 minutes from Genoa.
A winter station.
Though the place is rather dusty and windy, it has of late attracted many visitors, owing to the excellence of its hotel. It affords very comfortable accommodation, at moderate rates. The town possesses little of interest beyond its shipbuilding yards.

Hotels: The *Grand Hotel Sestri*, a first-class house, 15 minutes from Genoa; heated by caloriféres; moderate charges for a continuous winter stay; play-grounds specially arranged for children. *Meyer* and *Arrigoni*, also proprietors of *Hôtel de la Grande Bretagne*, at Bellagio.

Seville.—Spain, capital of the Province of same name, from Paris, viâ Bordeaux, Irun and Madrid.
As the town is said to have been founded by the Phœnicians, its great antiquity must be admitted. It is one of the most interesting towns in Spain, and is situated in a plain on the right bank of the Guadalquivir.
It was one of the most formidable of Roman strongholds, and became a flourishing city after its siege and capture by Julius Cæsar.

OBJECTS OF INTEREST.—The Cathedral, the Alcazar, the Giralda, the Triana, Museum, Churches of la Caridad, St. Nidors, San Miguel, Sta. Maria la Blanca, Town Hall, Lonja, Fabrica de Tabaco, **Picture** galleries, Palace of **St.** Telmo, Italica, Baños de Carmona, the walls around **the town**, Sta. Lorenzo, **University,** Colombian Library, the **Palace of** the Archbishop, **Hospital,** Mint, **C**asa de Pilatos, **Tower of Gold.**
EXCURSIONS: **Ruins of** Ithalica, Azualfarrache, Triana Cartujà.
CABS: One or two persons, **2 fr.**; three or four, 3 fr.; by hour, one or two persons, 4 fr.; three or four, **5 fr.**
Post-office: Close to Hotel Madrid **and de Paris.**
Telegraph-office: Calle de Bailen.
  Doctors: Dr. Hauser, de Uriarte.
  Hotels: **De las** *Cuatro Naciones,* **de** *Paris,* de *Madrid,* **de** *l'Europe.*

Shanklin.—England, **Isle of** Wight.
  Sea **baths, fine sandy** flat beach.
  Air **pure and healthy.**
  Season: June—October.
  Doctors: Drs. Dabbs and Moss.
  Hotel: The *Royal.*

Shap.—England, Cumberland, near Gisland.
  Alkaline **waters and sea baths.**

Sheerness-on-Sea.—England, Kent, on the Medway.
  Sea baths, shingly and sandy beach, climate bracing.
  Season: June—September.
  Doctors: Drs. Arrol and Swales.
  Hotel: The *Fountain.*

Shotley.—England, Northumberland.
  Chloride **of** calcium and ferruginous **waters.**

Sibitschudi-Suz.—Roumania.
  Ferruginous and sulphurous waters. Very primitive arrangements, **and the place** is **little** visited.

Sibó.—**Austria-Hungary,** comitat **of** Zsolnók.
  Sulphurous waters, little used.

Sid.—Austria-Hungary, comitat of Gömör, near Vargede.
  Alkaline ferruginous waters.
  The arrangements are **of** the most primitive kind, and very **little is** done for comfort.

Sidmouth.—England, Devonshire, **near Exeter.**
  Sea baths, sandy beach; climate equable.
  Season: June—September.
  Doctor: Dr. Hodge.
  Hotels: The *Royal York.*

Siekeriki.—Russia, Poland, near **Varsovia.**
  Ferruginous waters.
  The establishment is incomplete.

Sierk.—Germany, Lorraine, near Thionville.
  Cold saline waters, similar to those of Mondorf.

Sierra Alhamilla.—Spain, Province of Almeria, near Pechina; 1,800 feet above sea level.
  Alkaline effervescent waters, 120° F.; used in drinking.
  Special indications: Rheumatism, gout, uric diathesis, paralysis, white swelling, tendinous contractions, caries and necrosis, chronic ulcers, hyperæmia of liver and intestinal obstructions.
  The establishment is thoroughly well fitted up. The accommodation is very inferior, the rooms containing only chairs and tables; patients are expected to bring their own bedding and linen. The prices are moderate.
  Doctor: Dr. Hernandez.

Sierra Elvira.—Spain, Province of Granada, near Santa Fé.
  Sulphurous earthy waters, 60° to 65° F.
  Special indications: Skin diseases, rheumatism, stomachic disorders, vesical and uterine catarrhs, chorea and scrofula.
  The establishment is much behindhand, and the accommodation very inadequate.
  Doctor: Dr. Saez de Tejada.

Siete-Aguas.—Spain, Province of Valencia, near Chiva.
  Effervescent alcaline waters, 50° F., containing traces of iron.
  Special indications: Nervous complaints, dyspepsia, skin diseases, scrofula.
  The establishment and accommodation are fairly good. The situation is fine.
  Doctor: Dr. Gil.

Sigliano.—Italy, Romagna, valley of the Teverino.
  Alkaline ferruginous waters.
  The arrangements are still of a primitive character.

Silloth.—England, Cumberland, near Carlisle.
  Sea baths; sandy and gravel beach; climate mild, dry, equable.
  Season: June—September.
  Doctor: Dr. Leitsch.
  Hotel: The *Solway.*

Silvanès.—France, Department of Aveyron, near St. Affrique.
　　Ferruginous bicarbonate waters, 90° F.

Sinigaglia.—Italy, Province of Ancona, and near that town.
　　Sea baths and sulphurous spring, 110° to 130° F.
　　Special indications: Arthritis, rheumatism, gout, skin diseases, affections of the liver and spleen.
　　Much frequented by wealthy Romans, although there is no bathing establishment. The spring is half-an-hour distant from the town.

Sinzig.—Germany, Rhenish Prussia, near Bonn.
　　Alkaline muriatic waters; a climatic station in summer, with whey and grape cure.
　　A very old town, with picturesque surroundings.

Siradan.—France, Department of Hautes Pyrénées, near Bagnères de Bigorre.
　　Sulphate of lime and ferruginous waters.
　　Special indications: Affections of the urinary organs, chlorosis, and anæmia.

Sironabad.—Germany, Hessen, near Nierstein.
　　Sulphurous waters.
　　The ruins, said to be those of Roman buildings, serve to show the long period during which these waters have been used. The place is now little visited.

Sisso.—Austria-Hungary, comitat of Neutra, district of Gross-Tapolcza.
　　Alkaline-saline earthy waters.
　　Little visited.

Sitka.—Russian America.
　　Hyperthermal sulphurous waters.

Sixt.—France, Department of Upper Savoy, near La Croix de Pilly.
　　Effervescing ferruginous waters.

Skara-Chori.—Greece, Morea, Elis, near Katacolbi.
　　A sulphurous warm lake.
　　There are some ruins of ancient baths with mosaic floors. They are still used as baths.

Skegness.—England, Lincolnshire.
　　Sea baths, sandy beach; climate bracing.
　　Season: June—September.
　　Hotel: The *Sea View*.

Skinburness.—England, Cumberland, on the Solway.
  Sea baths, fine sandy beach; climate mild and dry.
  Season: June—September.

Slepzoff-Michailoff Springs.—Russia, Caucasus, near Wladiskawkas.
  Sulphurous waters, 70° F.
  There is a military hospital, but all other arrangements are as yet in a primitive state. The site is romantic and the scenery wild. Good hunting can be had.

Smokobe.—Turkey, on the Greek frontier.
  Sulphurous waters.
  Very primitive; holes are sunk in the ground.

Smyrna.—Asia Minor.
  Thermal sulphurous waters.

Sobron.—Spain, Province of Alava, near Vitoria.
  Alcali-saline waters, 60° F.
  Special indications: Chronic intestinal derangements, hyperæmia of liver and spleen, vesical catarrhs, menstrual difficulties.
  The place is situated in a charming valley. The establishment is not quite complete, but affords good accommodation at low prices.
  Doctor: Dr. Tejada y España.

Soden.—Germany, Nassau, at the foot of the Taunus, near Frankfort-on-Main.
  Effervescing and saline ferruginous water, 45° F., with laxative and alterative action.
  Special indications: Scrofula, abdominal plethora, pulmonary phthisis.
  Fine climate, beautiful situation.
  Doctors: Drs. Köhler, Thilenius and Fresenius.
  Hotels: Kurhaus, d'Europe, de Frankfort.

Soest.—Germany, Westphalia.
  Saline waters. Brine baths.
  Special indications: Scrofula.
  The establishments are good.

Sohl.—Germany, Saxony, near Elster; 1,418 feet above sea-level.
  Ferruginous waters.
  The arrangements are primitive.

Solan de Cabras.—Spain, Province of Cuenca, near Priego.
Bicarbonate of lime waters, 45° F.
Special indications: Nervous affections, skin diseases, tabes, intermittent fever and marasmus, vesical catarrhs.
The establishment is an indifferent one.
Doctor: Dr. Valdivieso.

Solares.—Spain, Province of Santander.
Saline waters, 70° F.
Special indications: Lymphatic affections.
Arrangements primitive; the accommodation is Spanish.
Doctor: Dr. Sarraci.

Sombór.—Austria-Hungary, comitat of Háromszek, near Torja.
Sulphurous waters, 45° F.
Very little used.

Sorrento.—Italy—Bay of Naples, one hour's drive from Castellammare; 7,500 inhabitants.
Climatic air station, with sea bathing.
Season: February till June.
The situation of Sorrento, facing Capri, is one of the most lovely in the world. The country around is in a state of high cultivation. Very good fruits and fish. Tasso was born at Sorrento, 1544. The manufacture of scarves and stockings is the chief industry, together with wood mosaic work. The environs are very beautiful.
English Church Service in the Hôtel Tramontana.
Banker and Exchange Office: A. Falangola.
Post and Telegraph Office: Piazza Tasso.
Chemist: Dr. Galano.
Doctors: Drs. Mayo, and Galano.
Hotels: The *Hotels Tramantano, Tasso and de la Sirène* are very comfortable, and afford splendid views over the whole bay. Large garden, with baths in sea. L. Tramontano, proprietor.

Sostó.—Austria-Hungary, comitat of Szabolcs Natron lake.
A thoroughly well arranged establishment.

Sotteville-lez-Rouen.—France, Department of Seine Inférieur, near Rouen.
Thermal saline bromo-iodurated waters, 55° F.

Soulac-les-Bains.—France, Department of Gironde.
Sea baths.
Very fine beach.

Soulieux.—France, Department Isère, near Grenoble.
Sulphate of magnesia and soda waters, 5 grammes of mineralisation.

Sousas and **Caldeliñas, or Aguas de Verin.**—Spain, Province of Orense, on the frontier of the Portuguese province Traz-oz-Montez.
Alkali saline waters, 49° F.; chemically superior to the waters of Vichy.
Special indications : Calculus, hyperæmia of liver and spleen, gastralgia.
There is a fairly good establishment at Souzas, but at Caldeliñas the most primitive arrangements still prevail. Comfort is conspicuous by its absence, though the efficacy of the waters would amply repay a large outlay. They are unrivalled in the treatment of calculus.
Doctor : Dr. Gavilanes.

Southampton.—England, Hants.
Sea baths, unsatisfactory beach; climate mild and healthy.
Season : June—September.
Doctors : Drs. Olliver and Ward.
Hotel : The *South Western.*

Southbourne-on-Sea.—England, Hants.
Sea baths, excellent sandy beach; climate healthy and invigorating
Season : July—September.

Southend-on-Sea.—England, Essex.
Sea baths, sandy beach; pure and bracing air.
Season : June—August.
Doctors : Drs. Jones and Phillips.
Hotel : The *Royal.*

Southport.—England, Lancashire.
Much frequented sea bathing-place, with sandy beach, and mild and pure air.
Season : June—July.
Doctor : Dr. Elias.

Southsea and Portsmouth.—England, Hants, opposite Isle of Wight.
Sea baths, sandy beach, with very good bathing; climate mild and steady.
Season : July—October.
Doctors : Drs. Axford and Simpson.
Hotel : The *Queen's.*

South Shields.—England, Durham, on the Tyne.
Sea baths, sandy beach; climate pure and bracing.
Season : June—September.
Doctors : Drs. Dalziel and Spear.

Southwold.—England, Suffolk.
  Sea baths, sandy beach; climate very equable and healthy.
  Season: June—September.
  Doctor: Dr. Vertue.
  Hotel: The *Swan*.

Spa.—Belgium, close to the German frontier, 17 miles from Liege and Aix-la-Chapelle, 1,050 feet above the sea level; nearly 7,000 inhabitants. For routes see table.
  Highly effervescent ferruginous and acidulous waters. Every description of bath.
  Season: May to October. Air healthy and bracing.
  Number of Visitors: More than 30,000 annually.
  Special indications: Anæmia, chlorosis, female complaints, hysteria, gastralgia, sterility, difficult menstruation, liver complaints, urinary disorders, cachexia, mucous catarrh of the uterus, &c.
  English and Scotch churches.
  Bookseller: Jos. Engel, opposite the Pouhon.
  Doctors: Drs. Thompson, Lezaack, De Damescau and Scheuer.
  Hotel: The *Hotel de Flandre*.

Spag.—Russia, Government of Pleskow, on the Düna.
  Sulphurous waters. Primitive arrangements.

Spalato.—Austria, Dalmatia.
  Saline and sulphurous waters, 28 grammes of mineralisation, of which 19 grammes are chloride of sodium.
  Special indication: Scrofula.

Spezia (La).—Italy, on the Riviera di Levante, an Italian maritime port, between Genoa and Pisa. Fine Bay.
  Winter air-cure and summer sea bathing station.
  Mean winter temperature, 50° F.; 74 rainy days, 64 cloudy days, and 63 sunny, during the winter season. The atmosphere is more humid than on the Riviera di Ponente. Spezia is especially recommended to people who find a difficulty in sleeping in other seaside places.
  The Piazza Vittorio Emmanuele is a very fine square, and the gardens are being extended every year. The walks are admirably clean, and this is a very favourite one.
  Very picturesquely situated; air fine and bracing. Many excurisons. The new road to the fortifications, commanding a magnificent view over the bay, is open to visitors.
  There is a good theatre and a casino.
  Vegetation is luxuriant, and composed chiefly of olive and vine trees. The water is good, but should always be mixed

with wine at the beginning of residence, to prevent diarrhœa. There is very little dust.
English church in winter.
Post office : Corso Cavour.
Telegraph : Via Facio.
Chemist : Fossati, Farmacia Regia.
Doctor : Dr. Sparks.
Hotel : The *Grand Hotel de Spezia,* a thoroughly well appointed house, comfortable, open all the year; excellent situation, overlooking bay. Well managed by Menetrey Hauser, the proprietor.

Spital.—England, Durham, near the mouth of the Tweed.
Cold sulphate of soda waters.
Laxative action.

Sprofondo.—Italy, Tuscany, near Asciano.
Nine springs. Sulphurous waters, 37° to 75° F., mud baths. The establishments are good.

Stachelberg.—Switzerland, Canton of Glarus; 2,800 feet above sea-level.
Cold sulphur waters.
Special indications : Skin diseases, chronic laryngitis.
A very well-appointed establishment, with comfortable accommodation. The situation is highly picturesque.

Staden.—Germany, Hessen-Nassau, near Nidda.
Earthy saline waters. Exported.
The establishments are incomplete, though the spring is one of the oldest in Germany.

Staraja-Rossa.—Russia, District of Nischni-Novgorod.
Cold saline iodo-bromurated waters.
The establishments are very good, and the place is one of the most frequented of Russian stations. Its situation is charming.

Starbeck.—England, Yorkshire, near Harrowgate.
One ferruginous and one sulphurous spring, both slightly mineralised.
There is a bathing establishment lately opened, which is beginning to be frequented.

Stavenhagen.—Germany, Mecklenburg.
Earthy alkaline ferruginous waters.
The arrangements leave room for improvement.

Steben.—Germany, Bavaria, district of Upper Franconia, near Hoff; 1,900 feet above sea level.

Five ferruginous effervescing springs, and mud baths.

Mountain air cure.

The situation is most picturesque. The establishments are fitted with all modern improvements. The accommodation is comfortable and moderate in price.

Doctor : Dr. Klinger.

Hotels : **Anchor,** Bavière.

Steinabad.—Germany, Baden, in the Black Forest.

A climatic station.

The place is situated in the midst of pine forests. The climate is mild and strengthening. Disorders of the digestive organs are those chiefly benefited by residence.

Doctor : **Dr. Wiel.**

Steinfurt.—Germany, Hessen-Nassau, near Wisselsheim.

Saline waters.

The place deserves to be better known. At present the waters are used only by the surrounding inhabitants.

Steinheyde.—Germany, Saxe-Meiningen.

Saline ferruginous waters.

The site is picturesque, but little has been done towards improving the place.

Steinwater.—Austria, Bohemia.

Sulphate of magnesia waters, 31 grammes of mineralisation, of which 28 grammes are sulphate of magnesia.

Purgative, as Püllna and Sedlitz.

Sternberg.—Austria, Bohemia, near Prague.

Chalybeate **waters. Whey cure station.**

The **establishment is very** efficient. The site is picturesque and much visited.

Stoika.—Austria-Hungary, comitat of Szolnok.

Alkali-saline waters, 45° F.

No establishment, and the waters are used only in the neighbourhood.

Stolypin.—Russia, on the Kuschum, one of the tributaries of the Volga.

Saline sulphurous waters.

The establishment is thoroughly well fitted up, is very elegant, and affords comfortable accommodation. Its situation is charming, and the place is rising rapidly in public favour.

Strathpeffer.—Scotland, Rosshire, near Dingwall, in a very picturesque situation.
 Strong sulphurous waters.
 Hydropathic establishment.
 Special indications: Skin diseases, rheumatism.
 Doctors: Drs. Manson and Middleton.

Strasburg.—Germany, Capital of Alsace-Lorraine, on the left bank of the Rhine.
 One of the strongest fortresses in Europe, captured in 1871 by the Germans, after a long and desperate resistance.
 OBJECTS OF INTEREST.—The Cathedral, the University (formerly the Episcopal Palace), Church of St. Thomas, Public library, Town Hall and old Prefecture, Theatre, Academy, Lyceum, Tobacco factory.
 EXCURSIONS: Mont St. Odile, Saverne, Greifenstein.
 CABS: A drive for one or two persons, 50 pfgs.; one hour 2 marks.
 Post and telegraph-office: Schlossplatz.
 Bookseller: Derivaux, Rue des Hallebardes.
 Dentist: Jacques Wiesner.
 Doctors: Drs. De Barry, Kussmaul and Kley.
 Hotels: The *Hôtel d'Angleterre*, first-class, nearest to the central station. Telegraph, post-offices and cathedral. Great comfort with moderate charges. Omnibus at the station. C. Mathys, proprietor.

Streatham.—England, near London.
 Sulphate of magnesia waters; purgative action.

Streitberg.—Germany, Bavaria, near Forchheim; 1,800 feet above sea level.
 A climatic air station with whey cure.
 The surrounding hills are covered with pine forests. The country is picturesque, and abounds in walks and excursions. The establishment is thoroughly fitted up and affords ample accommodation.

Stresa.—Italy, Lago Maggiore, in front of the Borromean Islands.
 Station of call. Air cure.

Stromstad.—Sweden, on the Skagerak, near Gothenburg.
 Sea baths; strong surf, stony beach.
 The most popular of Swedish sea-bathing places.

Stuben.—Austria-Hungary, comitat of Arva Turócz, on the Stubna.
 Earthy saline bitter waters, 110° F.
 The establishment and accommodation generally are very satisfactory. The place is romantically situated, and the park and gardens well kept.

Stuttgart.—Germany, Capital of Würtemberg; **from Paris by Strasburg and Kehl, 17¼ hours. 88 fr. 1st class.**

Stuttgart, most beautifully situated a **short** distance to the **west of the** Neckar, surrounded by hills, covered with orchards **and vineyards.** The town is of comparatively recent origin. It occupies the second place in Germany as **a centre of the** book trade.

OBJECTS OF INTEREST.—Old and new Royal Palace, **Rosenstein** and Wilhelmina, Public library, Museum of natural **history,** Commemorative Column, Museum of fine arts, **Königsbau,** Exchange, Barracks, Ministry of foreign affairs, **Post-office, Museum, Liederhalle, and Carlsschule.**

EXCURSIONS: Villa **Rosenstein, Palace** Gardens, Wilhelmina, Cannstadt and Solitude, and **Hohenheim.**

CABS: according to tariff.

Dentist :—Ferd. Young.
Doctors: Drs. Von Koch, **Von** Hausmann and Gärthner.
Hotels: *Marquardt, Royal.*

Suderode.—Germany, near Quedlinburg.
Saline waters.
Special indications: Scrofula, pulmonary **affections, anæmia,** &c.
Doctor: **Dr.** Barwinsky.
Hotel: *Kurhaus.*

Sujo.—Italy, **in the valley of** Garigliano.
Thermal sulphurous ferruginous and effervescent waters.
Doctor: **Dr. Rocco-Tagliata.**

Suliguli.—Austria-Hungary, **comitat of Mármaros, District** of Visso.
Alkali-saline waters.
The establishment **is incomplete, and the accommodation not such as** to attract visitors.

Sulz.—Hungary, comitat of Eisenburg, **near** Güssing.
Cold alkaline-earthy waters.
Very **celebrated in the district. The** establishment is good, **and the accommodation satisfactory.**

Sulza.—Germany, Saxe-Weimar.
Saline waters; brine baths; grape **cure.**
Special indications: Scrofula.
The site is charming, and the **establishment good.**
Doctor: **Dr. Sanger.**
Hotels: *Kurhaus* **and** *de Saxe.*

Sulzbach.—Germany, Alsatia, **near** Colmar.
Cold bicarbonate ferruginous waters, 4 grammes of minerali**sation.**
Special indications: Dyspepsia, anæmia.

Sulzbach.—Germany, Baden.
Ferruginous bicarbonate waters, 50° F., 2 grammes of mineralisation.
Special indications: **Dysmenorrhœa.**

Sulzbad.—Germany, Alsatia, near Weissenburg and Strasburg.
Saline iodo-bromurated waters, 4 grammes of mineralisation. The same therapeutic properties as Wildegg and Kreuznach.
Special indications: Scrofula and lymphatic affections.
The gardens are beautiful, and the establishment well fitted up.

Sulzmatt.—Germany, Alsatia, near Gebweiler and Colmar; 850 feet above sea level.
Effervescing alkaline waters, more gaseous than those of Seltzer.
Special indications: Catarrh of the bladder and **lungs.**
An excellent table water.

Suot-Sass.—Switzerland, Canton of the Grisons.
Strongly effervescing chalybeate waters.

Sutinsko.—Austria-Hungary, Croatia, near Agram.
Inert waters, 100° F.; climatic station.
This is a very favourite bathing-place. The situation is good, and the air mild and exhilarating.
The establishment is elegant, and thoroughly well fitted up. Comfortable accommodation at reasonable prices.

Sutton.—England, Lincolnshire, near Alford.
Sea baths, sandy beach; **climate** bracing and healthy.
Season: June—September.
Hotel: The *Jolly Bacchus.*

Swanage.—England, Dorsetshire, **Isle of Purbeck.**
Sea baths; sandy beach. **Climate very healthy.** Hydropathic establishment.
Season: July—September.
The surrounding **scenery is fine.**
Doctors: **Drs.** Geo. Delamott and Pearse.
Hotel: The *Victoria.*

Swanlibar.—Ireland, near Enniskillen.
Cold hydrosulphuric waters.

Swansea and the Mumbles.—Wales.
Sea baths; sandy beach. Climate good.
Rather hot in summer; **good** establishments.
Doctors: Drs. Bend and Couch.

Swinemünde.—Germany, Pomerania, on the Baltic Coast.
　Very well frequented sea baths, near Stettin; very fashionable.
　Doctors: Drs. Schulze and Wilhelm.
　Hotels: *Wilmelmsbad* and *Three Crowns*.

Sylt.—Island in the North Sea, on the coast of North Schleswig.
　Sea baths; sandy beach.
　Doctors: Drs. Ditmann and Mareus.
　Hotels: *Royal, Strand*, **German** *Emperor*.

Szalatnya.—Hungary, comitat of Hont, near Schemnitz.
　Cold saline earthy waters.
　Digestive and diuretic action.
　The situation is fine, and the waters are efficacious, but an utter want of energy on the part of the proprietors is gradually destroying the place.

Szaldobos.—Austria-Hungary, Transylvania, District of Udvarhély.
　Alkaline ferruginous waters.
　The arrangements are very primitive.

Szaploncza.—Austria-Hungary, comitat of Mármaros.
　Alkaline effervescent waters; for table use only.

Szator.—Austria-Hungary, near Rimaszecs.
　Alkali-saline waters.
　Only used by the surrounding inhabitants.

Szczawnicza.—Austria, Galicia, near Neusandek; 1,000 feet above sea level.
　Alkali-saline ferruginous waters; whey cure.
　Special indications: Chronic metritis, vesical catarrhs, kidney affections, fluor albus, vaginal and uterine catarrhs, goitre, rhachitis.
　Owing to its magnificent situation, and the elegance and comfort of its bathing establishments and accommodation, this place is the most important of the Galician bathing stations. Each year the number of its visitors increases. The efficiency of its waters contribute to this result in no small degree. The excursions are very varied and interesting, while the Kursaal supplies all that is required for making the time pass pleasantly.

Szent-Ivan.—Austria-Hungary, near Magyarfalva.
　Sulphurous, highly effervescent waters, with gaseous emanations.
　Little has been done to attract visitors, and the arrangements are very primitive.

Szinye-Lipócz.—Austria-Hungary, near Eperies.
Alkaline waters, containing iodine and lithium.
Special indications: Kidney and vesical disorders.

Szinyak.—Austria-Hungary, comitat Bereg, near Munkacz.
Sulphurous cold waters, containing alum and sulphuric acid.
Special indications: Gout and rheumatism.
Situation romantic; pure forest air. The establishment is well fitted up.

Szkleno.—Hungary, comitat of Bars, near Schemnitz.
Several thermal sulphate of lime springs, 50° to 135° F.
Special indications: Rheumatism, gout, skin diseases.
The establishments leave much room for improvement.
They are all very elegant, but not well suited to modern requirements. The situation is highly picturesque, and excursions are abundant. The parks and gardens are scrupulously kept, and the surrounding woods are laid out with walks. The sojourn is an agreeable one, and the place is gaining in favour steadily.

Szklo.—Austria, Galicia, 5 hours from Lemberg.
Sulphate of lime water.
Special indications: Skin diseases, rheumatic affections.
The establishments are good.

Szliacz.—Hungary, comitat of Sohl, near Kremnitz; 1,200 feet above sea-level.
Eight ferruginous effervescing springs, 25° to 80° F.
Diuretic, tonic, and fortifying.
Special indications: Anæmia, menstrual disorders, atony.
Well-appointed and well-frequented; beautiful situation.
Doctor: Dr. Hasenfeld.

Szmrdak.—Austria-Hungary, comitat of Oberneuthra, near Szenitz.
Sulphurous waters.
Special indications: Chronic skin diseases, paralysis, hæmorrhoids, disorders of liver and spleen, hypochondriasis, hysteria, scrofulous and rhachitic affections, dyscrasia, chronic blenorrhœa.
The situation is picturesque, and the concourse of visitors large.
The two bathing establishments are thoroughly well fitted up. The large Kurhaus, built in the style of an old feudal castle, affords very excellent accommodation. The establishments are situated in an extensive English-like park.

Szobrancz.—Hungary, near Kaschau.
Saline sulphurous waters, 35° F.
In a very picturesque position, with mild salubrious air and well-appointed arrangements. The list of visitors is composed principally of Russians, Moldavians and Hungarians of the upper classes. The whole place is one large park.

Szolyva.—Austria-Hungary, comitat Bereg, near Munkács.
Effervescent alkali-saline waters. The water is largely exported.

Szombat-Falva.—Austria-Hungary, Transylvania, 2 hours from Udvarhély.
Earthy ferruginous and sulphurous waters, the latter of a temperature of 212° F.
The site is picturesque, but little has been done to bring these excellent waters into notice.

Sztubicza.—Austria-Hungary, comitat of Agram.
Inert waters, 140°.
Special indications: Chronic, nervous and skin affections, rheumatism, gout, lymphatic and scrofulous disorders.
The establishment is incomplete.

Szulin.—Hungary, comitat of Saros.
Cold ferruginous and effervescing lime waters, 7 grammes of mineralisation, of which 3 grammes are bicarbonate of soda, and 2 grammes chloride of sodium.

Tabbiano.—Italy, near Piacenza, and Parma.
Sulphurous waters, 60° F.
There is a small establishment.

Talamonaccio.—Italy, Tuscany, Valley of Osa.
Saline ferruginous sulphurous springs, 80° F.
Special indications: Lymphatic affections, rheumatism.
There are some ruins of old bathing establishments, but the waters are now used only at the spring.

Talloires.—France, Department of Haute Savoie, on the banks of the lake of Annecy.
Cold sulphurous waters.

Tambach.—Germany, Saxe-Coburg, near Gotha.
Climatic station in summer.
One of the most picturesque of the valleys of Thuringia. The surrounding country is covered with pine forests.

Tannenbrunnen.—Germany, Bavaria, near Haagard Ampfing.
Indifferent waters.
The establishment is good, and affords comfortable accommodation.

Taormina.—Italy, Sicily, 30 miles from Messina; 1 mile by carriage, and ¼-mile by footpaths.
Winter station, and one of the most interesting places in Sicily.
The *Hôtel Belle Vue* truly deserves its name, and will be found very clean and comfortable. G. Kellermann, Proprietor.

Tapolcza.—Austria-Hungary, near Miskolcz.
Inert waters, 70° F.
There is a small establishment.

Tapolcz-Bisztra.—Austria-Hungary, comitat of Zemplén.
Effervescent ferruginous waters.

Tarascon.—France, Department of Ariège, near Foix, Ax and Ussat.
Bicarbonate of iron waters.

Tarasp.—Switzerland, Engadine; 4,500 feet above sea level.
Mild alkaline and ferruginous waters, climatic station.
Special indications: Affections of the bladder.
Doctors: Drs. Killias and Pernish.

Tardon.—Spain, near Sevilla.
Sulphate of magnesia waters.

Tarna.—Austria-Hungary, comitat of Ugocza, near Kiralyhaza.
Ferruginous waters.
Small bathing establishment.

Tatenhausen.—Germany, Westphalia, near Bielefeld.
Indifferent waters; mud baths.
Special indications: Nervous affections.
The establishment is good and the accommodation comfortable and moderate in price.

Tatzmannsdorf.—Austria-Hungary, near Güuz and Steinamauger.
Bromurated earthy, ferruginous waters; climatic station.
A well-appointed establishment, beautifully situated.
Well-frequented station; mild and equable air.
Doctor: Dr. Thomas.

Tegernsee.—Germany, Bavaria, near Holzkirchen; 2,490 above sea level.
Climatic mountain air station, with whey cure.
The site is charming and the accommodation comfortable.

Teignmouth.—England, on the Channel Coast.
Sea baths, smooth sandy beach; climate bracing.
Doctor: Dr. Lake.

Teinach.—Germany, Würtemberg, in the Black Forest, near Wildbad.
   Cold earthy alkaline waters.
   Special indications: Nervous debility.
   The situation is charming and residence agreeable.
   Doctor: Dr. Wurm.

Telese.—Italy, Province of Benevento, on the line from Naples to Foggia carriages and omnibus from the station to the establishment; 1,900 feet above sea level.
   Sulphurous waters, 68° F., containing much sulphuretted hydrogen gas. Used externally and internally.
   Seasons: June—September.
   Special indications: Herpes, affections of the glandular system, rheumatism and mercurialism.
   This place is much frequented and during the season, a special service of fast trains runs between it and Naples.
   The establishment is fitted with all modern requirements. Visitors and patients generally prefer, however, staying at San Salvatore Telesine, a fine village in a very picturesque position, quarter of an hour by carriage from the establishment. There are comfortable hotels with pensions, from 10 fcs. upwards a day. Beautiful walks and excursions.
   Post and Telegraph Office.
   Doctor: Dr. Abbamondi.

Telgart.—Austria-Hungary, Comitat of Gran.
   Alkaline-ferruginous waters.
   Special indications: Constipation, hypochondriasis, hysteria.
   Little accommodation, and the establishment is incomplete.

Temburg.—England, Worcestershire.
   Cold saline waters.
   Special indications: Scrofula.

Tenby.—Wales, near Pembroke.
   Chalybeate springs; tonic and fortifying action, sea baths, sandy beach.
   Special indications: Anæmia.
   Doctors: Drs. Chater and Dyster.
   Hotels: The *Albion*.

Teneke.—Austria-Hungary, comitat of Südbihar, near Grosswardein.
   Effervescent ferruginous waters.
   The small establishment is good, and the accommodation it affords satisfactory.

Tennstädt.—Germany, Prussia, Province of Saxony, near Langensalza.
   Sulphate of lime waters, 30° F.
   Special indications: Rheumatism and skin diseases.
   The establishment is efficient, but the country is uninteresting.

Tenos.—Greece, island in the Archipelago.
  Alkaline waters.
  There are no arrangements for visitors.

Teplitz.—Austria, Moravia, ¼-hour from Weisskirchen.
  Effervescent-ferruginous waters; climatic summer station.
  The air in the valley of Teplitz is very mild and steady. The scenery around is picturesque. The establishments are good and the accommodation comfortable and moderate. A quiet place of residence. It is much frequented.

Teplitz-Schönau.—Austria, Bohemia. 700 feet above sea level; 17,000 inhabitants. For routes, see table.
  Alkali-saline waters. The temperature of the several springs ranges between 95° and 125° F. The coolest is the Steinbath, and the hottest the Stadtbath. Refrigerating basins enable the waters to be used at any temperature.
  Season: 1st May to end of September, but open all the year.
  Number of visitors in 1883, 32,625.
  Special indications: Rheumatism, gout, neuralgias, paralysis, incipient spinal complaints, scrofulous tumours and sores, sequelæ of severe wounds, *fractures, anchylosis, and affections of the tendons.*
  Teplitz-Schönau, the oldest of the Bohemian baths, is a thermal mineral bathing station of first importance. It lies in a valley of the Erz mountains, by which it is sheltered from north winds. Against the hot south winds, the Mittel mountains provide efficient protection. The parks and gardens are magnificent, while the walks and drives offer every variety of scenery and vegetation. The excursions into the Erz and Mittel mountains are very interesting.
  The bathing establishments contain arrangements for mud baths. This mud is applied in most of the above-mentioned complaints in conjunction with thermal water baths. The thermal waters are applied in sitz baths, with very satisfactory results.
  Percussion, jet, shower and douche baths.
  The bathing establishments: Stadtbad, the Kaiserbad (with very luxurious arrangements), and Steinbad, belong to the municipality of Teplitz. The Schlangenbad is the property of the adjoining Schönau. The cure is taken in summer and winter, and with the exception of the Schlangenbad, the bathing establishments offer very comfortable accommodation. Teplitz has a very fine Cursaal, a magnificent theatre, and a municipal band of musicians; in Schönau a military band gives concerts several times a week. There is an elegant reading room in Teplitz, and one also in Schönau. Churches of various denominations, and two railway stations.
  Post: Schlossplatz, No. 2./60.
  Telegraph Office: Waisenhausgasse, No. 4./736.

For detailed information, apply to the Bath Inspector in Teplitz, and to the Mayor in Schönau.

There are a great many cafés, restaurants, confectionaries and wine-rooms; villas and private apartments.

Doctors: Drs. Seiche, Eberle, Ficker, Krauss, Von Krajewski, Musil, Saumely and Wantuch.

Hotels: *Grand Hôtel de la Poste, Hôtel Roi de Prusse, de Londres,* both 1st class. *Hôtel Habsburg,* good 2nd class.

Tercis.—France, Department of Landes.

Saline sulphurous springs, 80° F.

Special indications: Rheumatism, skin diseases, and disorders of the bladder.

Terme Luigiane.—Italy, Calabria, Province of Cosenza, on the line from Naples to Cosenza.

Two sulphurous (212° F.) and ferruginous (70° F.) springs.

Special indications: Rheumatism, skin diseases, anchylosis, contractions, anæmia, chlorosis, difficult menstruation.

Establishments not good; the accommodation is of a primitive kind.

Termignon.—France, Savoy.

Cold saline waters.

Slightly purgative.

Termini Castroreale.—Italy, Sicily, Province of Messina.

Ferruginous sulphurous waters, 90° F., slightly effervescent.

Special indications: Affections of the respiratory organs, uterine diseases, rheumatism, gout, herpes.

There is an establishment, and the town is gaining in popular favour.

Termini Imerese.—Italy, Sicily, Province of Palermo, 22,000 inhabitants.

Sulphate of sodium water, 115° F.

Special indications: Skin diseases and rheumatism, liver, kidney, vesical and intestinal disorders.

There is a good establishment.

Ternaut.—France, Puy de Dôme.

Cold ferruginous bicarbonate waters.

Special indications: Feeble digestion.

Terrasse (La).—France, Department of Isère, near Grenoble.

Sulphate of calcium waters.

Terrau.—France, Department of Cantal.

Cold ferruginous bicarbonate waters.

Teruel.—Spain, Aragon, on the Alhama.
Inert waters, 80° F.
The establishments are quite insufficient, and are almost entirely neglected.

Testa.—Italy, Province of Siena, valley of the Paglia.
Ferruginous effervescent **waters, 110° F.**
Known to the Romans **as the "Balnea Clusina,"** but now almost abandoned. The **establishments are not good.**

Teste (La).—France, Department of Gironde, near Arcachon.
Sea baths; sandy beach; mild climate, but heavy surf.

Teufen.—Switzerland, Canton of St. Gall; 2,610 feet above sea level.
Sulphurous waters.
Climatic station.
There are three bathing establishments, all alike antiquated.

Thalgut.—Switzerland, Canton of Berne, half-hour **from** Thun; 1,650 feet above sea-level.
Cold ferruginous sulphurous **waters.**
Special indications: Rheumatism **and nervous** affections.
The **establishment is** fairly good.

Tharandt.—Germany, Saxony, on the Weiseritz, near Dresden.
Cold ferruginous waters.
Very good arrangements; the place is much frequented.

Theissholz.—Austria-Hungary.
Earthy waters.
Very primitive arrangements, used **only by the** surrounding inhabitants.

Thera.—Greece, one of the Cyclades.
Chalybeate and sulphurous waters, 75° to 105° F.
Very potent waters, but there are no arrangements whatever for strangers.

Thermas (Las).—Spain, near Saragoza, on the frontier between Aragonia and Navarra.
Saline and sulphate of soda waters, 100° F.
Special indications: Rheumatism.
For Spain, very well appointed.

Thermia.—Greece, island in the Archipelago, anciently Dryopis.
Saline and saline ferruginous waters, 115° to 145° F.; mud baths.
No establishment, and used in a primitive fashion.

Thermopylæ.—Greece, Thessalia, also called Herculesbath.
Saline sulphurous waters, 140° to 145° F.
The arrangements are altogether behindhand.

Thesbis Spring.—Greece, Booetia, on the Helikon.
Ferruginous waters.
Renowned and much used in antiquity in cases of sterility; they are now altogether abandoned.

Theusserbad.—Germany, Würtemberg, 3 hours from Heilbronn.
Earthy waters.
The situation is romantic, but the arrangements inadequate.

Thingöe-Syssl.—Iceland, near Reikindal.
Alkaline waters, 195° F.
Not used therapeutically.

Thorp Arch, or Boston.—England, Yorkshire, West Riding, near Bramham.
Saline waters.
Very picturesque site; air and climate good.
Doctor: Dr. Seatchard.
Hotel: Crown Hôtel.

Thoune or Thun.—Switzerland, Canton Berne, on the lake of Thun; 4,600 inhabitants.
Climatic station.
Season: June—September.
Climate bracing and strengthening.
The surrounding country is covered with pine forests, and excursions are numerous. Boating and fishing can be had; the town is old and interesting, with a mediæval castle, and several good public buildings.
English church.
Banker: A. Knechtenhofer.
House and Estate Agent: A. Knechtenhofer.
Hotels: The *Hôtel and Pension Baumgarten* is situated in extensive gardens, and is a very comfortable and homelike place. J. H. Beilick, proprietor.

Thuez.—France, Department of Pyrénées Orientales, near Olette.
Sulphurous waters, 110° F.
Arrangements are imperfect.

Thurso Bay.—Scotland, Caithness, near Dunnet.
Sea baths, sandy beach; climate bracing.
Season: June—September.
Doctors: Drs. Craven and Smith.

Thusis.—Switzerland, Canton of Grisons, on the Via Mala, near Coire.
Mixed sulphurous waters.
Special indications: Skin diseases.
The establishment is old; accommodation comfortable.

Tiermas.—Spain, Province of Zaragoza, near Sos.
Sulphurous saline waters, 80° to 100° F.
 Special indications: Anchylosis, consequent upon injury, gravel, gout, rheumatism, paralysis, scrofula, menstrual derangements.
The establishment has been considerably enlarged of late, and is up to all modern requirements. The accommodation is good and cheap. There are large gardens, and the situation is charming.
Doctor: Dr. Pastor.

Tiflis.—Asia, Caucasus, Georgia.
Thermal sulphurous springs.

Tillerborn.—*See* Tönnistein.

Tissières (les Boules).—France, Department of Cantal.
Cold bicarbonate of soda waters.
Table water.

Tissington.—England, Derbyshire, near Ashbourne.
Climatic and holiday resort.

Titus (Banos de).—Spain, on the road from Barcelona to Gerona.
Thermal saline waters, 105° F.
Good arrangements.

Tivoli.—France, Department of Upper Savoy, near Evian.
Effervescing ferruginous waters.

Tobelbad.—Austria, Styria; 1,050 feet above the sea level, near Graz.
Inert waters, 80° F. A climatic station in summer.
The establishments are excellent; accommodation abundant and comfortable. The situation is picturesque, and residence very agreeable. There are large well-kept gardens and a park.
Doctor: Dr. von Kottowitz.

Tolpa.—Italy, Romagna, near Campaccio.
Effervescent ferruginous waters.
There are no arrangements whatever for bathers.

Tongeren.—Belgium, Province of Limburg, near Liege.
Cold bicarbonate of iron waters.

Tönnisstein.—Germany, Rhenish Prussia, near Neuwied and Andernach.
Bicarbonate table waters. Only used for exportation.
There is an establishment.

Tôplika.—Austria, Croatia, near Warasdin and Agram.
Several sulphate of **soda** and alkaline springs, 145° F.
Special indications: Arthritis and skin diseases.
Several establishments and mud baths.
The **Emperor Constan**tine rebuilt the thermæ after their destruction by fire in the 3rd century.
Doctor: Dr. A. Fodor.

Topolschiz.—Austria, Styria, near Schönstein.
Indifferent waters, 80° F.
The arrangements are somewhat antiquated, but the accommodation is good, and the situation very fine.

Topuczko.—Austria, Croatia, near Sissek.
Several inert springs, 115° to 140° F.
Mud baths; climatic air station.
Special indications: Arthritism, scrofula, menstrual disorders.
Good arrangement; military hospital; large parks.
Doctor: Dr. Hinterberger.

Torda.—Austria, Transylvania.
Bromo-iodurated saline waters.
Special indications: Scrofula.
Arrangements pretty good.

Toropetz.—Russia, Government of Pleskow.
Effervescent ferruginous waters.
There are **no proper** arrangements.

Torpa.—Sweden, Province of Gothenburg.
Bromo-iodurated saline waters. Exported.
A good establishment.

Torquay.—England, Devonshire, Channel Coast.
Sea baths, small beach; cool climate in summer, warm in winter.
Special indications: Consumption and rheumatism.
Very much frequented.
Doctors: Drs. Dally, Huxley, Pollard and Thompson.
Hotel: The *Imperial Hotel*, in a most magnificient position, well appointed, and **historically** interesting through the sojourns of imperial and royal personages; southern aspect and facing the sea; one of the most popular houses in the West of England; homelike. G. Hussey, manager.

Torre-Anunziata.—Italy, Naples, on the rail from Naples to Salerno.
Bicarbonate of soda and effervescent acidulated waters, 85° F. Mild and healthy climate.

Season: All the year round, but principally June—September.

Special indications: Feeble digestion, incipient hepatitis, hypochondriasis, hæmorrhoids, leucorrhœa and skin diseases.

A bathing establishment is in course of construction. Daily concerts in summer on the beach.

Torre del Greceo.—Italy, near Naples.
Natural gaseous baths.

The situation is very lonely, and the baths are consequently little used.

Torre de San Miguel.—Spain, Province of Sarragozza.
Cold sulphurous waters.

Special indications: Nervous affections, pulmonary catarrh.

Torres.—Spain, Province of Madrid, near Alcala de Henares.
Sulphurous magnesian waters, 45° F.

The establishment is fairly complete; the visitors lodge in the village; gardens, park and environs are charming.

Doctor: Dr. Jaime.

Torres-Vedras.—Portugal, near Lisbon.
Saline ferruginous waters, 130° F.

The bathing establishment affords only the most modest accommodation.

Torretta (La).—Italy, Tuscany, valley of the Nievole.
Saline waters.

The arrangements are very primitive.

Totland Bay.—England, Isle of Wight, 2 miles from the Needles.
Sea baths; fine sandy beach. Climate bracing and healthy.

Season: June—September.

The place is rapidly coming into fashion and favour.

Toulon.—France, on the Mediterranean.
Sea baths, sandy beach.

Very good arrangements; somewhat neglected latterly.

Toulouse.—France, Capital of Department of Haute Garonne,; by O. Railway, 15 hours. 89 fr. 10 c. 1st class.

An old Roman town, taken by the Visigoths and afterwards by the Saracens, full of antiquities and objects of interest. A busy commercial town, also trading more especially in fruit, vegetables, wines, truffles, and pâté de foie gras. There are likewise manufactories of cabinet work, hats, and steel goods.

OBJECTS OF INTEREST.—Cathedral and 10 other Churches, the most interesting of which is St. Severin and Jacobins; Town Hall, Lyceum, Library, Law Courts, Hospitals, Botanic Gardens, Artillery schools, Arsenal, Observatory, and Cemetery.

Hotel: The *Hôtel Tivollier*. Celebrated for its cellars of fine wines, and Perigord truffled pies. A very comfortable house. Tivollier, proprietor.

Towyn.—Wales, Merionethshire.
Sea baths; hard sandy beach. Climate mild and bracing.
Doctor: Dr. Jones.
Hotel: The *Cambrian*.

Tramore.—Ireland, **Waterford, on the Atlantic coast.**
Sea baths.

Trani.—Italy, 15¼ hours rail from Bologna, on **the Adriatic.**
Sea baths; good beach.
The establishments are indifferent.

Traishorloff.—Germany, Hessen-Nassau, near Hungen.
Saline waters.
The establishments are old-fashioned.
The situation is charming.

Travemünde.—Germany, Mecklenburg, on the Baltic.
Sea baths; sandy beach. Very fashionable.
The establishment and accommodation are good.

Trebas.—France, Department of Tarn, near Alby.
Bicarbonate of iron waters, 45° F.
Special indications: Liver complaints, **stomachic disorders, chlorosis.**
There is an establishment.

Tremblade (La).—France, Department **of Charente** Inférieure, on the Bay of Biscay.
Sea baths; a very fine sandy beach.

Tremiseau.—France, Department of Cantal.
Ferruginous bicarbonate waters.

Trenczin-Teplitz.—**Hungary,** near Schemnitz and Pressburg.
Bicarbonate of lime **and** sulphurous waters, 90° to 100° F.
Mud baths, climatic station, **whey cure.**
One of the principal Hungarian establishments.
Special indications: Rheumatism, skin diseases, affections of the respiratory organs.
Doctors: Drs. **J. N. von Heinrich,** Nagel and Ventura.
Hotels: *Establishments*.

Treport (Le).—France, Department of Seine Inférieure, near Dieppe.
Sea baths; sandy and shingly beach.
Very picturesque scenery.
Season: July—September.

Tresclaix.—France, Department of Hautes Alpes, near Gap.
Cold alkaline springs, containing silicate. (Silicate of aluminium, 0·112 gr.).

Trescore.—Italy, near Bergamo.
Saline sulphurous waters, 55° F.; mud baths.
The establishments are inferior. The site is pretty.

Triberg.—Germany, Baden, in the Black Forest; railway station on the Baden line of the Black Forest; 2,500 inhabitants, and 1,957 feet above the sea level.
Climatic station; whey and milk cure.
Season: May till October.
Triberg is mostly visited on account of its beautiful surroundings and as a centre for excursions. The small town contains many manufactories, and during the season there is a permanent exhibition in the upper town of "articles de souvenir." The air is particularly mild and bracing, owing to the pine forests on the surrounding hills.
Doctor: Dr. Feederle.
Hôtel: The *Black Forest Hôtel* is situated on a height overlooking the town and valley, within five minutes' walk of the waterfall (the Giessbach of the Black Forest). The waterfall is illuminated during the season by electric light and Bengal fires. The hotel is well managed, and exceedingly comfortable for a prolonged stay. Omnibus meets all trains. L. Bieringer, proprietor.

Trieste.—Austria, Istria, on the Adriatic, the most important Austrian seaport.
Sea baths, fine sandy beach. The mean temperature of the sea is 80° F. in June, July, and August.
The arrangements are good. The town contains many buildings and objects of interest. It is the centre of Austrian commerce, and has a naval dockyard. There is a disposition to make a winter station of Trieste, but all attempts to do so have failed so far. The "Bora" is the natural enemy. A splendid view is obtained by ascending the Optschina.
Bookseller: J. H. Schimpff, Pinzza della Borsa.
Hotels: The *Hotel de la Ville* is a first-class house, well situated on the sea. Large bathing establishment, with fresh and salt water baths; moderate prices. F. Rosslacher, proprietor.

Trillo.—Spain, Province of Guadalaxara, near Cifuentes.
Ferruginous nitrogen, saline, and sulphurous waters, 55° to 75° F.
Special indications : Arthritis, scrofula and skin diseases.
A very complete balneal establishment. The situation is fine. Good gardens and well kept promenades on the banks of the Tajo.
An hospital for the poor.
A well-frequented station.
Doctor : Dr. Taboada.

Tritoli.—Italy, near the ruins of Cumæ, Bay of Naples.
Saline waters, 140° F.; brine baths.
Used in primitive fashion.

Trois-Torrens.—Switzerland, Grisons, near St. Moritz; 2,570 feet above sea level.
Earthy ferruginous **waters, 80° F**.
The establishment is **inadequate**.

Trollière (La.)—France, Department of Allier, near St. Pardoux.
Effervescing, ferruginous bicarbonate waters, containing more carbonic acid than the waters of St. Pardoux.

Troutbeck.—England, Yorkshire, Ilkley, near Leeds.
Hydropathic establishment.
Doctor : Dr. Thomas Scott.

Trouville-sur-Mer.—Department of Calvados, near Pont l'Evêque.
Very fashionable and much frequented **sea baths**; sandy beach.
There is a Kurhaus. The **establishment, and all** appertaining to it, is of the best.
Dentist : A. Preterre.
Hotel : The *Hôtel de la Roche Noire*, a really splendid house, situated in an unrivalled position, and under new management; has been entirely refurnished. It is the only hotel with a lift. Kitchen and cellar excellent. Ebensperg and Ritz, same proprietors as of *Hotel Splendid*, in Paris.

Truskowice.—Austria, Galicia, District of Sambór.
Three cold springs of saline, ferruginous, and sulphate of lime waters; mud baths, whey cure.
Special indications : Arthritis, chlorosis, skin diseases.
Fairly arranged establishments. The accommodation is comfortable and the site picturesque.

Tschawitz.—Austria, Bohemia, near Kaaden.
Earthy alkaline waters.
There is **a small establishment, with good** accommodation

Tuebingen.—Germany, Würtemberg.
Earthy alkaline waters.
The establishment is good. There is a university, which is one of the best in South Germany. The situation is very good, and the accommodation ample, comfortable, and moderate in price.

Tüffer.—Austria, Styria, station on the railway line from Vienna o Trieste.
Thermal indifferent springs, 100° F. Drinkhall. Climatic station.
Special indications: Nervous disorders, gout, rheumatism, tendinous retractions, articular pains, and female complaints.
The situation is amongst the finest in Styria.
Season: Open all the year round.
Doctor: Dr. von Schön-Perlashof.
The Kaiser Franz Josef's baths is one of the best establishments of its kind in Austria. Theod. Gunkel, proprietor.

Tunbridge Wells.—England, Kent.
Cold ferruginous bicarbonate waters, containing 0·32 gr. of protoxide of iron.
Climate bracing and healthy.
Season: June–September.
Special indications: Anæmia, chlorosis.
Excellent arrangements.
Doctors: Drs. Ranking, and Rix.
Hotels: The *Calverley*, The *Castle*.

Tunis.—Africa.
The regency of Tunis has several thermal springs, which are in great repute with the natives.

Tür.—Austria-Hungary, Transylvania, near Blasendorf.
Bitter waters. Exported.
A difficult place to reach.
The town is situated in a plain, and owing to flooding of the rivers, is sometimes inaccessible for weeks. Bathing arrangements old-fashioned.

Turin.—Italy, capital of Piedmont.
Turin is situated between the Dora and Po rivers, is very regularly built, and does not contain a single poor-looking house.
The climate is changeable and on the whole unpleasant.
OBJECTS OF INTEREST.—The Duomo or Cathedral, Churches of La Consolata, Corpus Domini, Royal palace, Palazzo

Madama, Royal Armoury, Academy of Science, Royal Gallery of Paintings, Museum of Antiquities, Natural History, Royal University, Arsenal, Royal Theatre, Public Garden.

EXCURSIONS: To La Superga, to Nice, *via* the Col di Tenda, Stupianigi, Cuneo, and to the Chartreuse de Pesio.

Cabs: Course 1 fr.; per hour, 1½ fr.; two horses by hour, 2 fr.

Post-office: Palazzo Carignano.

Telegraph: Piazza Castello.

Bookseller: H. Loesher, 19, Via di Po.

Hotels: The *Hotel d'Europe*. A large, excellent, first-class house, facing the King's palace and in the best situation in the town; railway ticket office in the hôtel. Borgo and Gagliardi, proprietors.

Tusnád.—Austria-Hungary, Transylvania, near Kronstadt.

Saline ferruginous waters; climatic station.

The town is surrounded by very extensive pine forests. The climatic conditions are most favourable, and the establishment good.

Tuy.—*See* Caldas de Tuy.

Twér.—Russia, capital of the Government of the same name on the Tmak.

Ferruginous waters. Exported.

The arrangements are primitive, and the waters used only by the surrounding inhabitants.

Tynemouth.—England, Northumberland, at the mouth of the Tyne, on the North Sea.

Cold ferruginous waters and sea baths; fine sandy beach. Climate very mild.

Season: June—September.

Doctors: Drs. Bramwell and Gibbson.

Hotel: The *Bath*.

Ueberkingen.—Germany, Würtemberg, near Ulm.

Cold effervescing alkali-saline and ferruginous waters.

Special indications: Atony, catarrh of the respiratory organs, anæmia, rhachitis.

Pretty good establishment and accommodation. The situation charming and scenery romantic.

Ueberlingen.—Germany, Baden, near Schaffhausen; 1,233 feet above the sea level, on the north-west shore of the Lake of Constance.

Effervescent ferruginous waters, bathing in the lake; mild climate.

Ugod.—Hungary, comitat of Vesprim, near Pápa.
Earthy-saline ferruginous waters.
Special indications: Uterine affections and vesical disorders.
The establishment is antiquated; situation charming; accommodation indifferent.

Uhlmühle.—Germany, Hanover.
Cold ferruginous waters.

Uleaborg.—Russia, Finland.
Hydromineral station.
Sea baths in the Bay of Bothnia.

Ullersdorff.—Austria, Moravia, district of Ollmütz, near Hohenstadt.
Sulphurous waters, 75° F. Whey cure and climatic station.
The situation is romantic, and the surrounding country very mountainous. The establishments and accommodation are good.

Ullswater.—England, Westmorland, in the Lake District.
Hydropathic establishment, and summer holiday resort.
Doctor: Dr. MacNalty.

Undary.—Russia, Government of Simbirsk.
Effervescent ferruginous waters.
The arrangements are most primitive, and the water is used only in the immediate neighbourhood.

Unterbad.—Switzerland, Canton of Appenzell, on the Sitter; 2,130 feet above sea level.
Earthy alkaline waters; whey cure.
The situation is pretty, but the establishment leaves room for improvement.

Unterhallau.—Switzerland, near Schaffhausen; 1,530 feet above sea level.
Sulphurous waters.
The bathing establishment is old fashioned, and the accommodation not good.

Unter-Micsinye.—Austria-Hungary, near Neusohl.
Earthy ferruginous waters; exported.
The arrangements are intended to facilitate exportation only.

Unterseen.—Switzerland, Canton Berne.
Whey cure station.

Upton.—England, on the south coast.
Sea baths, and sandy beach, with very mild climate and sheltered position.

Urbalacone.—France, Corsica, near Ajaccio.
Sulphurous arsenical waters, 90° F.
Special indications: Phthisis of the larynx, skin diseases.

Urberoaga de Alzola.—Spain, Province of Guipuzcoa, near Plasenzia.
Bicarbonate of lime waters, 75° F.; also two ferruginous springs.
Special indications: Affections of the uterus and urinary organs, anæmia, chlorosis, spleen and liver complaints.
The establishment is complete, and contains in its newer portions assembly-rooms, ball, concert and billiard rooms. The accommodation is good and moderate.
Doctor: Dr. de la Pedruesa.

Urberoaga de Ubilla.—Spain, Biscaya, near Villa de Marquina.
Sulphurous alkaline waters, 82° F. Sitz, steam, shower, and spray baths. Inhalations.
Special indications: Catarrh and other affections of the mucous membranes, laryngitis, pharyngitis.
The establishment is elegant and the accommodation comfortable and moderate in price. Scenery picturesque.
Doctor: Dr. Jimenez de Pedro.

Uriage.—France, Department of Isère, near Grenoble; 1,250 feet above sea level. For routes see table.
Saline sulphurous waters, 81° F. There is also a muriatic ferruginous spring.
The former, the important water of the place, contains 7 grammes of chloride of sodium, and 0·0113 grammes hydrosulphuric acid per litre. The waters likewise contain a considerable quantity of lime, soda and magnesia, with an appreciable percentage of arsenic.
Season: From 15th May till 15th October.
The thermal establishment, situated in one of the most picturesque valleys of the Dauphinese Alps, 400 m. above sea level, is one of the best in Europe, and can accommodate 2,000 bathers at a time; from 60,000 to 70,000 baths or douches are administered here annually. Taken to the amount of from three to six glasses per diem, the water acts as a purgative; in from one to two glasses, it is alterative, and stimulates the digestive functions; used externally in baths, douches, &c., it acts as a tonic and reconstituent; in the form of spray it is frequently used with great success in chronic affections, especially those of the eyelids (blepharitis), of the pharynx (granular pharyngitis), and of the face (eczema, acne, &c.).

The *special indications* for the use of the waters are very numerous. They are employed with great success in scrofula and its various manifestations, especially in children; in chronic affections of the skin, notably eczema, impetigo, acne, &c.; in chronic rheumatism, in nervous affections, in convalescence after severe illnesses, long nursing, and in various pathological conditions resulting from anæmia and debility. Their efficacy has also been undeniably established in a certain class of cases which are met with chiefly among young girls. With such it is often necessary to encourage or regulate the menstrual functions. At the period of the menopause in elderly persons they are also highly beneficial, especially in cases where derivative action is desirable in order to prevent cutaneous or other localisations of disease, which either the family history or an acquired taint, render probable. In cases of incipient consumption, they often arrest the progress of the disease, by modifying the whole system; also in constitutional syphilis, and especially in its tertiary and more inveterate forms they constitute an active auxiliary to mercurial treatment.

The baths are situated in the midst of a large and magnificent park. Ample accommodation may be found in the hotels and furnished villas. Milk and whey cure.

The walks and excursions are numerous, and the scenery most picturesque. Concerts every day; balls every Thursday and Sunday. Theatre; Casino.

Post and Telegraph. English Church. Carriages, horses and mules for excursions.

Doctors: Drs. Doyou (medical inspector of the baths) and Tenlon-Valio.

Hotels: Are all under the supervision of the Bath Committee.

Information from the Director of the Establishment free of charge.

Urnäschen.—Switzerland, Canton of Appenzell; 2,514 feet above sea-level.

Alkaline earthy waters; whey cure.

The site is romantic and the establishment fairly good.

Ussat.—France, Department of Ariège, near Foix; 1,300 feet above sea-level. For routes see table.

Saline and sulphurous waters, 80° to 100° F.

Special indications: Affections of the uterus, with a neuropathic tendency; neuroses.

The scenery is mountainous, the establishment very good, and the accommodation abundant and comfortable.

Usson.—France, Department of Ariège.

Sulphate of soda waters, 50° to 75° F.

Ustrom.—Austria, Silesia, near Teschen.
Hydropathic and whey cure station.
The establishment is thorough, and the accommodation comfortable.

Utzera.—Russia, Caucasus, near Kutais.
Effervescent ferruginous waters.
The arrangements are very incomplete.

Uzsok.—Austria-Hungary, comitat of Ung, on the frontier of Ung.
Ferruginous waters.
A small but well-arranged establishment.

Vacia-Madrid.—Spain, near Madrid, on the borders of the Tarama.
Purgative waters.

Vaisse.—France, Department of Allier, near the thermal establishment of Vichy.
Composition of the waters similar to that of Vichy; they are more effervescing.

Val-André (Le).—France, Department of Cotes-du-Nord, *via* Lamballe.
Sea baths; sandy beach.

Valdeganga.—Spain, Province of Cuenca, and near that town.
Effervescent ferruginous waters, 60° F.
Special indications: Anæmia, chlorosis, dyspepsia, intestinal obstructions, hyperæmia of liver.
The establishment is bad, everything being in a backward condition.
Doctor: Dr. Yebenes.

Valdieri.—Italy, Piedmont, Province of Coni; 4,100 feet above sea-level.
Several hyperthermal sulphate of soda springs, 180° F.
Special indications: Skin diseases, rheumatism, old sores, ulcers.
The establishment is incomplete, the site mountainous, but the climate rather humid. The garden and parks are extensive. Accommodation good and moderate in price.

Valdorf.—Germany, Westphalia.
Cold sulphate of lime waters.

Vale-Szkragye.—Austria-Hungary, comitat of Mármaros, near Felsö-Visső.
Sulphurous ferruginous waters; alum waters.
This place would amply repay more attention; at present the water is only used in the immediate neighbourhood.

Valencia.—Spain, capital of Province of the same name, on the Guadalaviar, about one mile from the Mediterranean.

Winter station and sea baths. The beach is sandy.

As a sea bathing station Valencia is one of the most important in Spain.

As a winter station, owing to the marshes in the vicinity, it is less frequented than its climate merits. Opinions vary, however, considerably as to this. The sky is almost always clear, the rainy days amounting to only 38 in the year. The mean winter temperature is 57° F., and never falls below freezing point. There is a heavy dewfall in the evening; snow is extremely rare. Relative humidity of the air, 66 %. The drinking water contains a good deal of lime, as on the Riviera. The autumn months is the season for sea baths. Drainage comparatively good, being aided by the river.

Doctor: Dr. Serrano.

Hotels: De la Ciudad de Madrid.

Valenza.—Italy, Piedmont, Province of Alexandria.

Sulphurous iodurated waters.

Vale-Vinului.—Austria-Hungary, Transylvania, comitat Bistritz-Naszod, near Ródna.

Alkali-saline waters.

Very romantically situated.

Valle de la Cueva.—Spain, Province of Madrid.

Cold sulphate of soda waters; nine grammes of sulphate of sodium.

Valle de Rivas.—Spain, Province of Gerona, near Bruguera.

Bitter and ferruginous effervescent waters, 45″ F.

Neither the establishment nor the accommodation encourage visitors to remain.

Doctor: Dr. Mir.

Vals.—France, Ardèche, near Aubenas.

Various cold alkaline springs, which may be classed as weak, medium, and strong, according to the amount of mineralisation.

Principally used as table waters.

The exportation of Vals water reaches several million bottles annually.

Special indications: Gravel, liver complaints, and disorders of the spleen.

Doctor: Dr. Charvet.

Vals.—Switzerland, Grisons, near Peiden; 2,450 feet above sea level.

Ferruginous waters, 70° F.; very little used.

Varennes.—France, Department of Maine and Loire, near Angers.
  Ferruginous arsenical waters.

Várgéde.—Austria-Hungary, comitat of Gömör.
  Effervescent ferruginous waters, 45° F.
  The establishment is an old-fashioned one.

Vaugnières.—France, Department of Drôme, Arrondissement of Die.
  Mixed bicarbonate of iron, and iodurated waters.

Vaxholm.—Sweden, near Stockholm.
  Sea baths, with sandy beach.
  A favourite resort with the inhabitants of Stockholm.

Veierbach.—Germany, Baden, near Offenburg.
  Cold ferruginous bicarbonate waters.
  Establishment very incomplete.

Veldes.—Austria, Carinthia.
  Climatic station; hydropathic establishment, with whey cure.
  The establishment is new and thoroughly well fitted up.
  The cure consists chiefly in the influence of sun and air on the skin, the whole body being uncovered during the time the patient is undergoing treatment; patients walk in the park as lightly clad as in a Turkish bath. The site is charming and quiet; the air pure and healthy. Good fishing and hunting.
  Doctor: Dr. Germonik.
  Hotel: *Mallner's*.

Velejte.—Austria-Hungary, comitat of Zemplén.
  Alkali saline waters.
  A bathing establishment with large parks.

Vellebro (Acqua di).—Italy, Province of Bari, stop at Fasano, between Foggia and Brindisi.
  Earthy alkaline waters.
  Special indications: Constipation, hepatic complaints, vesical calculus and gravel.
  Primitive establishments, with inadequate accommodation.

Velleminfroy.—France, Department of Haute Saône, near Vesoul, station of Creveney.
  Cold sulphate of lime and magnesian waters; two grammes of mineralisation.
  Special indications: Affections of the uterus and bladder.

Velleron.—France, Department of Vaucluse, near Carpentras.
Bicarbonate of soda water.

Velmont.—France, Department of Seine Inférieure, near Fecamp.
Cold ferruginous waters.

Venafro—Italy, Province of Naples, near Teano, on the Volturno.
Saline waters. Exported. Used only in the locality as drinking waters. The arrangements are very imperfect.

Venice.—Italy, Province of Venetia, a seaport on the Adriatic.
Sea baths and winter station; sedative climate.
The mean winter temperature is 45° F. The climate is somewhat variable, and windy days are frequent. Mean barometric pressure, 760·07 $^m/_m$., with daily changes ranging from 2·23 $^{in}/_m$; mean relative humidity of the air, 75½ °/₀ ; 50 rainy and snowy (five of latter) days during season are recorded. Fogs are very rare. The climate of Venice may thus be said to assimilate somewhat to that of Pau and Pisa, but chiefly as regards atmospheric moisture. The drinking water is not good.
The most favoured and most sheltered quarters are the Riva and the Zattere. Here patients can enjoy outdoor exercise without being obliged to hire a gondola. The choice of an apartment, &c., should, however, always depend on the advice of a resident medical man.

English Church : Palazzo Contarini San Trovaso.
Optician and Photographs : Carlo Ponti, 52, Place St. Marco.
Chemist : Zampironi, St. Moisé.
Bookseller : Münster, Piazza San Marco.
Bankers : S. & A. Blumenthal & Co., on the Traghetto.
Doctors : Drs. Kurz, Keppler, Lewi and Minich.
Hotels : Danieli's *Royal Hotel*, first, both by its position and management.
The *Grand Hôtel*, splendidly situated on the Grand Canal; magnificent apartments for families; private gondolas. Warms and Melano, managers.
The *Hôtel d'Angleterre*, situated full south on the Quay des Esclavons, near St. Mark's Square. Reduced prices for a prolonged winter stay. Modest and comfortable. F. Venturini, proprietor.

Ventnor.—England, Isle of Wight.
Sea baths; sandy and pebbly beach; climate mild and steady.
Very fashionable.
Seasons : June—September, and November—February.
Doctor : Dr. Bell.
Hôtel : The *Royal*.

Vernet (Le).—France, **Department** of Pyrénées Orientales, near Prades; 1,900 feet above sea level.
    Mountain air cure, **and** winter station.
    Sulphurous saline waters, 45° to 140° F.
    Eleven springs, two establishments.
    Special indications: Rheumatism, scrofula, syphilis and pulmonary affections.
    The establishments are elegant, and the accommodation comfortable. The situation is charming, and the park and gardens extensive.

Verrazano.—Italy, Tuscany, valley of the Teverino.
    Effervescent alkaline waters, 50° F. Little used.

Vescovo (Bagno del).—Italy, near Grosseto, Val d'Ombrone.
    Bitter waters, slightly mineralised, 70° F.
    There is a small and incomplete establishment.

Vetzel.—Austria-Hungary, Transylvania, Valley of Kalamar.
    Alkaline waters, 50° F.
    There are no arrangements to accommodate visitors.

Veules.—France, Department of Seine Inférieure, near St. Valéry.
    Sea baths; sandy beach.

Veulettes.—France, near Dieppe.
    Sea baths; sandy beach. Air bracing.
    Season: July—October.
    Residence here is quiet.

Vevey.—Switzerland, Canton Vaud, on the Lake Leman.
    Air cure station.
    Bookseller: Jacot, Guillarmod, 18, Rue du Simplon.
    Doctor: Dr. Muret.

Veytaux.—Switzerland, on the north-east shore of Lake Leman, near Montreux.
    Climatic station; chiefly visited as a station of call.
    Season: Spring and Autumn.

Vialla.—Italy, Tuscany, near Firenzuola.
    Sulphurous waters, 40° F.
    The establishment is incomplete.

Viareggio.—Italy, between Spezia and Leghorn.
    Sea baths; sandy beach; climate very mild.
    Season: August—October.
    Coming into favour as a winter station, though rather foggy.
    Doctor: Dr. Hirschl.
    Hotel: De la Grande Bretagne.

Vicarello.—Italy, near Rome.
   Alkali-saline waters, 112° F.
   These waters are supposed to be the Apollinaris waters of the ancient Romans. Arrangements poor, and very much neglected.

Vicarsbridge.—Scotland, near Dollar.
   Sulphate of iron waters.
   Some of the richest iron waters known.

Vichy.—France, Department of Allier. For routes see table.
   Several bicarbonate of soda alkaline springs, ranging from 35° to 105° F.
   Three of the springs are ferruginous. Vichy may be taken as the type of alkaline waters generally. Any disease due to excess of acidity is benefited by waters such as those of Vichy.
   Special indications: Dyspepsia, splenic and hepatic disorders, uric acid diathesis, catarrh of the urinary organs, diabetes.
   Vichy has been well called "the Queen of thermal watering places;" it is visited by as many as 40,000 persons yearly.
   Bookseller: Vve. César, Passage du Casino.
   Doctors: Drs. Durand-Fardel, Souligoux and Nicolas (Rue de Nimes).
   Hotels: The *Hôtel Mombrun and du Casino*, first-class house, situated in the park, facing the casino and baths. M. Giboin, proprietor.
   The *Nouvel Hotel* is exceptionally the best appointed, and is a first-rate house in all respects. Situated on the park near the bathing establishment. Large central garden and lift. Mr. and Mrs. Gentil, managers.

Vic-le-Comte.—France, Puy du Dôme, near Isoire.
   Ferruginous waters.
   Used only in the immediate neighbourhood as table waters.

Vicoigne.—France, Department of Nord.
   Cold sulphate of sodium waters.

Vic-sur-Cer.—France, Department of Cantal, near Aurillac.
   Bicarbonate of iron waters.
   Doctor: Dr. Soubeyran.

Victoire.—Italy, near Aosta, Val d'Entréves; 3,800 feet above sea level.
   Effervescent ferruginous waters.
   Very mild climate.
   The arrangements are not satisfactory.

Victoria Spa.—England, Warwickshire.
   Sulphate of sodium waters.
   Special indications: Rheumatism, dyspepsia, liver complaints.

Vidago.—Portugal, near Oporto.
Alkali-saline and ferruginous waters.
Season: September—November.
The establishments and arrangements generally are modern, and provided with all the latest and most scientific improvements. Trout fishing can be had. The situation is picturesque, and residence pleasant.

Vienna.—Capital of Austria, from Paris, *viâ* Strasburg, Munich, and Passau, 48¼ hours. 190 fr. 1st class.
The town is situated near the southern bank of the Danube, and is of very ancient origin. It has been the residence of the German Emperors from the 13th to the 19th centuries, and is full of historical interest.
It was unsuccessfully besieged by the Turks in 1683.
OBJECTS OF INTEREST.—The Cathedral of St. Stephen, Churches of the Capucines, Latin, Carmelites, St. Michael, St. Charles Borromeo, Imperial Palace or Burg, Imperial Library, Cabinet of antiquities, Cabinet of minerals, Museum, Coach-house and riding school, Imperial Jewel office, People's garden, Picture gallery, the Lower Belvedere with Ambras collection of armour, Esterhazy gallery, Liechtenstein gallery, Imperial arsenal, Town arsenal, University, General and Lying-in Hospitals, Opera-house, and Hofburg Theatre, the Prater, the Ring Strassen, Dreher's and Geyer's beer-houses, Aquarium and Exhibition Palace of 1873, many private gardens.
Amateurs of the bath will be amply repaid for travelling any distance by a visit to the Turkish baths of Dr. Von Heinrich.
The Royal Baths are well worth inspection, as being very complete and approaching more nearly to the ancient Roman Baths in luxury than any other in Europe.
EXCURSIONS Schönbrunn, Hütteldorf, Hietzing, Brühl Laxenburg, Mödling, Baden, warm springs; Leopoldsberg, Kahlenberg, and Klosterneuburg.
CABS: Half hour or less, 1 fl., and each succeeding ¼ hour, 50 kr.; one-horse carriage, ¼ hour, 50 kr., and not exceeding ½ hour, 60 kr.; each succeeding ¼ hour, 20 kr.
Post-office: Alte Postgasse, 10.
Telegraph: I. Börsenplatz 1; Hotel National.
Doctors: Drs. Bamberger, Nothnagel, · von Heinrich, Bettelheim and Kaposi.
Hotels: Imperial, Golden Lamb, Grand, Metropole, and National.

Vignale.—Italy, Piedmont.
Cold saline sulphurous waters.

Vignolles.—France, Department of Vienne.
Cold saline waters.

Vignone.—Italy, Tuscany, Siena, near San Quirico.
Ferruginous alkaline waters, 100° F. Natural steam baths.
The establishments are good, as is also the accommodation.
Vegetation luxurious, and situation charming.

Vihnye.—Austria-Hungary, near Schemnitz.
Earthy, ferruginous effervescent waters, 100° F. Climatic station in summer. Very mild climate.
The establishment might easily be improved. Accommodation indifferent. The situation is fine and well sheltered. Vegetation luxurious. An extensive park has recently been laid out.

Villacarillo.—Spain, Province of Jaen.
Sulphurous waters, 30° to 45° F.

Villa delle Caselle.—Italy, near Arezzo.
Alkaline waters, 55° F.
There is nothing here except the spring.

Villa d'Este.—Italy, Lake of Como, near Cernobbio.
Hydropathic establishments; arranged to admit a limited number of indoor patients.

Villafafila.—Spain, Province of Zamora.
Nitrate of potash to the extent of 1·80 gr. is found in these waters.

Villaharta.—Spain, Province of Córdoba, near Alhondiguilla.
Acidulous bitter waters containing iron.
Special indications: Chlorosis, anæmia, menorrhagia, amenorrhœa, dysmenorrhœa, vesical catarrh.
The site is very beautiful. The establishment is situated in the midst of large gardens. Accommodation abundant.
Doctor: Dr. Illescas.

Villar del Pozo.—Spain, Province of Cindad Real, near that town.
Acidulous ferruginous waters, 70° F.
Neither the establishment nor the accommodation offer any attractions.
Doctor: Dr. Ocaña y Pazos.

Villaro.—Spain, Province of Bizcaya, near Bilbao.
Sulphurous earthy waters.
Special indications: Skin diseases, lymphatism, syphilis, pharyngeal and laryngeal disorders, vesical and uterine catarrhs.
The site is very picturesque, and the establishment is surrounded by a large garden and very complete. The accommodation is good and prices low.
Doctor: Dr. Palacios Tomás.

Villatoya.—Spain, Province of Albacete, and near that town.
Alkaline waters, 70° F.
Special indications: Rheumatism; **one spring is** used in cases of parasitical affections, diabetis, scrofula.
The establishment is well fitted, and accommodation suitable to all purses.
Doctor: Dr. Garcia Teresa.

Villavieja de Nules.—Spain, Province of Castellon de la Plana, six hours from Valencia.
Sulphate of magnesia waters, 75° to 110° F.
Special indications: Rheumatism, neuralgia, paralysis.
A well-frequented and well-appointed bathing establishment. The hotels are good, and the accommodation within the reach of all. There are assembly, reading, smoking, ball and billiard-rooms.
Doctor: Dr. Barraca.

Villefranche.—France, Department of Alpes Maritimes, near Nice.
Sea baths and winter station; little frequented, but in a very sheltered position.
The temperature is in the mean 5° higher in winter than at Nice and Cannes.

Villefranche.—France, Department of Aveyron.
Alkaline waters, 27° F.

Villeneuve.—Switzerland, Canton Vaud, on the Lake of Geneva, seven miles from Vevey, near the Castle of Chillon.
Climatic station.
Season: All the year round.
English Church: Service at the Hôtel Byron.
Hotel: The *Hôtel Byron*, splendidly situated on the hills, is a first-class house with a fine garden and extensive farm connected with it. Here Byron composed his "Prisoner of Chillon." Exceptionally reduced tariff for winter season; suitable for large families, where children can have the best masters. Address the manager.

Villers-sur-Mer.—France, near Trouville.
Sea baths; fine sandy beach. Climate bracing.
Season: May—September.

Villerville.—France, near Deauville.
Sea baths; fine sandy beach. Climate bracing.
Season: May—September.
The scenery is picturesque, and Villerville is rising rapidly in public favour.

Vilo or Rosas.—Spain, Province of Malaga, near Colmenar.
Bitter and ferruginous effervescent waters, 50° F.
Special indications: Skin diseases, anæmia, chlorosis, glandular swellings, hyperæmia of liver and spleen.
The construction of an establishment with an hotel is in contemplation. Until this is finished, residence is scarcely possible.
Doctor: Dr. Alvarez.

Vilsbiburg.—Germany, Bavaria.
Bicarbonate of lime waters. There is a small bathing establishment with good accommodation.

Vinadio.—Italy, Piedmont, near Coni.
Saline and sulphurous waters, 40° to 150° F. Diuretic and purgative action. In baths they are very invigorating and tonic.
Special indications: Rheumatism, and vesical disorders.
A good establishment. The accommodation is comfortable; the society very select.

Vinca.—France, Department of Pyrénées Orientales, on the railway from Perpignan to Prades.
Saline sulphurous waters, 55° F.
Special indications: Rheumatism and skin diseases.
The establishment is good, and the situation picturesque.

Vippach-Edelhausen.—Germany, Saxe-Weimar, near Weimar.
Saline waters.
There is an establishment, but little has been done to develop the resources of the place.

Visibachbad.—Switzerland, between Schaffhausen and Baden.
Inert waters, 45° F.
The site is fine. There is a small establishment.

Visk.—Hungary, comitat of Marmarós, near Tecsö.
Ferruginous bicarbonate of soda waters.
Special indications: Gravel, dyspepsia.
The site is romantic—on the slope of a hill in the Theiss valley—and the establishment excellent. Much frequented.

Visone.—Italy, district of Rivalto d'Acqui.
Sulphurous waters, 60° to 75° F.
The site is unhealthy, and the establishment old-fashioned.

Visos.—France, Department of Hautes Pyrénées.
Sulphate of lime waters.
Special indications: Old sores and wounds.

Viterbo.—Italy, Province of Rome.
Sulphurous thermal waters, 145° F., **and cold** ferruginous waters.
Special indications: Rheumatism, skin **diseases,** and **syphilis.**

Vitry-sur-Marne.—France, Department of Marne.
Ferruginous waters.

Vittel.—France, Vosges. For routes see table.
Alkaline ferruginous waters, 30° F.
Laxative action.
Special indications: Dyspepsia, vesical disorders, gout, gravel.
Well-appointed establishment.
Doctor: Dr. Bouloumié.

Vizella.—Portugal, Province of Minho, five miles from Guimaraes.
Chalybeate and sulphurous waters, 110° F.
Season: June—September.
Special indications: Skin diseases, gout, scrofula, rheumatism, old sores.
The establishment is situated in an extensive park, and is well arranged, and the town is held in much favour. The scenery is grand.

Voltaggio.—Italy, Province of Nori, on the Morcione.
Sulphurous waters, 70° F.
Formerly very celebrated, and much visited.

Volterra.—Italy, Tuscany, valley of the Cecina.
Saline waters and brine baths.
The establishment is incomplete and the situation barren.

Voltri.—Italy, on the Riviera di Ponente.
Sulphurous waters, 70° F. Little used.

Vonitza.—Greece, Arcania.
Saline waters, 90° F.
There is no accommodation, and the country around being flat and swampy, the place is little visited.

Vöslau.—Austria, Province of Lower Austria, one hour from Vienna by rail. 700 feet above sea-level. For routes, see table.
Indifferent waters 75° F.; cold and warm baths; whey and grape cure; climatic station; hydropathic establishment.
Season: 15th May—30th September; grape cure, September and October.
Special indications: Female affections; hysteria, nervous complaints, anæmia, disorders of the lower intestines, convalescence after prolonged illness.

The annual list of **strangers** shows over 4,000 visitors.

Vöslau is a very favourite spot with the Viennese (who are not included in the above number), on **account of its** remarkably pure and healthy air, and charming situation on **the vine-clad** slopes of the Vienna forest. The waters of its **springs belong** to the class of "Acrato Thermæ."

The bathing appliances consist of three large **swimming** baths, with douches and shower baths, open alternately to ladies and gentlemen. Warm and tepid waters are supplied in wooden and deep earthenware baths. Brine, resinous and extract of pine-cone baths are likewise to be had.

The characteristics of Vöslau as a "*climatic station*" are a charming, healthy and picturesque situation on the well drained borders of the "Viennese Forest," extensive pine woods by which the town is surrounded, and last, but not least, vine-clad hills. From these latter comes its celebrated grape cure in autumn.

The **Curhaus** contains large **concert rooms, billiard, card** and reading rooms; balls and private **theatricals are also given**. The orchestra plays three times daily—in the morning **at the baths, and** in the afternoon and evening in front **of the Curhaus**.

The environs of Vöslau **are** noted for their attractive walks and drives. The forest begins, so to speak, in the very town, and **is** intersected by innumerable walks, roads and paths. In fact everything is arranged in such a way as to make **a** sojourn here as pleasant as possible. The constant return of visitors who have once passed **a season** here, is the best evidence that the authorities **have succeeded in their** endeavours to please.

Villas and private apartments to suit all tastes and purses.

Doctors: Drs. Venninger, Krischke.

Hotels: *Hôtel Hallmayer, Back, Comunal, and Tägerhorn.*

Vrécourt.—France, Department **of** Vosges.
Sulphate of soda **waters**.

Wachenheim.—Germany, Palatinate.
Grape cure station.
The accommodation is good and moderate in price.

Wahlberga.—Sweden, Wermland, near the Lake **of Wenner**.
Mineral mud baths. Good establishment.

Waidhaldenbad.—Switzerland, near Basle and Rheinfelden; 1,340 feet above sea level.
Whey cure station.
The accommodation is good.

Walby-Brunnen.—Sweden, near Upsala.
Effervescent alkaline, sulphurous and ferruginous waters.
The establishments are good, and the place much visited.

Waldbad.—Germany, Württemberg, near Kloster Weingarten.
Inert waters.
There is a small bathing establishment.

Waldstatt.—Switzerland, near Appenzell; 2,410 feet above sea level.
Earthy ferruginous waters.
The bathing establishment is built of wood, but affords good accommodation.

Walton-on-the-Naze.—England, Essex, near Harwich.
Sea-baths, good beach; climate bracing and healthy.
Doctor: Dr. Hayman.
Hotel: The *Albion*.

Wangeroog.—Germany, Oldenburg, an island in the North Sea.
Sea-baths, sandy beach; climate bracing and healthy.

Warasdin-Töplitz (Kroatic-Toplitz).—*See* Töplika.

Warmbrunn.—Germany, Silesia, near Hirschberg; 1,034 feet above sea level.
Sulphate of soda waters, 90° to 100° F.
Special indications: Rheumatism, paralysis, and affections of the respiratory organs.
One of the oldest of Silesian bathing places. The scenery is very picturesque, and the gardens, parks, &c., are scrupulously kept.
Establishments efficient and accommodation comfortable.
Doctors: Drs. Franz, Herzog and Höhne.
Hotels: De *Prusse*, *Thomas*, *Victoria*.

Warm Springs.—United States of America, Arkansas.
Thermal springs, used against rheumatism by the Indians.

Warm Springs.—United States of America, Georgia.
Thermal waters, 80° F.
A well frequented station.

Warm Springs.—United States of America, Virginia.
Thermal effervescent alkaline waters, containing magnesia and chloride of lime.
A very well-appointed establishment, and much frequented.

Warnemünde.—Germany, Mecklenberg-Schwerin, near Dobberan, on the Baltic sea coast.
Sea baths. Very much frequented; a first-class station.
The establishments and accommodation are excellent and very comfortable.
Doctor: Dr. Uterhard.
Hôtel: Kurhaus.

Warnicken.—Germany, Prussia, on the Baltic.
Sea baths, fine sandy beach.
The situation is very fine and the country around picturesque and romantic, with luxuriant vegetation.

Warrenspoint.—Ireland, County Down, near Newry.
Sea baths, pebbly beach; pure and healthy air.
Season: June—September.
Doctor: Dr. Douglas.
Hotel: The *Crown*.

Wartenberg.—Austria, Bohemia, near Königsgrätz.
Hydropathic establishments, climatic station in summer.
The country around is very charming.
The establishments, walks, gardens and parks, are well kept, and residence is very pleasant.

Wassacherberg.—Germany, Württemberg; 1,700 feet above sea level.
Effervescent ferruginous waters.
The establishment is thoroughly well fitted up.

Watchet.—England, Somersetshire.
Sea baths; bad beach; climate pure and bracing.
Season: June—September.

Wattenweiller.—Germany, Alsatia, near Belfort.
Bicarbonate of iron waters. Mud baths.
Special indications: Chlorosis and anæmia.

Wattwyl.—Switzerland, Canton of St. Gall.
A whey cure station.
There is a drink-hall, with a variety of natural mineral waters, which can be mixed with the milk.

Weggis.—Switzerland, Lake of Lucerne; 1,350 feet above sea level; 1,500 inhabitants.
Climatic station; whey cure.
The situation is picturesque and sheltered, on the south bank of lake of IV. Cantons between Lucern and Vitznau, it is a good point of departure for excursions. Exceedingly quiet.
There is a Doctor in attendance.
Hotel The *Hôtel and Pension du Lac*, centrally situated for excursions to Righi and other places; good table; pension from 5 to 8 francs. F. Faulstich, proprietor.

Weilbach.—Germany, Nassau. For routes see table.
Sulphurous alkaline iodurated waters.
Special indications: Catarrh of the respiratory organs.
A well-appointed bathing establishment.
Doctor: Dr. Stift.
Hotel: *Kurhaus* and Dependencies.

Weissbad.—Switzerland, Canton of Appenzell; 2,524 feet above sea-level.
    Cold bicarbonate of lime waters.
    Whey cure; good arrangements; fine situation.

Weissenburg.—Switzerland, Canton of Berne, near Thune.
    Alkali-saline waters, 50° F. Sedative action.
    Special indications; Pulmonary catarrh, chronic bronchitis.
    The bathing establishment and the accommodation are good.

Weissenstein.—Switzerland, near Solothurn; 3,050 feet above sea-level.
    Whey cure, fine situation, in a picturesque country.
    Good accommodation.

Wells-next-the-Sea.—England, Norfolk.
    Sea baths, good sandy beach.
    Season: June—September.
    Doctor: Dr. Fort.
    Hotel: The *Crown*.

Wemyss-Bay, and Skelmorlie.—Scotland, Renfrewshire.
    Hydropathic establishment. Sea baths, good beach.
    Climate healthy and air pure.
    Season: July—September.
    Doctor: Dr. Wyllie.

Wenzelsbad.—Austria, Bohemia, near Tschawitz and Komotau.
    Alkali-saline waters, 65° F.
    The bathing establishment is good, though small.

Westerland-Sylt.—Germany, on the west coast of Schleswig.
    Sea baths; sandy beach with heavy surf.
    The establishments are good, accommodation is comfortable and moderate in price.

Westgate-on-Sea.—England, Kent.
    Sea baths; firm sandy beach; bracing air.
    Season: June—September.
    Doctor: Dr. Flint.
    Hotel: Royal.

Weston-Super-Mare.—England, Somersetshire, near Bristol.
    Sea baths; bad beach; pure and bracing air.
    Season: May—October; but principally July—September.
    Doctors: Drs. Martin and Wallis.
    Hotel: Bath Hotel.

Westward Ho.—England, Devonshire, near Northam Burrows.
Sea baths; sandy beach. Climate bracing, and air pure.
Season: June—September.
Hotel: The *Westward Ho!*

Weymouth.—England, on the Channel coast, near Portland.
Sea baths, firm sandy beach, with mineral water springs in the neighbourhood.
Doctors: Drs. Colmer and Smith.
Hotel: Bourdon.

Whitburn.—England, Durham, near Sunderland.
Sea-baths; in the environs are several mineral springs; sandy beach; bracing climate.

Whitby.—England, Yorkshire.
Sea-baths; level sandy beach; climate bracing.
Season: June—September.
Doctor: Dr. Hayward.
Hotel: The *Duke of Cumberland*.

White Sulphur Springs.—United States, Delaware.
Sulphurous and ferruginous waters.

White Sulphur Springs.—United States, Virginia.
Sulphurous waters.
A much frequented station.
Resorted to by invalids, but not to the same extent as by pleasure-seekers.

Wieliczka.—Austria, Galicia.
Saline waters, 139 grammes of mineralisation, of which 137 grammes are chloride of sodium; brine baths.
Special indications: Scrofula.
The establishment is thoroughly well fitted up.

Wiesau.—Germany, Bavaria.
Two effervescing ferruginous springs, analogous to the Pyrmont waters.

Wiesbaden.—Germany, Hessen-Nassau, capital of district of same name. 53,500 inhabitants; 371 feet above sea level. For routes, see table.
Thermal alkaline-saline waters, 155° to 160° F.; climatic air and winter station; whey, koumiss, and milk cure; a very important grape cure station.
Season: May till November; but open all the year round.
Number of Visitors: At any given time, between 8,000 and 10,000; during the whole season, above 70,000.
Special indications: Chronic gout, chronic catarrh of the stomach and intestines, syphilis, diabetes, chronic rheumatism,

arthritis, affections of the respiratory organs, laryngitis, pharyngitis, skin diseases, such as eczéma, urticaria, etc.

The waters of Wiesbaden were known to the ancient Romans, but their extended use dates from the present century, and more especially the last twenty years. The situation of the town is charming—on the southern slopes of the well-wooded Taunus mountains, an hour from the Rhine. Being thus well sheltered against all northerly, north-east, and north-west winds, the climate, considering the latitude, is exceptionally mild. Magnolias, chestnut, and almond trees thrive in the open air all the year round. Mean summer temperature, 65° F.; mean winter temperature, 35° F.; mean rainy days during the year, 138; mean rainfall, 26·55 inches; death rate, 19 per 1,000; relative humidity, 80 %. The natural heat of the soil is remarkable; the temperature in the centre of the town is relatively high, consequently snow never lies long. The excellent drinking water, the perfect drainage, as also the scrupulous cleanliness of the streets and roads are worth mentioning.

The bathing establishments are complete, and fitted up with all modern requirements. Pine cone sap, Russian, Roman, Irish, or Turkish, mud, steam and swimming baths; mineral, electric and medicinal baths of every description; pneumatic and electric apparatuses; two hydropathic establishments, and three cliniks for eye diseases. The mineral water is exported and keeps well. The establishments are all suitably arranged for a winter cure.

The surroundings of Wiesbaden offer an endless variety of very interesting and romantic excursions, and embrace the Taunus and the most interesting part of the Rhine Valley.

The Curhaus is a very elegant building, situated in an extensive park. It contains—being open all the year round—restaurant, concert, card, conversation, and a very ample reading-room, etc., etc.

The amusements are very various, and the Directors of the baths and Curhaus spare no expense or trouble to render a sojourn agreeable. The orchestra, composed of 50 musicians, amongst which are always several well known soloists, plays three times a day. Vocal and instrumental concerts are given, with star artists; balls, matinées dansantes, fancy dress balls, etc., illuminations, fireworks, pic-nics, races, lectures, etc., etc. There is a very fine theatre, museum, art exhibitions, fishing and hunting, football, cricket, croquet and lawn-tennis grounds, rowing and sailing.

English Church: Frankfurterstrasse.
Post Office: Rheinstrasse No. 9.
Telegraph Office: Rheinstrasse No. 9.
Bank and Exchange Office: Pfeiffer & Co., Langgasse 16.
House and Estate Agent: L. Rettenmeier, Rheinstrasse.
Dentist: Dr. Hofmann, Taunusstrasse 18.
Doctors: Drs. Von Langenbeck, Seitz, Cohn, Roth, Heymann, Schirm, Velten, Ziemssen, Pagenstecher, Mordhoust, Maerklin, Clouth.

Hotels: The *Hôtel and Bains de Nassau* with dependence Villa Nassau, facing the Kurhaus, the best position in the town. Goetz Bros., proprietors.

The *Four Seasons Hotel and Baths* (Vier Jahreszeiten Hotel) is a first-class establishment, with fine open position, facing the Kurhaus and park; 120 front rooms; 30 airy and comfortable bathing cabins, supplied by mineral spring, at rate of 120 litres a minute, 145° F. Arrangements made for winter residence. W. Zais, proprietor.

The *Taunus Hotel* is the nearest to the three railway stations, post and telegraph office. The omnibus for Schwalbach starts from the post next door. Pension all the year round; well-appointed and managed by the proprietor, J. Schmitz-Volkmuth.

Wiesenbad.—Germany, Saxony, near Annaberg.

Bicarbonate of soda waters, 50° F.

Special indications: Neuropathic conditions, female diseases, and calculous affections.

Hotel: Kurhaus.

Wiesloch.—Germany, Baden, near Langenbrücken.

Sulphurous waters, not used.

Wildbad.—Germany, Württemberg, Black Forest, in the Enz Valley; 3,000 inhabitants; 1,320 feet above sea level. For routes, see table.

Indifferent waters from 80° to 95° F.; whey and milk cure; climatic air station.

Season: 1st May to 1st October, but open all the year round.

Number of Visitors: At any given time about 1,200, but more than 8,000 during the season.

Special indications: Uric acid diathesis, gout, rheumatism, neuroses, all affections of the spinal chord, neuralgia, hyperaesthesia, anæsthesia, cramps, partial and general paralysis, anchylosis, debility from overwork (mental or bodily), convalescence, scrofula, rhachitis, chronic derangements of the digestive organs, catarrh of the respiratory and urinary organs; female complaints, fractures and laxations.

The town is situated in a very picturesque valley of the Black Forest, through which passes the Enz, a clear mountain stream. Wildbad is surrounded by mountains, clad with pine forests, which protect it from high winds, and at the same time renders the air buoyant and bracing in ozone, a very fortifying quality.

The environs and the scrupulously well kept garden and parks offer great variety of promenades and excursions.

The climate is very strengthening and stimulating. The atmospheric changes are never rapid. The mean temperature in summer is 65° F. The nights are rather cool, which is considered an advantage.

The bathing establishments are replete with every modern appliance, and afford every comfort.

The amusements are well provided for, and an orchestra plays three times a day. There is also a good theatre. Curtax for each person for the season, 10 shillings. Good fishing and hunting.
English Church: Rev. W. Ludlow, M.A.
Post Office: In the Hôtel Frey.
Telegraphic Office: In the railway station.
Bank and Exchange: Klumpp and Gewerbe bank.
Dentist: Partik.
Chemist: Umgelter.
Doctors: Drs. Von Renz, Inspector of the Baths, Von Burkhardt, Haussman sen., and Haussman jun., De Ponte, and S. Wagner.
Hotels: Royal Baths' Hotel, Klumpp.
The *Belle Vue Hôtel* is a first-class house for families; it has a covered way leading to the bath-house (new). The hotel faces the promenade, and is near the English Church. J. Stockinger, proprietor.

Wildegg.—Switzerland, Canton of Argau. For routes see table.
Saline iodo-bromurated waters used at Schinznach.
Special indications: Lymphatic affections and scrofula.

Wildhausbad.—Switzerland, Canton St. Gall; 3,430 feet above sea level.
A whey cure station; sulphurous spring, not used.
The establishment is good.

Wildungen.—Germany, Waldeck, near Waldeck and Cassel. For routes see table.
Cold alkaline ferruginous waters.
Special indications: Diseases of the genito-urinary organs, and dyspepsia.
Nature and art have made of Wildungen a very charming place, and residence there is very pleasant.
Doctors: Drs. Muller, Kruger, Röhrig and Von Lingelsheim.
Hôtels: Post, Waldeck.

Wilhelmsbad.—Germany, Hessen, near Hanau.
Cold chalybeate waters.
Special indications: Scrofula, anæmia, female complaints.

Wilhelmsbad.—Germany, Saxony, near Aschersleben.
Brine baths.
The establishment is incomplete.

Wilhelmsbad.—Germany, Silesia, near Ratibor.
Cold sulphurous waters.
The establishments are good, the accommodation comfortable and moderate in price, and the scenery very picturesque.

Willoughby.—England, Warwickshire, near Rugby.
  Sulphurous and saline waters.
  Climate healthy.
  Season : June—September.

Wimpfen.—Germany, Hessen-Darmstadt, on the road from Heidelberg to Heilbronn.
  Saline water, brine baths.
  The establishment is good, and the scenery picturesque.

Windermere.—England, Westmorland, Lake District.
  Summer air cure station, and holiday resort.
  Doctor : Dr. Hamilton.

Windsor Forest.—England, Berkshire.
  Sulphate of magnesia and soda effervescing waters; 8 grammes of mineralisation. Laxative action.

Winterbach.—Germany, Württemberg, near Stuttgart; 760 feet above sea level.
  Sulphurous waters.
  The establishment is good, the accommodation abundant and comfortable, and the scenery charming.

Wipfeld.—Germany, Bavaria, near Kissingen.
  Cold sulphate of lime waters; mud baths.
  The establishment is sufficiently good.

Withernsea.—England, Yorkshire, near Seathorne.
  Sea baths; shingly beach; climate bracing.
  Season : July—September.
  Hotel : The *Queen's*.

Wittekind.—Germany, Prussian Province of Saxony, near Halle.
  Brine baths, Russian and sulphurous baths.
  The establishment is thoroughly well fitted up, and is very comfortable and elegant. The gardens are very beautiful and admirably laid out. The situation is charming.
  There is a Casino, with assembly, concert, ball, and reading rooms.

Wolfach.—Germany, Baden, in the valley of the Kinzig; 875 feet above sea level.
  Cold ferruginous waters.
  Wolfach is better known for its pine-cone sap baths and resinous inhalations, than for its ferruginous spring.

Wolfs.—Austria-Hungary, near Oedenburg, on the Lake of Neusiedel.
  Earthy alkaline sulphurous waters.
  The establishment might be improved.

Wolfsegg.—Austria, near Linz.
    Alkaline waters.
    The establishment is old fashioned.

Woodhall.—England, 3 hours from London, and 1 hour from Lincoln.
    Cold saline bromo-iodurated waters.
    An imperfectly arranged establishment; in a flat country.

Wörth.—Germany, Bavaria, near Regensburg.
    Effervescent ferruginous waters.
    No establishment and very imperfect appliances.

Worthing.—England, Sussex, near Brighton.
    Sea baths; hard sandy beach. Climate mild.
    Season: July—October.
    Doctors: Drs. Collet and Smith.
    Hotel: Egremont Hôtel.

Wyk.—On the Island of Föhr, in the Baltic, steamboat from Husum.
    Sea baths, much frequented, sandy beach; climate mild.
    The bathing arrangements are very good.
    Doctors: Drs. Gerber and Hitscher.
    Hotels: Kurhaus, Redlefsen.

Wyk aan Zee.—Holland, 1¼ hours from Amsterdam, on the North Sea.
    Sea baths. Sandy beach. The surf is strong.
    Season: 15th June—1st October.
    Hôtel: The *Hôtels Reunis*.

Yarmouth.—England, Norfolk.
    Sea baths.

Yeuzet.—France, Department of Gard.
    Sulphurous waters.
    The establishment is good.

Yverdon.—Switzerland, Canton of Vaud.
    Mixed bicarbonate waters, 55° F.
    Special indications: Affections of the pulmonary and vesical mucous membranes.
    Very agreeably situated, salubrious climate, and well organised bathing establishment.

Yport.—France, Department of Seine Inférieure, *via* Fécamp.
    Sea baths, sandy beach.

Zafarana.—Italy, Sicily, at the foot of Ætna.
    Effervescent ferruginous waters.
    No arrangements.

Zagwera.—Russia, Caucasus, near Borschom.
Alkaline ferruginous waters. Exported.
No establishment; owing to insufficient communication the place is not visited.

Zaisenhausen.—Germany, Baden, near Carlsruhe.
Cold sulphate of lime waters, with slightly laxative action.
Establishment incomplete.

Zajzon.—Austria, Hungary, Transylvania, near Kronstadt; 1,790 feet above sea level.
Ferruginous iodurated waters. Good accommodation. Very changeable climate.
The establishments are ample and much frequented.

Zaldivar.—Spain, Province of Biscaya, near Durango.
Sulphate of lime waters, 50° F.
Special indications: Skin diseases, scrofula, leucorrhœa, chronic ulcers, caries, necroses, white swelling.
The site is very picturesque and well sheltered. The establishment and hotel form different buildings. Both are surrounded by large parks with extensive woods. The establishment is fitted with all modern appliances and the hotel is comfortable.
Doctor: Dr. Castañon.

Zandvoort.—Holland, near Haarlem.
Sea baths, sandy beach, climate bracing.
The arrangements are good. Accommodation abundant and comfortable, but rather dear.

Znyka.—Austria, Hungary, comitat of Bereg.
Ferruginous waters.
No establishment.

Zante.—Greece, Ionian Islands, near Chieri.
Possesses a sulphurous and saline spring, which contains petroleum and bitumen.

Zavelstein.—Germany, Baden, near Teinach; 1,800 feet above sea level.
A whey cure station; well appointed.

Zea.—Greece, island in the Archipelago.
Alkaline waters; used in primitive fashion.

Zerbst.—Germany, Saxe-Anhalt.
Effervescent ferruginous waters.
The establishments are good.

Zermatt.—Switzerland, Canton Valais; 5,315 feet above sea level. Visp railway station 3 miles from Brigue, 9 miles from St. Maurice.

Climatic station; with bracing climate.

The scenery is very grand. The place is secluded, but is a good centre for excursions. The situation is one of the best wooded and most picturesque in the Swiss Alps. Zermatt is reached in 7 hours from Visp, mule path and carriage included, the first $3\frac{1}{2}$ hours are by mule to St. Nicolas.

Mr. Seiler, proprietor of the following hotels, is unceasing in his endeavours to render every assistance to all who visit the locality.

Hotel Cervin, 180 rooms, first hotel on entering Zermatt.
Hotel Monte Rosa, 70 rooms.
Hotel Zermatt, 130 rooms.

The *Hotel Seiler*, at **Riffel Alp**, $2\frac{1}{2}$ hours distant, 150 rooms. For all information, address Mr. A. Seiler.

Zinnowitz.—Germany, on the island of Usedom in the Baltic.
Very fashionable and much-frequented sea bathing station.
Doctor: Dr. Sachse.
Hotel: Belvedere.

Zittau.—Germany, Saxony, Oberlausitz.
Pine-cone sap and resinous vapour baths.
The establishment is thorough and the situation romantic.

Zögg.—Austria, Tyrol, valley of Passeyer.
Effervescent ferruginous waters.
Special indications: Gout, chronic nervous disorders, chronic skin diseases.

Zoppot.—Germany, near Danzig, on the Baltic.
Sea baths, much frequented by Germans and Poles.
There is a thoroughly well-appointed establishment.
Doctor: Dr. Benzler.
Hotels: Kurhaus, Schulz.

Zovány.—Austria-Hungary, Transylvania, near Szilagy-Sombyo.
Ferruginous waters.
Special indications: Chronic catarrh, blenorrhœa, disorders of the uterus, and chronic sores.
The establishment is antiquated.

Zug.—Switzerland, Canton Zug, near Zurich; 2,967 feet above sea level.
Climatic mountain air station; whey, milk and grape cure.
Special indications: Exhaustion from mental overwork, anæmia, chlorosis, failing nutrition and asthma.

The situation is well sheltered, and the hills covered with pine-woods. The views are very fine, embracing the Lake of

Lucerne and the Bernese Oberland. The air is particularly bracing, and the sojourn, though quiet, is pleasant.

Doctors: Drs. Keiser and Steiger.

Hôtels: *Hôtel and Pension Schönfels*, a first-class house, with baths; very popular and home-like. Chas. Boderner, manager. In winter, at *Hôtel Continental*, Cannes.

Zujar.—Spain, Province of Granada, near Baza.

Thermal **sulphato of lime springs, 100° F.**

Special indications: Rheumatism, paralysis, gout, articular swellings, tendinous contractions, skin diseases, old wounds and bronchial catarrhs.

The establishment contains tanks and sweating rooms, but is much neglected.

Doctor: Dr. Quesada.

Zürich.—Switzerland; 1,280 feet above sea level.

Earthy alkaline waters.

There are three establishments, all equally good—the **Rösli** baths, the Drahtschmidli baths, and the sulphur baths.

Capital of the Canton of the same name, is a very thriving commercial town of much interest. Formerly a fortress, but the walls have now been pulled down.

OBJECTS OF INTEREST.—Gross Münster, or the Cathedral; Churches of St. Peter, St. Augustine, and Frauen Münster, Arsenal and Armoury, Cantonal buildings, Polytechnic school, University, Great Hospital and Observatory, Hall of Art and Paintings, Asylums, Library.

EXCURSIONS: Uetliberg, Kappel, Albisbrunn, Zürichberg, Rapperschwyl, Horden, Wadensweil, and up the lake.

CABS: One horse, two persons, 80 cts.; two horses, four persons, 1 fr. 20 cts. Post and Telegraph office: Bahnhofsstrasse.

Bookseller: Cesar Schmidt, Munsterberg.

Doctors: Drs. Rahn-Escher and Peters.

Hotels: The *Hotel Victoria*, facing the railway station. A well-appointed house, exceedingly comfortable, cooking first-class. J. Boller and Sons, proprietors.

Zwieselsalpe.—Austria, near Abtenau, on the Lammer.

Bitter waters. Exported.

The situation is charming. The establishments are very good, and the accommodation comfortable and abundant.

Zwolle.—Holland, on the Zuider Zee.

Sea baths, sandy beach; the waters are considerably poorer in brine than those of the North Sea.

St. Moritz—Kurhaus, "Hôtel du Lac," "Hôtel Engadiner Kulm," "Hôtel Victoria," *see* page 275.

Campfer—"Hôtel d'Angleterre, *see* page 67.

Silvaplana—

Sils—

Maloja—The Kurhaus, *see* page 196.

Bormio—

Tiefenkasten—

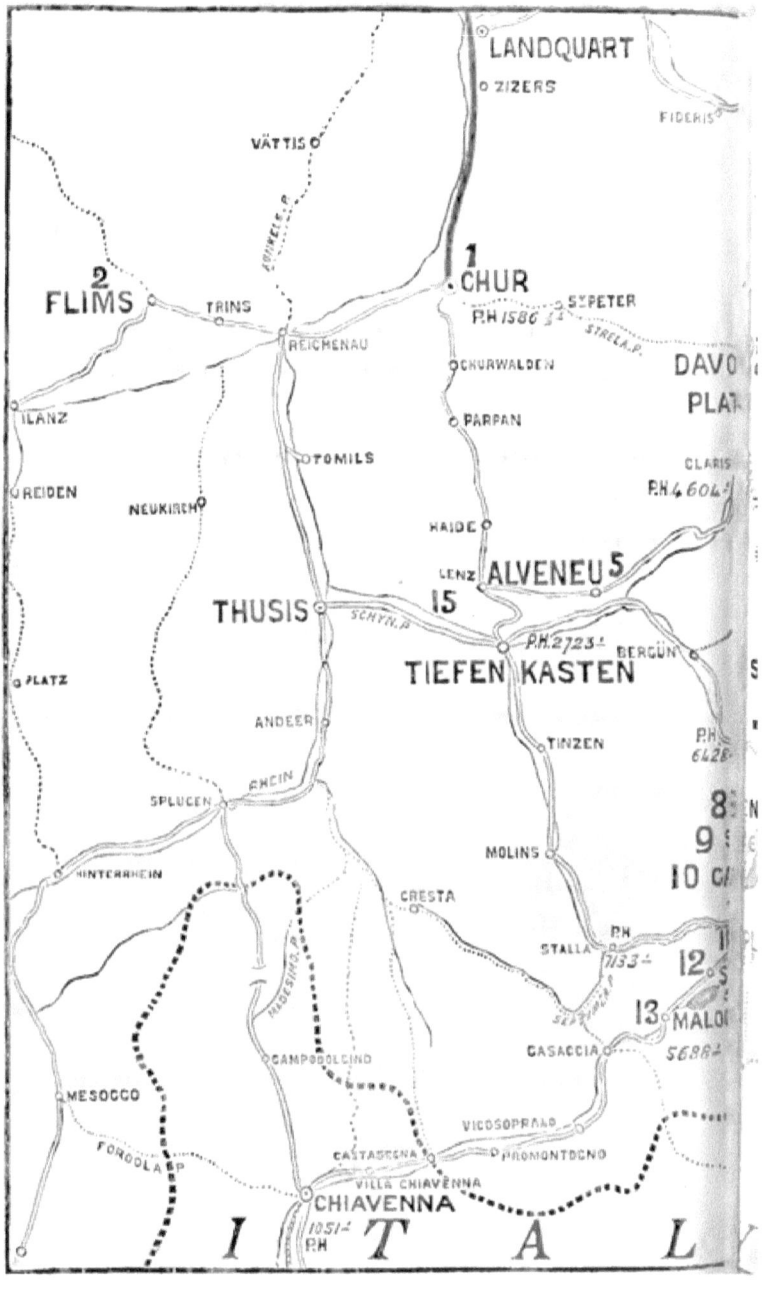

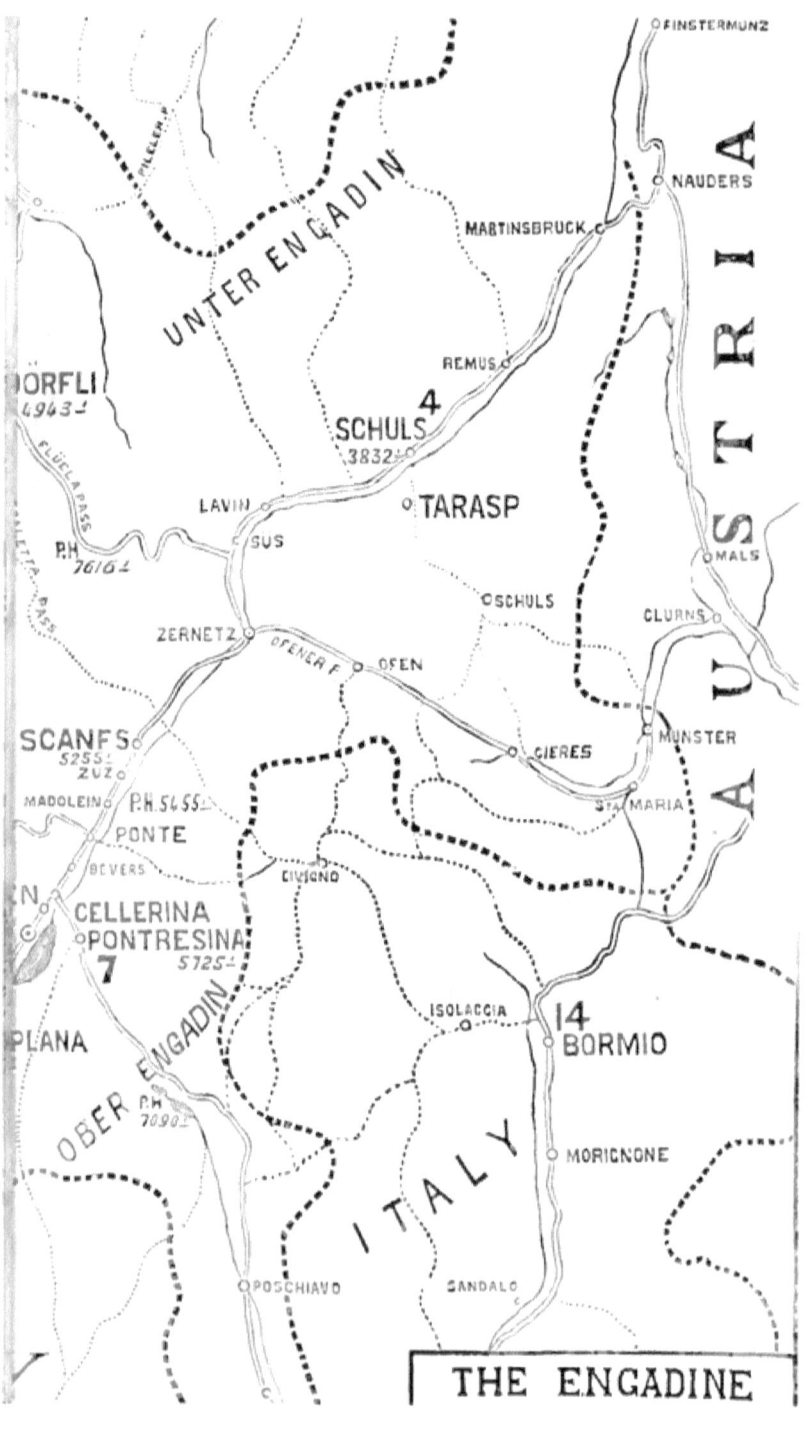

1. COIRE—"Hôtel Steinbock," *see* page 87.

2. FLIMS—The Kurhaus, *see* page 116.

3. DAVOS—

4. TARASP—

5. ALVENAU—The Kurhaus, *see* page 17.

6. SCANFS—

7. PONTRESINA—"Hôtel Enderlin," *see* page 248.

8. SAMADEN—"Hôtel Bernina," *see* page 282.

3
REIBURG

1. STRASBURG—"Hôtel d'Angleterre," see page 310.

2. BASEL—"Hôtel Hofer," see page 37.

3. FREIBURG—"Hôtel Sommer Zähingerhof," see page 121.

4. BADENWEILER—"Hôtel Sommer," see page 30.

5. RIPPOLDSAU—Hôtel and Baths Gœringer, see page 262.

6. WOLFACH—

7. TRIBERG—"Hôtel Black Forrest," see page 326.

8. GRIESBACH—

9. FREUDENSTADT—

10. WILDBAD—"Hôtel Belle Vue," see page 351.

11. BADEN-BADEN—"Hôtel de Russie," Hôtel de France," see page 30.

12. TEINACH—

13. ST. BLASIEN—

14. SCHAFFHAUSEN—

LIST OF DOCTORS

NAMED IN THE TEXT, AND THEIR SUMMER AND WINTER RESIDENCES.

Doctors practising in Great Britain and Ireland, *and mentioned in text, are not included in this list.*

Name.	Page.	Winter Station.	Page.	Summer Station.
Abbamondi	317	Teleze	317	Teleze.
Adam	116	Flinsberg	116	Flinsberg.
Adhéma	45	Biarritz.		
Aguglia	82	Chiatamone	82	Chiatamone.
Aguilera			135	Gravulos.
Aitken	265	Rome	265	Rome.
Almendariz			219	Nanclares de Oca.
Alrig	25	Aulus	25	Aulus.
Alsina			35	Bañolas.
Althammer			22	Arco.
Altherr			144	Heiden.
Alvarez				Vilo ó Rosas.
Andres			146	La Hermida.
Anglada			105	Eaux-Chaudes.
Anjel	133	Gräfenburg	133	Gräfenberg.
Armieux			36	Bareges.
Arnal	17	Amélie-les-Bains	17	Amélie-les-Bains.
Arnesto				St. Bartolomé de a Cuadra.
Arnold	144	Heidelberg	144	Heidelberg.
Arntz	84	Cleves	84	Cleves.
Arone	132	Görz	132	Görz.
Aschenbach			185	Lobenstein.
Asenjo y Cáceres			234	Ormaiztegui.
Audual			125	La Garde de Bio.
Augey	45	Barritz.		
Averbeck	175	Laubbach	175	Laubbach.
Avila			187	Loujo.
Ayegui			50	Borines.
Aymé			89	Contrexéville.
Baer	42	Berlin	42	Berlin.
Bagnell	241	Pau	31	Bagns. de Bigorre.
Bailly	33	Bains-les-Bains		Bains-les-Bains.
Balbas			285	San Gregorio de Brozas.
Balzer	17	Alveneu	17	Alveneu.

Name.	Page.	Winter Station.	Page.	Summer Station.
Bamberger	...	Vienna	...	Vienna.
Bauning	61	Bushey	61	Bushey.
Baradue	...	...	80	Chatel-Guyon.
Bardenherier	87	Cologne	87	Cologne.
Barraca	...	...	...	Villavieja de Nules.
Barráca	...	...	120	Frailes y la Rivera.
Barrio	...	...	70	Carballino.
Barth	28	Baden (Austr.)	28	Baden (Austr.)
Barrols	...	...	22	Argentona.
Barwinsky	...	...	311	Suderode.
Bary, De	310	Strasburg	310	Strasburg.
Battersby	69	Cannes	50	Bormio.
Battlehner	71	Carlsruhe	71	Carlsruhe.
Baumann	293	Schlangenbad	293	Schlangenbad.
Becker	141	Hanover	141	Hanover.
Beeli	...	...	95	Davos-Platz.
Beissel	7	Aix-la-Chapelle	7	Aix-la-Chapelle.
Bellini	...	...	4	Acqua Santa.
Belvin	213	Monte Carlo.		
Beneke	...	...	220	Nauheim.
Benuet	204	Mentone	—	London.
Benzler	...	...	355	Zoppot.
Bernabéu	...	...	124	Fuente Sta. de Lorca
Berry	...	...	275	St. Moritz-Bad.
Bertholet	...	...	215	Montreux.
Bertier	...	...	273	St. Gervais.
Bertou	30	Baden-Baden	30	Baden-Baden.
Bertrand	...	...	293	Schlangenbad.
Betons	...	...	36	Barèges-Barsun.
Bettelheim	339	Vienna.		
Biden	...	...	153	Hyères.
Biefel	...	...	280	Salzbrunn.
Biermann	...	...	257	St. Moritz-Bad.
Bilfinger	140	Hall (Würt.)	140	Hall (Würt.).
Birnbaum	87	Cologne.		
Blanc	...	...	9	Aix-les-Bains.
Blezinger	69	Cannstadt	69	Cannstadt.
Blom-Coster	139	Hague	139	Hague.
Blondin	...	...	277	St. Sauveur.
Bochin	...	...	296	Schwalbach.
Bode	...	...	220	Nauheim.
Bodenstein	...	...	87	Colberg.
Boell	...	...	226	Niederbronn.
Boggs	239	Paris	239	Paris.
Bonjau	...	...	277	St. Simon.
Bottentint	...	...	246	Plombières.
Boucomont	...	...	...	Rouzat.
Boner	...	...	95	Davos-Platz.
Bonilla	...	...	...	Caldas de Oviedo.
Bongard	...	...	52	Bourbonne.
Boulonmié	...	...	243	Vittel.
Bourgeois	42	Berne	42	Berne.
Boutemps	...	...	277	St. Simon.
Bowes	...	Herne Bay	...	Herne Bay.
Boyer	...	...	291	Saxon.
Brabant	97	Dinan	97	Dinan.
Brachet	...	...	9	Aix-les-Bains.
Braudes	7	Aix-la-Chapelle	7	Aix-la-Chapelle.
Brandstetter	28	Baden (**Austr.**)	28	Baden (Austr.)

LIST OF DOCTORS.

Name.	Page.	Winter Station.	Page.	Summer Station.
Brandt	...	...	269	Royat.
Brauley	...	...	138	Guillon.
Breen	49	Bordeaux	49	Bordeaux.
Brehmer	132	Görbersdorf	132	Görbersdorf.
Breiting	127	Genoa	127	Genoa.
Breñosa	...	...	22	Archevaleta.
Breunig	25	Augustusbad	25	Augustusbad.
Bright	69	Cannes.		
Broichox	...	...	89	Contrexéville.
Brongnait	...	...	89	Contrexéville.
Brockmann	...	...	137	Grund.
Bruce	241	Pau.		
Brück	101	Driburg	101	Driburg.
Bruckner	37	Basel	37	Basel.
Brüggelmann	155	Inselbad	155	Inselbad.
Brügger	...	...	282	Samaden.
Brunner	...	...	186	Loëche les Bains.
Brunner	...	...	109	Engelberg.
Bütonan	...	...	87	Colberg.
Bujakowsky	79	Charlottenbrun	79	Charlottenbrun.
Bundsen	195	Malaga.		
Burdach	...	...	258	Reichenhall.
Calderon	...	...	78	Cestona Guezalaya.
Calvo	...	...	64	Caldas de Estrac.
Camaran	119	Forges-les-Eaux	119	Forges-les-Eaux.
Cantani	219	Naples	...	
Carrio y Grifol	...	...	...	Fuente Sta de Gayanga.
Carrar	...	...	215	Montreux.
Carretero	...	...	190	Lugo.
Carretero	194	Madrid	194	Madrid.
Caspari	...	...	202	Meinberg.
Castell	...	...	...	Ntra. Sa de Abellá.
Castells	...	...	70	Carballo.
Castañon	...	...	...	Zaldivar.
Cazalis	69	Cannes	213	Mont-Dore.
Cazaux	...	...	104	Eaux-Bonnes.
Clay	241	Pau.		
Cazenave de la Roche	...	...	104	Eaux-Bonnes.
Cerdo y Oliver	...	...	200	Martos.
Cérerville	177	Lausanne	177	Lausanne.
Cerezo	...	...	40	Bellus.
Cerio	69	Capri	69	Capri.
Cervera	...	...	153	Ibero.
Cessens	153	Hyères	9	Aix-les-Bains.
Cessens	...	...	9	Aix-les-Bains.
Chacel	...	...	119	Fortuna.
Challand	177	Lausanne	177	Lausanne.
Champtier	...	...	77	Cauvalat.
Chapman	239	Paris	239	Paris.
Charvet	...	...	334	Vals.
Chlorin	...	...	202	Meluadia.
Clemens	...	...	269	Rudolstadt.
Clouth	349	Wiesbaden	349	Wiesbaden.
Coche	...	...	89	Condillac.
Cohn	...	...	349	Wiesbaden.
Colonna	10	Ajaccio.		

Name.	Page.	Winter Station.	Page.	Summer Station.
Collignon	58	Brussels	58	Brussels.
Coll y Amo	...	...	115	Fitero.
Commandré	...	...	33	Bagnole.
Companyo	...	Estramer	110	Les Escaldas.
Cordes	...	...	12	Alexandersbad.
Cornilo	189	Lugano	189	Lugano.
Costina	...	...	82	Chiclana.
Cosyn	46	Blankenberghe	46	Blankenberghe.
Cramer	108	Elster	108	Elster.
Crespo y Escoriaza	...	...	123	Fuencaliente.
Cros	...	...	173	La Malou.
Crucero	...	...	211	Mondariz.
Cumine	141	Hanover	141	Hanover.
Cüppers	—	Trèves	42	Bertrich.
Czerny	144	Heidelberg	144	Heidelberg.
Daina	289	Saragoza	289	Saragoza.
Dantaud	...	...	112	Evian-les-Bains.
Danjoy	...	...	53	La Bourboule.
Daremberg	204	Mentone	204	Mentone.
Daubeney	287	San Remo	109	Engelberg.
Davids	19	Amsterdam	19	Amsterdam.
Davidson	117	Florence	117	Florence.
Debont d'Estrée	...	...	89	Contrexéville.
De Alessandri	...	...	5	Aequi.
De Gregorio	...	...	159	Jaruba.
De Daineseau	...	...	307	Spa.
Deetz	149	Homburg	149	Homburg.
Deetz	...	...	190	Luisenbad.
De Rugama	...	...	253	Puente Viesgo.
De la Harpe	177	Lausanne	177	Lausanne.
De la Pedruesa	...	...	...	Urberoaga de Alzola.
Del Rio	...	...	212	Montanejos.
Delfau	70	Capvern	70	Capvern.
Deputowsky	...	...	224	Neyrac.
Desprez	...	...	280	Salins (Savoy).
Desprez	...	...	55	Brides.
Dettweiler	112	Falkenstein	112	Falkenstein.
De Uriarte	301	Seville	301	Seville.
De Wette	37	Basel	37	Basel.
Diaz	...	...	160	San Juan de Azcoitia
Dickinson	11	Alassio.		
Diez	...	...	110	Escoriaza.
Diruf	...	...	166	Kissingen.
Ditmann	...	...	313	Sylt.
Dlauhy	...	...	224	Neudorf.
Döbner	...	...	182	Liebenstein.
Doiz	...	...	195	Malahá.
Dorner	25	Augustusbad	25	Augustusbad.
Donné	233	Orezza	233	Orezza.
Dörr	201	Mayence	201	Mayence.
Doyon	...	...	322	Uriage.
Drummond	225	Nice	275	St. Moritz-Bad.
Dubouloz	...	...	87	Coise.
Duchosal	...	...	79	Chamounix.
Ducoux	...	...	34	Balaruc-les-Bains.
Duffy	117	Florence	117	Florence.
Duhoureau	76	Cauterets		Cauterets.

Name.	Page.	Winter Station.	Page.	Summer Station.
Dumuz	...	...	111	Evians-les-Bains.
Dupeyron	74	Castrera Verduzan	55	Barbotan.
Duran	...	...	...	Ntra. Sa. de las Mercedes.
Dupont	177	Lausanne	177	Lausanne.
Dürr	140	Hall (Würt.)	140	Hall (Wurt).
Durran-Fardel	...	...	338	Vichy.
Duwez	...	London	196	Maloga.
Dyes	141	Hanover	141	Hanover.
Ebert	41	Berka	41	Berka.
Eberle	...	Prague	318	Teplitz-Schönau.
Eichler	...	...	318	Teplitz.
Elb	101	Dresden	101	Dresden.
Emond	...	...	213	Mont-Dore.
Engelmann	170	Kreuznach	170	Kreuznach.
Erb	144	Heidelberg	144	Heidelberg
Erfurth	113	Feldberg	113	Feldberg.
Escudero	...	...	78	Cervera.
Estrada	...	...	133	Graena.
Enriquez	...	...	108	Elorrio.
Ewe	...	...	220	Neundorf.
Exsell	141	Hanover	141	Hanover.
Fanti	...	...	32	Bagno-in-Romagna.
Farral	143	Le Hâvre	143	Le Hâvre.
Fedeli	...	...	213	Monte Catini
Feederle	...	...	...	Triberg.
Fernandez	...	...	147	Hervideros de uent Santa.
Ferrer	...	...	14	Alicum.
Ferretti	...	Bologna	154	Imola.
Ficker	...	...	319	Teplitz-Schönau.
Finkelnburg	131	Godesberg	131	Godesberg.
Fitz-Henry	204	Mentone	...	London.
Fleehsig	108	Elster	108	Elster.
Fleischanderl	...	...	170	Kreuzen.
Fleury	...	...	296	Schwalheim.
Flottard	...	...	112	Evian-les-Bains.
Fodor	...	...	...	Töplika.
Fontau	...	...	63	Cadéac.
Fontini	...	...	14	Alhama de Murcia.
Förster	...	...	23	Artern.
Fraisse	...	...	207	Miers.
Frank	69	Cannes	...	London.
Franke	143	Harzburg	143	Harzburg.
Franz	...	...	345	Warmbrunn.
Frédet	...	...	269	Royat.
Freeman	287	San Remo.	...	
Fresenius	...	...	304	Soden.
Freuler	...	...	128	Gersau.
Frey	...	...	184	Lippspringe.
Freygang	41	Berka	41	Berka.
Frickhöffer	...	...	296	Schwalbach.
Fromm	...	...	227	Norderney.
Fukel	...	...	294	Schmalkalden.
Fürstenberg	157	Ischl	157	Ischl.

LIST OF DOCTORS.

Name.	Page.	Winter Station.	Page.	Summer Station.
Gactschenberger	...	...	166	Kissengen.
Galano	...	...	170	Sorrento.
Gallus	190	Luhatschowitz	190	Luhatschowitz.
Gracia Lopez	...	...	179	Ledesma.
Garcia Teresa	...	...	341	Villatoya.
Garrigon	26	Ax	...	...
Gason	265	Rome	188	Lucca.
Gastl	129	Giesshübl	129	Giesshübl.
Gatti	—	Bologna	74	Castel St. Pietro.
Gaulejac, de	177	Fonfrede	177	Fonfrede.
Gavilanes	...	...	61	Buyères de Nava.
Gavilanes	...	...	306	Souzas y Caldeliñas.
Gazert	...	...	227	Norderney.
Geissé	108	Ems	108	Ems.
Genicys	17	Amélie-les-Bains	17	Amélie-les-Bains.
Genoves y Tio	...	...	40	Beninarfull.
Genth	...	...	296	Schwalbach.
Gerber	...	...	353	Wyk.
Gerber	131	Godesberg	131	Godesberg.
Germonik	...	...	335	Veldes.
Gil	...	...	302	Siete-Aguas.
Gimenez de Pedro	...	...	115	Fitero.
Gimeno	...	...	279	Sacedon.
Girdlestone	45	Biarritz.		
Giserius	121	Freienwalde	121	Freienwalde.
Glaesgen	217	Münster am Stein	217	Münster am Stein.
Glax	...	...	263	Rohitsch.
Goldschmidt	194	Madeira	194	Madeira.
Goltz	287	San Remo	108	Ems.
Gongona y Joanico	...	...	252	La Puda.
Goodchild	49	Bordighera.		
Gourdon	12	Aled	12	Aled.
Goyeneche	273	St. Jean de Luz	273	St. Jean de Luz.
Grabham	194	Madeira	194	Madeira.
Graef	129	Frankenhausen	129	Frankenhausen.
Grandvilliers	194	Nice.		
Graux	...	...	89	Contrexéville.
Grebert	...	...	296	Schwalbach.
Green	17	Anacapri.		
Griffiths	153	Hyères.		
Grigor	265	Rome	265	Rome.
Grimmaud	...	...	36	Barèges.
Grimmand	...	...	38	La Baneke.
Grossmann	...	...	293	Schlangenbad.
Gubian	...	...	173	Lamotte.
Guedea	...	...	20	Aranmyona.
Guetz	...	...	199	Marienbad.
Guilland	...	...	9	Aix-les-Bains.
Guimbert	69	Cannes.		
Gurrucherri	...	...	43	Betelú.
Gustorf	42	Berlin	42	Berlin.
Gutschke	71	Carlsruhe	71	Carlsruhe.
Guycnot	...	...	279	Salins (Jura).
Haase	...	...	149	Homburg.
Huberer	136	Griesbach	136	Griesbach.
Hahn	42	Berlin	108	Elster.
Hail	87	Cologne	87	Cologne.

Name.	Page.	Winter Station.	Page.	Summer Station.
Hamburger	121	Franzenbad	121	Franzensbad.
Hamean	21	Arcachon	21	Arcachon.
Harvey	...	...	51	Boulogne-sur-M
Hasenfeld	...	...	314	Szliácz.
Hassall	287	San Remo	...	...
Hassenstein	154	Ilmenau	154	Ilmenau.
Haueisen	141	Hall (Wurt.)	141	Hall (Würt.).
Haufe	...	Strasburg	272	St. Blasien.
Hauser	301	Seville	301	Seville.
Heer	177	Lausanne	177	Lausanne
Heinemann	157	Ischl	157	Ischl.
Heller	...	...	318	Teplitz.
Helligenthal	...	...	30	Baden-Baden.
Hellwig	201	Mayence	201	Mayence.
Hernandez	...	...	11	Alange.
Hernandez	...	...	302	Sierra Alhamilla.
Herzog	...	...	345	Warmbrunn.
Hesse	...	...	181	Liebenstein.
Heusinger	...	Harzgerode	13	Alexisbad.
Heymann	349	Wiesbaden	349	Wiesbaden.
Hinterberger	...	...	323	Topusko.
Hirschl	...	Pisa	337	Viareggio.
Hirschfeld	85	Colberg	85	Colberg.
Hitscher	...	...	353	Wyk.
Hœber	149	Homburg	149	Homburg.
Höffer	...	...	170	Krankenheil.
Höhne	...	...	345	Warmbrunn.
Hoisel	...	...	263	Rohitsch.
Höring	...	...	205	Mergentheim.
Hugemann	10	Ajaccio.		
Hüller	101	Driburg	101	Driburg.
Hunt	241	Pau.		
Hysem	194	Madrid	194	Madrid.
Iborra	...	...	255	Quinto.
Illescas	...	...	...	Villaharta.
Ingrish	...	...	199	Marienbad.
Ionard	...	...	271	St. Amand.
Jabob	...	...	...	Cudowa.
Jaegerschmied	136	Griesbach	136	Griesbach.
Jaime	...	...	...	Torres.
Janssens	...	...	234	Ostende.
Janvier	244	Pierrefonds	244	Pierrefonds.
Japhet	109	Enghien	109	Enghien.
Jarmay	294	Schmécksz	294	Schmécksz.
Jelly	194	Madrid	194	Madrid.
Jimenez de Pedro	...	Madrid	331	Urberonga de Ubilla.
Joël	177	Lausanne	177	Lausanne.
Joseph	...	...	173	Landeck.
Joubert	118	Gréoulx	33	Bagnolles.
Jungmayer	...	...	170	Krankenheil.
Junnie	...	...	42	Ostende.
Kaatzer	...	...	258	Rohburg.
Kafka	...	...	162	Karlsbad.

Name.	Page.	Winter Station.	Page.	Summer Station.
Kallay	121	Franzensbad	121	Franzensbad
Kaposi	339	Vienna.		
Kartüm	99	Doberan	99	Doberan.
Kastan	108	Ems	108	Ems.
Katzer	139	Hall (Austr.)	139	Hall (Austr.)
Kaufmann	103	Dürkheim	103	Dürkheim.
Keil	123	Friedrichroda	123	Friedrichroda.
Keiser	...	...	...	Zug.
Keith	10	Ajaccio	—	Edinburgh.
Kelling	...	...	48	Boltenhagen.
Kelly	58	Brussels	58	Brussels.
Keppler	336	Venice	336	Venice.
Kern	...	...	183	Lipik.
Kiel	...	...	256	Rastenberg.
Killias	...	...	316	Tarasp.
Kirchheim	108	Elmen	108	Elmen.
Kisch	...	...	199	Marienbad.
Kispert	194	Madrid	194	Madrid.
Klein	121	Franzensbad	121	Franzensbad.
Klein	...	...	256	Niederbronn.
Kley	310	Strasburg	310	Strasburg.
Klinger	...	...	309	Steben.
Knorr	...	...	169	Kösen.
Koblovsky	...	...	269	Roznau.
Köhler	...	...	254	Pyrmont.
Köhler	...	...	394	Soden.
Kohn	...	...	168	Königswart.
Kopf	160	Johannisbad	160	Johannisbad.
Körbl	139	Hall (Austr.)	139	Hall (Austr.).
Kostens	...	...	15	Allevard-les-Bains.
Krauss	...	...	205	Mergentheim.
Krauss	162	Karlsbad	162	Karlsbad.
Krischke	174	Kreuzen	344	Vöslau.
Kruger	351	Wildungen	351	Wildungen.
Küchler	...	...	190	Luhatschowitz.
Kuhnemann	189	Monaco	189	Luchon.
Kurz	336	Venice.		
Kussmaul	310	Strasburg	310	Strasburg.
Lacoste	172	Labassère	31	Bagnères de Bigorre.
Lacort	...	...	158	Jabalcúz.
Lafont	...	...	277	St. Sauveur.
Lahs	...	...	254	Pyrmont.
Lamboon	...	...	189	Luchon.
Langaudin	...	Paris	269	Royat.
Lange	99	Doberan	99	Doberan.
Lange	72	Cassel	72	Cassel.
Langner	...	...	173	Landeck.
Lanjurjo	...	...	251	Prelo.
Larguier	177	Lausanne	177	Lausanne.
Larisa	194	Madeira	194	Madeira.
Laspales	...	...	189	Luchon.
Laudry	...	...	250	Pouillon.
Lauffs	7	Aix-la-Chapelle	7	Aix-la-Chapelle.
Laugandin	...	...	269	Royat.
Laza	...	...	287	Sau Vincens.
Lazaro	...	...	...	{ St. Filomena de Gomillaz.

LIST OF DOCTORS. 365

Name.	Page.	Winter Station.	Page.	Summer Station.
Leclerc	...	...	246	Plombières.
Lehmann	...	...	190	Luisenbad.
Lehmann	...	...	230	Oeynhausen.
Lersch	7	Aix-la-Chapelle	7	Aix-la-Chapelle.
Lescamel	66	Cambo	66	Cambo.
Lesi	...	Bologna	154	Imola.
Leudi	...	...	282	Samaden.
Leukardt	179	Leipzig	179	Leipzig.
Lewi	336	Venice	80	Chatenois.
Lewis	149	Homburg	149	Homburg.
Ley	...	...	282	Salzungen.
Leznack	307	Spa	307	Spa.
Lietard	...	...	246	Plombières.
Lindemann	...	...	205	Mergentheim.
Lippert	225	Nice.		
Lleget y Caila	...	...	64	Caldas de Mombuy
Löchner	103	Dürckheim	103	Dürckheim.
Logerais	...	...	250	Pougues.
London	...	...	162	Karlsbad.
Lopez	...	...	111	Estadilla.
Lopez	...	...	62	Buzot.
Lopez	...	...	209	El Molar.
Lorent	112	Falkenstein	112	Falkenstein.
Louzet	7	Aix (Provence)	7	Aix (Provence).
Lozano y Rubio	...	...	297	Salinetes de Novelda.
Lucientes y Pueyo	...	...	186	Loëches.
Ludwig	...	...	248	Pontresina.
Lynker	...	...	254	Pyrmont.
Macario	...	...	93	Croisic.
Macé	...	...	9	Aix-les-Bains.
Macher	277	St. Radegund	277	St. Radegund.
Maclean	199	Luxo	...	London.
Maerklin	349	Wiesbaden	349	Wiesbaden.
Mahr	24	Assmannshausen	24	Assmannshausen.
Maier	71	Carlsruhe	71	Carlsruhe.
Malpas	45	Biarritz	45	Biarritz.
Mangold	124	Balaton Füred	124	Balaton Füred.
Mannheim	155	Inowrazlaw	155	Iwrazlaw.
Manzi	...	...	156	Ischia.
Marchal	...	...	211	Mondorf.
Marcus	...	...	313	Sylt.
Marquez	153	Hyères.		
Marriott	204	Mentone.		
Martinez	...	...	...	Annedillo.
Martinez	...	...	35	Barambio.
Martinez	...	...	118	Fonté.
Martin	77	Cayla	19	Andabre.
Mascarel	...	...	213	Mont-Dore.
Massot	...	...	51	Le Boulou.
Mast	310	Strasburg	20	Antogast.
Maury	260	Rotterdam	268	Rotterdam.
May	170	Kreuth	170	Kreuth.
Mayer	7	Aix-la-Chapelle	7	Aix-la-Chapelle.
Mayo	...	...	305	Sorrento.
Mayoral	...	...	23	Arteio.
Mayrhofer	...	Gries	265	Römerbad (Austr).
Meissen	112	Falkenstein	112	Falkenstein.

LIST OF DOCTORS.

Name.	Page.	Winter Station.	Page.	Summer Station.
Melgrani...	...	Ajaccio.		
Mendez Tejo	...	...	263	Riva los Baños.
Mengis ...	...	...	186	Loëche-les-Bains.
Menzies ...	...	Hyères	7	Aix-les-Bains.
Mermagen	198	Mannheim	198	Mannheim.
Merreguer	...	...	82	Chublla.
Mestre y Marzal	...	...	253	Puertolano.
Metzger ...	19	Amsterdam	19	Amsterdam.
Meyr ...	243	Pesth ...	108	Elöpatak.
Mezzini ...	...	Bologna	261	Riolo.
Michaelis	...	...	258	Rehburg.
Miesko ...	...	...	202	Mehadia.
Millet ...	...	...	186	Montmirail.
Million ...	112	Evian-les-Bains	112	Evian-les-Bains.
Mir ...	...	...	...	Valle de Rivas.
Möller ...	107	Eilsen	107	Eilsen.
Monnier ...	...	...	215	Montreux.
Morales ...	...	...	281	Salvatora.
Morin ...	204	Mentone	53	La Bourboule.
Mordhoust	349	Wiesbaden	349	Wiesbaden.
Müller ...	...	...	351	Wildungen.
Müller ...	...	...	230	Oeynhausen.
Müller ...	...	...	293	Schinznach.
Munk ...	...	...	202	Mehadia.
Müützel ...	...	...	192	Neuenahr.
Murret ...	337	Vevey	337	Vevey.
Nagel ...	...	...	325	Trenczin-Teplitz.
Nager ...	180	Lucerne	180	Lucerne.
Nath ...	121	Freienwalde	121	Freienwalde.
Neisser ...	79	Charlottenbrunn	79	Charlottenbrunn.
Nella ...	...	Bologna	74	Castel San Pietro.
Neubauer	...	...	162	Karlsbad.
Neus ...	...	...	220	Nenndorf.
Niebergall	...	...	23	Arnstadt.
Nieto ...	...	...	221	Navalpino.
Niehaus ...	42	Berne	42	Berne.
Niépce, fils	...	...	15	Allevard-les-Bains.
Notebaert	46	Blankenberghe	46	Blankenberghe.
Nothnagel	339	Vienna	339	Vienna.
Nuñez ...	194	Madrid	194	Madrid.
Ocaña y Pazos	...	...	...	Villar del Pozo.
Odin ...	...	...	273	St. Honoré.
Oliphant	241	Pau.		
Onetti ...	287	San Remo.		
Opitz ...	...	...	199	Marienbad.
Ortega ...	...	...	39	Belascoin.
Orth ...	108	Ems ...	108	Ems.
Ortíz ...	...	...	123	Fuente Amargosa
Osuna ...	...	...	88	Concepcion de Peralta
Ostrowitz	...	...	173	Landeck.
Oswald ...	...	...	23	Arnstadt.
Oton ...	...	...	199	Marmolejo.
Ott ...	...	...	199	Marienbad.
Otte ...	16	Alt Haide	16	Alt Haide.

LIST OF DOCTORS.

Name.	Page.	Winter Station.	Page.	Summer Station.
Pachmayr	...	...	258	Reichenhall.
Padrals	...	...	126	La Garriga.
Paesn	...	...	202	Mehadia.
Pagenstecher	349	Wiesbaden	349	Wiesbaden.
Palacios Tomás	...	...	...	Villaro.
Paltauf	...	...	222	Neuhaus.
Pantlen	...	...	69	Cannstadt.
Pardo	...	...	126	Gaviria.
Parraverde	...	...	13	Alhama de Aragon.
Pastor	...	...	...	Tiermas.
Pauer	160	Johannisbad	160	Johannisbad.
Peliäzus	107	Elgersburg	107	Elgersburg.
Penkert	...	...	23	Artern.
Penn	...	...	284	Sangerberg.
Pepere	33	Bagnoli	33	Bagnoli.
Perales Churt	...	...	13	Alhama de Granada.
Peters	356	Zürich	95	Davos-Platz.
Petit	...	Clermont	269	Royat.
Petrequin	...	Fernand	145	Heilbrunn.
Pfeuffer	...	...	46	Birmansdorff.
Pflug	120	Frankenhausen	120	Frankenhausen.
Philbert	...	...	55	Brides.
Pickering	210	Monte Carlo	...	...
Pierre	...	...	89	Contrexéville.
Pierson	101	Dresden	101	Dresden.
Pingler	...	...	168	Königstein.
Planat	...	...	220	Néris.
Pleschner	...	...	162	Karlsbad.
Plunert	...	...	181	Liebwerda.
Pohl	25	Aussee	25	Aussee.
Pozzioli	109	Enghien	109	Enghien.
Präsil	130	...	130	Gleichenber.
Preiss	...	...	162	Karlsbad.
Prochnow	...	...	218	Müskau.
Pröll	225	Nice	226	Gastein.
Prompt	225	Nice.		
Prussian	230	Odessa.		
Quesada	...	...	356	Zujar.
Rabel	...	...	64	Caldas de Malavella.
Rabl	139	Hall (Austr.)	139	Hall (Austr.).
Rahn-Escher	...	...	356	Zürich.
Raillard	95	Dax.		
Rak	170	Krapina-Toplitz	170	Krapina-Teplitz.
Rakowski	155	Inowrazlaw	155	Inowrazlaw.
Rapin	177	Lausanne	177	Lausanne.
Rapp	...	...	258	Reichenhall.
Recordon	177	Lausanne	177	Lausanne.
Regnault	...	...	52	{ Bourbon l'Archambault.
Regaern	...	...	22	Arcnosillo.
Rehm	...	...	294	Schmaekalden.
Reichenbach	...	...	291	Saxon.
Reinvillier	...	...	80	Chateauneuf.
Reitemeyer	71	Casamicciola	71	Casamicciola.
Reumont	7	Aix-la-Chapelle	7	Aix-la-Chapelle.
Reuss	45	Bilin	45	Bilin.

LIST OF DOCTORS.

Name.	Page.	Winter Station.	Page.	Summer Station
Richter ...	...	...	318	Teplitz.
Riefenstahl	101	Driburg	101	Driburg.
Rieken ...	...	...	45	Birkenfeld.
Rinteln ...	...	...	230	Oeynhausen.
Ritscher ...	...	...	177	Lauterberg.
Roberts ...	36	Barcelona	36	Barcelona
Rocca-Tagliata	...	...	311	Sujo.
Rocca-Tartarini	63	Caldaniccia	63	Caldaniccia.
Roden ...	...	...	184	Lippspringe.
Röhrig ...	...	...	351	Wildungen.
Rompler ...	132	Gobersdorf	132	Gobersdorf.
Roscher ...	...	...	291	Schandau.
Rosenberger ...	...	...	169	Kösen.
Roth ...	349	Wiesbaden	349	Wiesbaden.
Rouge ...	177	Lausanne	177	Lausanne.
Royer ...	...	...	78	Challes.
Rudifanchas ...	...	...	283	San Adrian y La Losilla.
Rühle ...	...	...	69	Cannstadt.
Sachse ...	...	...	355	Zinnowitz.
Saez de Tejada...	...	...	302	Sierra Elviria.
Salafranca	...	...	123	Fuen-Alamo.
Salazar ...	...	...	182	Lierganes.
Salazar ...	...	...	232	Ontaneda y Alceda.
Salzado y Guillermo ...	...	...	71	Carratraca.
Sammret, J. ...	196	Malta...	196	Malta.
Sänger ...	...	...	311	Sulza.
Sanchez y Jabra	...	...	285	San Juan de Compos.
Sarraci ...	...	...	305	Solares.
Sastre y Domingulz ...	...	...	214	Montemayor y Bejar.
Säuerwald ...	...	...	230	Oeynhausen.
Saulmann	...	...	234	Ostende.
Savill ...	141	Hammam R'Irha	...	London.
Schoeder ...	42	Berne...	42	Berne.
Scharrenbroich	236	Pallanza.	...	
Schembri	196	Malta...	196	Malta.
Schenk ...	...	...	...	Königsdorff-Jastrzemb.
Scherpfert	...	...	166	Kissingen.
Schetelig	221	Nervi.	...	
Scheuer ...	...	...	307	Spa.
Schider ...	126	Gastein	126	Gastein.
Schieder...	...	...	22	Arco.
Schielp ...	30	Baden-Baden	30	Baden-Baden.
Schildbach	179	Leipzig	179	Leipzig.
Schimpf...	...	...	95	Davos-Platz.
Schindler	133	Gräfenberg	133	Gräfenberg.
Schirm ...	349	Wiesbaden	349	Wiesbaden.
Schläger...	151	Hubertusrunnen	151	Hubertusbrunnen
Schliep ...	30	Baden-Baden	30	Baden-Baden.
Schmidt ...	50	Borkum	50	Borkum.
Schmidt ...	72	Cassel	72	Cassel.
Schmitz ...	87	Cologne	87	Cologne.
Schmitz ...	287	San Remo	221	Neuenahr.
Schnee ...	225	Nice ...	162	Karlsbad.

LIST OF DOCTORS.

Name.	Page.	Winter Station.	Page.	Summer Station.
Schneer	11	Alassio.		
Schneider	130	Gleisweiler	130	Gleisweiler.
Schoeder	42	Berne	42	Berne.
Scholz	...	...	93	Cudowa.
Schötensack	...	...	270	Sachsa.
Schotten	72	Cassel	72	Cassel.
Schreiber	22	Arco	25	Aussee.
Schrieder	...	...	22	Arco.
Schrön	...	...	219	Naples.
Schücking	254	Pyrmont	254	Pyrmont
Schulze	...	...	313	Swinemünde.
Schumacher	7	Aix-la-Chapelle	7	Aix-la-Chapelle.
Schuster	7	Aix-la Chapelle	7	Aix-la-Chapelle.
Schwann	131	Godesberg	131	Godesberg.
Schwarz	...	...	28	Baden (Austr.).
Secchi	...	...	258	Reinerz.
Sécrétan	177	Lausanne	177	Lausanne.
Seegen	...	Karlsbad	162	Karlsbad.
Seitz	349	Wiesbaden	349	Wiesbaden.
Seltzner	101	Dresden	101	Dresden.
Serrano	...	...	138	Guarda-Vieja.
Serrano y Sanchez	...	...	13	Alfaro.
Servajan	...	...	271	St. Alban.
Settari	...	...	205	Meran.
Sevillana	...	...	...	Alcantuz.
Seyd	...	...	269	Ruhla.
Shiel	30	Baden-Baden	30	Baden-Baden.
Shuttleworth	287	San Remo.		
Siegel	...	...	30	Badenweiler.
Simrock	121	Frankfurt-on-Main	121	Frankfurt-on-Main.
Siordet	204	Mentone.		
Smith	287	San Sebastian	287	San Sebastian.
Sommer	121	Franzensbad	121	Franzensbad.
Sotier	...	...	166	Kissingen.
Soubeyran	...	...	338	Vic-sur-Cère.
Souligoux	...	...	338	Vichy.
Sparks	308	...	308	La Spezia.
Spengler	...	...	95	Davos-Platz.
Spies	...	...	206	Michelstadt.
Spurway	241	Pegli.		
Stabel	170	Kreuznach	170	Kreuznach.
Steiger	189	Lucerne	189	Lucerne.
Steiger	215	Montreux	...	
Stein	121	Frankfurt-on-Main	121	Frankfurt-on-Main.
Steinschneider	...	...	70	Carlsbrunn.
Stephani	198	Mannheim	108	Mannheim.
Stephens	69	Cannes.		
Stift	...	...	346	Weilbach.
Stilon	196	Malta	196	Malta.
Stöcker	189	Lucerne	189	Lucerne.
Stöhr	...	...	166	Kissingen.
Stollenkamp	41	Bentheim	41	Bentheim.
Storer	136	Ischia.		
Strahl	170	Kreuznach	170	Kreuznach.
Strasser	...	...	155	Interlaken.
Strehalano	...	...	172	Lacanne.
Stroth	41	Bentheim	41	Bentheim.
Sturge	225	Nice	...	
Sturm	...	...	169	Köstritz.

A A

Names.	Page.	Winter Station.	Page.	Summer Station.
Suñer	...	...	288	Santa Ana.
Taberlet	...	...	112	Evian-les-Bains.
Taboada	...	...	327	Trillo.
Taverney	...	...	275	St. Moritz.
Tanb	329	Vienna	26	Baden (Aust.).
Tejada y España	...	...	304	Sobron.
Tenner	71	Carlsruhe	71	Carlsruhe.
Tenlon-Valio	...	...	332	Uriage.
Teschemacher	...	...	221	Neuenahr,
Thewalt	...	...	168	Königstein.
Thiersch	179	Leipzig	179	Leipzig
Thiery	...	...	89	Contrexéville.
Thilenius	...	...	308	Soden.
Thomas	287	San Remo	30	Badenweiler.
Thomas	...	...	146	Herlein.
Thomas	...	...	316	Tatzmannsdorf
Thompson	13	Algiers	309	Spa.
Thompson	58	Brussels	58	Brussels.
Tillot	...	...	272	St. Christau.
Toca	194	Madrid	194	Madrid.
Tomasi	219	Naples	219	Naples.
Trion	...	...	181	Liebenzell.
Trost	7	Aix-la-Chapelle	7	Aix-la-Chapelle.
Tucci	...	...	4	Acqua-Santa.
Tweedie-Stodart	219	Naples.		
Unschuld	...	...	221	Neuenahr.
Urbaschik	170	Krenzen	170	Krenzen.
Urdapilleta	...	...	150	Horcajo.
Uterhard	...	...	345	Warnemünde.
Valcourt, de	69	Cannes.		
Valdivieso	...	...	305	Solan de Cabras.
Valentiner	10	Ajaccio	280	Salzbrunn.
Valentini	...	...	52	Bourbon-Lancy.
Valenzuela	...	...	174	Lanjarron.
Van Multern	46	Blankenberghe	46	Blankenberghe.
Varenhorst	...	...	220	Nenndorf.
Varrentrop	121	Frankfurt-on Main	121	Frankfurt-on-Main.
Vasquez	...	...	240	Paterna y Gigonza.
Vaust	101	Dresden	101	Dresden.
Vecino Villar	...	...	188	Lucaineua.
Veiel	...	...	69	Cannstadt.
Velten	349	Wiesbaden	349	Wiesbaden.
Venninger	...	...	344	Vöslau.
Ventura	...	...	325	Trenczin Töplitz
Verdallo	...	...	189	Luchon.
Verey	7	Aigle	7	Aigle.
Verhaeghe	...	...	57	Brughéas
Vidal	230	Paris	230	Paris.
Vidart	98	Divonne-les-Bains.	98	Divonne-les-Bains.
Vioira	194	Madeira	194	Madeira.
Viejo	...	...	237	{ Paracuellos de Gilloen.
Vigneau	...	...	272	St.-Christau.
Villafranca	...	...	63	Caldas de Besaya.
Viñolas	...	...	200	Segura de Aragon.

LIST OF DOCTORS. 371

Names.	Page.	Winter Station.	Page.	Summer Station.
Vittorelli	...	...	219	Naples.
Vogel	...	...	169	Korytnica.
Volger	108	Ems	108	Ems.
Von Aigner	...	...	170	Krapina Töplitz.
Von Althammer	...	...	22	Arco.
Von Brunn	...	...	184	Lippspringe.
Von Bühnau	...	...	87	Colberg.
Von Burkhardt	...	...	350	Wildbad.
Von Corval	...	...	295	Schöueck.
Von Feder	217	Munich	217	Munich.
Von Gärttner	311	Stuttgart	311	Stuttgart.
Von Gietl	217	Munich	217	Munich.
Von Hausen	130	Abbazia	130	Gleichenberg.
Von Haussman	311	Stuttgart	311	Stuttgart.
Von Haussman, sen.	...	...	351	Wildbad.
Von Haussman, jun.	...	...	351	Wildbad.
Von Heilbronn	...	...	199	Marienbad.
Von Heinrich	339	Vienna	325	Trenczin.
Von Hochberger	...	...	162	Karlsbad.
Von Kaan	...	...	205	Meran.
Von Koch	311	Stuggart	311	Stuttgart.
Von Kottowitz	...	...	322	Tobelbad.
Von Langenbeck	349	Wiesbaden	349	Wiesbaden.
Von Liebig, F.	217	Munich	258	Reichenhall.
Von Liebig, J.	...	...	123	Friedrichshall.
Von Lingelsheim	351	Wildungen	351	Wildungen.
Von Mühlleitner	28	Baden (Aust.)	28	Baden (Aust.)
Von Renz	...	...	351	Wildbad.
Von Schön-Perlashof	328	...	328	Tüffer.
Von Szontagh	222	Neuschmécks	222	Neuschmécks.
Von Waldhäusl	...	...	322	Tobelbad.
Wagner	10	Ajaccio	...	...
Wagner	...	...	282	Salzungen.
Wahn	...	...	169	Kösen.
Wakefield	...	...	9	Aix-les-Bains.
Walker	...	...	51	Boulogne-sur-Mer.
Weber	140	Homburg	140	Homburg.
Weber	...	Chur	17	Alvenau.
Wegener	107	Eilsen	107	Eilsen.
Wehner	57	Brückenau	57	Brückenau.
Wehse	...	...	173	Landeck.
Weidner	123	Friedsrichroda	123	Friedsrichroda.
Weitz	...	...	254	Pyrmont.
Welby	45	Biarritz.	...	...
Welsh	...	...	166	Kissingen.
Welsch	217	Münster am Stein	217	Münster am Stein.
Wendel	42	Berlin	42	Berlin.
West	...	Nice.	...	...
Whiteley	69	Cannes	109	Engelberg.
Wilhelmi	...	...	313	Swinemünde.
Wilkinson	127	Geneva	127	Geneva.
Williams	63	Cairo	96	Dieppe.
Williams	69	Cannes	...	...

Names.	Page.	Winter Station.	Page.	Summer Station.
Wilson ...	117	Florence ...	117	Florence.
Winkler ...	155	Inowrazlaw...	155	Inowrazlaw.
Winternitz ...	...	...	161	Kaltenleutgeben.
Wise, Tucker ...	...	London ...	196	Maloja.
Wittmann	201	Mayence ...	201	Mayence.
Wolf ...	...	...	293	Schlangenbad.
Wurm ...	...	...	317	Teinach.
Wyatt ...	219	Naples ...	...	
Yebenes ...	...	...	333	Valdeganga.
Young ...	117	Florence ...	117	Florence.
Zabala ...	...	...	21	Archéna.
Zapater ...	...	...	209	Molinar de Carranza.
Zdun ...	...	...	171	Krynica.
Zeroni ...	108	Mannheim ...	108	Mannheim.
Ziegelmeyer ...	...	...	174	Langenbrücken.
Zielemewsky ...	...	...	171	Krynica.
Zaemssen ...	349	Wiesbaden ...	349	Wiesbaden.
Zimmermann ...	145	Heligoland ...	145	Heligoland.
Zurbach ...	149	Homburg ...	149	Homburg.
Zürcher ...	...	...	155	Interlaken.

EXPLANATIONS AND TRANSLATIONS OF TECHNICAL TERMS AND PHRASES.

Amenorahœa.—Suppression of the menses.
Anœmia.—Poverty of the blood.
Anchylosis.—Stiffening of the joints.
Antispasmodic.—Against spasms.
Aphonia.—Loss of voice.
Apnœa.—Difficulty of breathing.
Articulation.—A joint.
Articular.—Concerning the joints.
Atonic.—Want of tone, debility.
Atrophy.—Lack of nourishment, emaciation.

Bronchi.—Small air passages in lungs.
Bronchial-catarrh.—Catarrh of the air-passages.
Bronchitis.—Inflammation of the bronchi.
Bronchus.—Windpipe.

Cachexia.—An unhealthy habit of body.
Calcium.—Lime.
Calciferous.—Lime-bearing.
Carbonate.—Containing carbonic acid.
Cervical.—Belonging to the neck.
Chalybeate.—Iron.
Chlorosis.—Green sickness.
Cholagogues.—Medicines promoting a flow of bile.
Climacteric.—Critical period of life.
Congestion.—Undue fulness of blood vessels, lungs, &c.
Coryza.—Inflammation of the lining membrane of the nostrils.
Cutis.—Skin.
Cutaneous.—Concerning the skin.

Diabetes.—Inordinate flow of urine containing sugar.
Diagnosis.—Recognition of a disease.
Diathesis.—Peculiar constitutional state.
Diphtheria.—A disease of throat.
Dislocation.—Out of joint.
Diuresis.—Flow of urine.
Diuretic.—Causing a flow of urine.
Douche.—A form of bath in which the water comes from above or the sides by mechanical power.
Dysmenorrhœa.—Painful menstruation.
Dyspepsia.—Difficult digestion.

Eczema.—A skin disease.
Effusion.—Escape of a fluid from vessels within the body into other parts.
Empyema.—Internal collections of pus.
Endemic.—Diseases of particular countries or localities.
Expectorants.—A class of medicines which aid in clearing the throat and chest.

Febrifuges.—Medicines which dispel feverish conditions.

Granulations.—A stage in the process of healing.
Gravel.—A disease of the bladder or kidneys.

Hæmorrhage.—Bleeding.
Hæmorrhoids.—Piles.
Hepatic.—Belonging to the liver.
Hernia.—Rupture.
Herpes.—A skin disease.
Homœopathy.—A school of medicine.
Hydatids.—Minute parasites which form collections within the human body.
Hydropathy.—Water cure.
Hypertrophy.—Over-growth.
Hypochondriasis.—A form of melancholy.

Idiopathic.—Occuring without apparent cause.
Idiosyncrasy.—Individual peculiarity.
Influenza.—A severe form of feverish cold.
Intermittent.—Recurring at intervals.
Iritis.—A disease of the eye.

Laryngitis.—Inflammation of the larynx.
Larynx.—Portion of the throat in which the **voice** is formed.
Leucorrhœa.—A white discharge in females.
Leprosis.—A skin disease.
Lithontriptics.—Medicine supposed to dissolve stone in the bladder.
Lumbago.—A rheumatic pain in the loins.

Menorrhagia.—Excessive menstruation.
Miasma.—An infectious element in the air.

Neuralgia.—A nervous pain.
Neurosis.—Nervous affection.
Nephritis.—Inflammation of the kidney.

Obesity.—Fatness.
Ophthalmia.—An eye affection.

Panacea.—A universal remedy.
Paralysis.—Palsy, loss of power.
Percussion.—A shaking.
Pharynx.—The back part of the mouth.
Phlebitis.—Inflammation of the veins.
Phrenitis.—Frenzy.
Phthisis.—Consumption.
Plethora.—Full habit of body.
Pleurisy.—Inflammations of lining membrane of lungs.
Pneumonia.—Inflammation of the lung.
Pruritus.—Itching.
Pulmonary.—Concerning the lungs.

Rachitis.—Rickets.
Remittent.—Which diminishes and increases febrile action periodically.

Saccharine.—Sugary.
Sciatica.—Pain in sciatic nerve (back of hip).
Scrofula.—A certain unhealthy condition of body.
Styptics.—Arresting flow of blood.
Suppuration.—Discharge of matter (pus).
Synocha.—Continuous fever.

Tendinous.—Belonging to the tendons.
Tuberculous or Tubercles.—A diseased **condition** of the lungs or other organs.

Uterus.—Womb.
Uterine.—Concerning the womb.

Varicose.—Enlarged and tortuous veins.
Viscera.—Bowels and internal organs.

CLASSIFICATION

ACCORDING TO THE DIFFERENT COUNTRIES OF THE STATIONS NAMED IN TEXT.

AFRICA.

Aboukir
Ain-Nonicy
Algiers
Bains de la Reine
Ben Haroun
Cairo
Cèdres (Source des)
Chazam
Hamma

Hammam Berda
Hammam de Gabes
Hammam el Eux
Hammam Melouane
Hammam Meskutin
Hammam R'irha
Helouan-les-Bains
Luksor
Madagascar

Madeira
Mara
Mogador
Monzaio-les-Mines
Mustapha Supérieure
Okme
Ramleh
Salah Bey

AUSTRIA AND HUNGARY.

Abazzia
Adelsberg
Ajnacskö
Akna Szlatina
Alàp
Al-Gyógy
Almamezö
Almás
Alsó-Sebes
Altsohl
Alt-Tura
Apataka
Arco
Arpad
Atya
Aussee
Aussen
Baden
Bocskó-Rahó
Bocznch
Bajfalú
Balsócz
Baracza
Bari
Bartfeld
Bekecs
Bela
Bellus
Benedekfalva
Benyus
Berencze
Bernstein
Besenyöfalva
Bikszád
Bén
Bilin
Bistritz
Bisztrú
Bodajk
Bocsing
Borkut
Borsa

Borsaros
Borszek
Botzen
Bréb
Brixen
Bruszno
Budimir
Budis
Budos
Budosko
Bujak
Buzias
Carlsbrunn
Cormus
Czacko
Czigelka
Daruvar
Deutsch-Kreutz
Dios-Györ
Dios-Jenö
Disznopatak
Doktorka
Dolka
Dombhát
Dorna
Dotit
Dragomerfalva
Drahowa
Dreykirchen
Drohobycz
Dubowa
Dubrawa
Ebed
Ebrinch
Egartbad
Egbell
Eghegh
Egerdach
Egelhof
Eichwald
Einöd
Eisenbach

Elisabeth Salzbaths
Erdöbenye
Erlachbad
Erlau
Ernsdorf
Falu-Szlatina
Farkas-Mesö
Feletckut
Fellathal
Felsö-Alap
Felsö-Apsá
Felsö Bajom
Felsö Nercsznicze
Felsö Rusbach
Felsö Visso
Fortyógo
Franzensbad
Füred
Gaberneg
Gainfahrn
Galthof
Gastein
Gath
Geltschberg
Giesshübl-Puchstein
Gleichenberg
Gleisslibergerbad
Gmünd
Gmunden
Gräfenberg
Gran
Grasnawawoda
Gries
Grodeck
Gross-Schlagendorf
Grosswardein
Gross-Wunitz
Gyüzy
Haj-Stubna
Hall, nr. Linz
Hall, nr. Innspruck

AUSTRIA AND HUNGARY—continued.

Hallein
Harkányi
Harsfalva
Heiligekreuzbath
Heiligenstadt
Herculesbaths
Herlein
Hevitz
Hildegarde-Brunnen
Homok
Homorod
Hozumező
Hradiszko
Huniady-Yános
Innichen
Ischl
Ivanda
Ivanyi
Ivonicz
Jacobfalva
Jahoduika
Jamnicza
Jaróslaw
Jehu
Johannisbad, nr. Pardubitz
Johannisbad, nr. Melnik
Jood
Josephsbad
Joszo
Jurowla
Kabolapolyána
Kács
Kaltenleutgeben
Kammer
Karlsbad
Karpfen
Keked
Kemend
Kerö
Keruly
Kierling
Kiralyi
Kiralymező
Kis-Czeg
Kis-Kalnu
Kis-Saros
Klanssen
Klein-Chocholna
Klieningen
Klokocs
Kobersdort
Königswarth
Kouopkowka
Korond
Koroud
Korsow
Korytnica
Kostreinitz
Köstritz

Kovaszva
Krapinu-Töplitz
Kreuzen
Krynica
Krzessow
Kugelbad
La Caldare
Lesina
Leszina
Levico
Liebwerda
Lienzmühl
Lipik
Lubien
Lucsky
Luhatschowitz
Luhi
Magyar-Szent-Lazlo
Malnas
Marienbad
Mastinecz
Mattigbad
Mauer
Meidling
Meran
Mitterbad
Moha
Monfalcone
Mscheno
Nelefina
Neudorf
Neuhaus
Neu-Lublau
Neuschmecks
Neusiedel
Neusohl
Nook
Oberladis
Oberuhaus
Obernau
Ober-Rauschenbach
Oelves
Ofen
Olahfazlú
Olah St. Györzgy
Olenyova
Ollmütz
Oroslau
Osztrovsk
Pality
Parád
Pesolina
Pesth
Pilsen
Pistyán
Po-Cseviczc
Pojau
Polena
Pongyelok
Pötschnig

Preblau
Prodersdorf
Prützerbad
Püllna
Pyrawarth
Rabbi
Rabka
Radein
Rajecz
Rakós
Ranigsdorf
Ratzes
Reichenau
Rima-Brezó
Riva
Rodna
Roggendorf
Rohitsch
Römerbad
Ronya
Rosenau
Roznau
Ruzspolyana
Sadschutz
Saidschutz
St. Peter
St. Radegund
Salzburg (Salzkammergut)
Salzburg (nr. Hermannstadt)
Sangerberg
San Georgen
San Lorenzo
Schmecks
Schönau
Schums
Schwarzenberg
Schwefelbad
Sesavnik
Sedlitz
Sibó
Sid
Sisso
Sombór
Sostó
Spalato
Steinwasser
Sternberg
Stoika
Stuben
Suligul
Sulz
Sutinsko
Szalatnya
Szaldobos
Szaploncza
Szator
Szczawnicza
Szent-Ivan

AUSTRIA—continued.

Szinyak
Szinye-Lipocz
Szkleno
Szkló
Szliácz
Szmrdak
Szobrancz
Szombhát-Falva
Szubicza
Szulin
Tapolcza
Tapolz Bisztrá
Tarna
Tatzmannsdorf
Telgart
Teneke
Teplitz
Teplitz-Schönau

Theissholz
Tobelbad
Töplika
Topolschitz
Topuczko
Torda
Trenczin-Teplitz
Trieste
Truskowicze
Tschawitz
Tüffer
Tür
Tusnád
Ugód
Ullersdorf
Unter-Micsnye
Ustrom
Uszók

Vale Szkragye
Várgéde
Veldes
Velejte
Vetzel
Vihnye
Visk
Vöslau
Wartenberg
Wenzelsbath
Wieliczka
Wolfs
Wolfsegg
Zajzon
Zanyka
Zögg
Zovány
Zwieselsalpe

BELGIUM.

Blanchemont
Blankenbergh
Chaudefontaine

Heyst
Middlekerke
Ostende

Spa
Tongeren

DENMARK.

Barestrand-Syssel
Iceland
Kirstenpils

Kopenhagen
Marienlust

Randamel
Thingöe-Syssel

FRANCE.

Abbecourt
Absac
Agon
Aincelle
Aix (Provence)
AIX-LES-BAINS
Ajaccio
Albens
Alais
Aled
Allevard-les-Bains
Amélie-les-Bains
Amphion
Andabre
Annecy
Antibes
Arborme
Arachon
Arcs, Les
Argentieres
Arlum
Arromanches
Asnelles
Audinac
Angnat
Aulus
Availles
Avesne
Ax
Bagnères de **Bigorre**

Bagnet, Le
Bagnoles
Bagnols
Bains
Bains, near Arles
Baluruc
Banyuls
Baran
Barbazan
Barberie
Barbotan
Barèges
Barzun
Bastennes
Bauche La
Beaucent
Beaulieu, nr. Issoire
Beaulieu, nr. Nice
Beauprean
Bellesme
Belleville
Belloc
Bernières
Bernos
Berthemont
Besse
Bétaille
Benzeval
Biarritz
Biluzai

Bleville
Boisse, La
Bondonneau
Bonnefontaine, La
Bonneval
Boules
Boulogne-sur-Mer
Boulou, Le
Bourasol
Bouquéron
Bourbou l'Archambault
Bourbon Lancy
Bourbonne
Bourboule, La
Bourg d'Ault
Bourg d'Oisans
Brides
Brugheas
Bruneval
Bulgneville
Bullycome, Le
Busignargues
Bussang
Cabourg
Cadéac
Caille, La
Calais
Caldaniccia
Camarès

FRANCE—continued.

- **Cambo**
- Camoins
- Campagne
- Candé
- **Cannes**
- **Cap d'Ant**ibes
- Capveru
- Carconières
- Carteret
- Casnefouls
- Castel-Jaloux
- Castera-Verduzan
- **Cauterets**
- Canvalat-le-Vigan
- Cayeux
- Cayla, La
- Celles
- Chabetout
- **Chaldette, La**
- Challes
- **Chamounix**
- Champoleon
- Chapelle-Godefroy, La
- Charbonnières
- Chateau Gontier
- Chateauneuf
- Chateldon
- **Chatel Guyon**
- **Chaudes-Aigues**
- Chaumbon
- Chaumout
- Chemille
- Cherbourg
- Choranche
- Clermont
- Coise
- Collioure
- **Cont**rexéville
- Cordéac
- Corenc
- Corneille
- Couchon
- Coudes
- Coudillac
- Courpière
- Cours
- Courtomer
- Courseuilles
- Coutainville
- **Cransac**
- **Crèche**
- Criel
- Croisic, Le
- Crol, Le
- Crotoy, Le
- **Deauville**
- Désaigues
- Desvres
- **Deux Lots**
- **Dieppe**
- Dieu-le-fit
- Digue
- Dinan
- **Dives**
- **Divonne-les-Bains**
- Dinard
- Donéue
- Domeray
- Domèvre
- **Dorres**
- **Douarnez**
- **Douville**
- **Douai**
- **Duivon**
- **Dunkerque**
- Durtal
- **Eaux-Bonnes**
- Eaux-Chaudes
- Ebeaupin
- Echaillon
- **Ecquevilly**
- **Ecuillé**
- Encausse
- Enghien
- Enn
- Erquy
- Escaldas, Las
- Escouloubre
- Etretat
- Enzet
- **Evaux**
- **Evian-les-Bains**
- Farctte
- **Fécamp**
- Felines
- Ferranche
- Ferrière, La
- Féron
- Finmorbo
- **Fonciergue**
- **Fonfrede**
- **Fonsaiute**
- **Fonsalada**
- **Fousanches**
- **Fonsrouilleuse**
- **Fontagre or Sorède**
- **Fontenelle**
- **Foradade**
- **Forceral**
- Forges-les-Bains
- Forges-les-Eaux
- Foucaude
- Fourchambault
- **Fumades**
- Gabinu
- **Gamarde**
- **Garde de Bio, La**
- **Gardinière**
- **Garris**
- **Gazost**
- Genestelle
- Gigondas
- Ginoles
- Glaine-Montaigu
- Gournay
- Gramat
- Grandcamp
- **Grand**eyrol
- Graudril
- **Gran**ville
- Grasville-l'heure
- Gréoulx
- Grinneaux
- Guaguo
- Guillon
- Guitera
- **Havre, Le**
- Henuebon
- **Hermonville**
- Herse, La
- Heucheloup
- Home-Varaville, Le
- **Honfleur**
- **Houches, Les**
- **Houlgate**
- Hourdel
- **Hyères**
- Jalleyrac
- **Joanette**
- **Job**
- **Jose**
- Jouche
- Labassère
- Labestz-Biscaya
- Lacanne
- Laifour
- La Malou
- Lamotte
- Langrune
- Larivière
- Laroche
- Lassère
- **Lavardens**
- **Legué St. Breuil**
- **Lepouliguen**
- **Les Salins**
- **Lion-sur-Mer**
- **Llo**
- **Lons-le-Saulnier**
- **Luc-sur-Mer**
- **Luchon**
- **Luxeuil**
- **Macon**
- Magnac
- Marat
- Marcols
- Marlioz
- **Marseilles**
- Martigné-Briant
- Martiguy

FRANCE—continued.

Martres
Mayres
Mazel, Le
Medagues
Mentone
Mereus
Mers
Mezières
Miers
Moingt
Molitg
Monestier de Briançon
Monestier de Clermont
Montbrison
Montbrun
Montchanson
Mont-Dore
Montégut-Ségla
Moutlignon
Mont Louis
Montmirail
Montner
Montol
Montpellier
Montpensier
Mortefontaine
Nebouzat
Netlinch
Neris
Neuville
Neyrac
Nice
Nohanend
Nouvelle, La
Olette
Olliergues
Orezza
Origny
Oriol
Outraneourd
Paimpal
Panassan
Paramé
Paris
Passy
Pau
Pandraux
Perrière, La
Perruches
Petites Dalles, Les
Pierrefonds
Pietrapola
Plaine, La
Plan, Le
Planchamps
Plan de Phazy
Plombières
Poggeto Theniers
Pontaillac
Pontamafrey

Pont à Mousson
Pont de Bared
Pontrieux
Pornic
Port-Bail
Porte
Port-en-Bessin
Port Thareau
Pougues
Pouillon
Pouroille
Prades Vernet
Prechac
Preste, La
Prompsad
Propiac
Provins
Prugnes
Puzziehello
Quez
Quinéville
Regneville
Remollion
Renaisson
Renlaigue
Reunes
Reyrieux
Rieumajou
Roccabigliera
Romeyer
Roucevaux
Roscoff
Roueas Blanc
Rouzat
Royan
Royat
Roye
Sables d'Olonne
Sail-les-Bains
Sail-sous-Couzan
St. Alban
St. Amand
St. André
St. Boes
St. Bonnet
St. Briac
St. Christau
St. Denis
St. Diery
St. Galmier
St. Gervais
St. Gildas
St. Honoré
St. Icaire
St. Jean d'Aulph
St. Jean de Luz
St. Laurent
St. Louhouer
St. Lunaire
St. Malo

St. Marie de Mont
St. Martin de Fenouille
St. Martin Lantosquo
St. Maurice
St. Myon
St. Nectaire
St. Ours
St. Pair
St. Pardoux
St. Parize
St. Pierre d'Argenson
St. Prieste des Champs
St. Prieste-la-Roehe
St. Quentin
St. Raphael
St. Remy
St. Roman
St. Sautin
St. Sauveur
St. Simon
St. Thomas
St. Vaast
St. Valery-en-Caux
St. Valery-sur-Somme
St. Vallier
St. Yorre
Ste. Adresse
Ste. Helene
Ste. Madeleine
Ste. Marie (Cantal)
Ste. Marie (Hautes Pyrenées)
Salees
Salies
Salies de Béarn
Salins
Salins Moutiers
Salles
Salz
Santenay
Sarcey
Saubuse
Saucats
Saulee
Sault
Sausillonges
Saute-Vaux
Savergnolles
Saxe, La
Secy
Segray
Segre
Semur
Senteiu
Sermaise
Silvanès
Siradan
Sixt
Sotteville

FRANCE—continued.

Soulac
Soulieux
Talloires
Tarascon
Tercis
Termignon
Termaut
Ternau
Terrasie, La
Teste, La
Thuez
Tissières
Tivoli
Toulon
Trebas
Tremblade, La
Tremiscau
Treport, Le

Tresclaix
Trollière, La
Trouville
Urbalacone
Uriage
Ussat
Usson
Vaisse
Val André, Le
Vals
Varennes
Vaugnières
Velleminfroy
Velleron
Velmont
Vernet, Le
Veules
Veulettes

Vichy
Vie-sur-Cer
Vicoigne
Vic-le-Comte
Viguolles
Villefranche (Aveyron)
Villefranche (Alpes Maritimes)
Villers-sur-Mer
Villerville
Vinça
Visos
Vitry
Vittel
Vrécourt
Yeuzet
Yport

GERMANY.

Abach
Abensberg
Adelheidsquelle
Adelholzen
Adorf
Aix-la-Chapelle
Alexanderbath
Alexisbath
Alt-Haide
Altwasser
Antogast
Apollinaris
Arnstadt
Artern
Assmannshausen
Auerbach
Augustusbath
Baden-Baden
Badenweiler
Bebra
Bentheim
Berg
Berg-Gilfshuebel
Beringerbath
Berka
Bertrich
Beuren
Birkenfeld
Birlenbach
Birresborn
Blankenburg
Bocklet
Bodendorf
Bodenfelde
Böll
Boltenhagen
Borbyc
Borkum

Bourtscheidt
Bramstadt
Braunfels
Brückenau
Burgbergheim
Buschbath
Calw
Camstadt
Charlottenbrunn
Charlottenburg
Chatenois
Christenhofsbad
Cleve
Colberg
Crailsheim
Cudowa
Cuxhaven
Czarskow
Dangast
Dicdenow
Diemeringen
Dievenow
Dingolfing
Dinkhold
Dirsdorf
Dizenbach
Dobberan
Dorfgeismar
Draitschbrunnen
Driburg
Dürkheim
Dürrheim
Durrwangen
Düsternbrook
Eberbach
Ebningen
Echzell
Eckartsbrunnen

Eckerberg
Eckernförde
Edenkoben
Ehrenbreitstein
Eilsen
Eimbeck
Elizabethbath
Elmen
Elster
Empfing
Ems
Eppenhausen
Ernabrunnen
Eschelloh
Fachingen
Falkenberg
Falkenstein
Feldafing
Feldberg
Fiestel
Flinsberg
Forbach
Frankfort-on-Maine
Freienwalde
Freyersbach
Friedrichshafen
Friedrichshall
Friedrichsroda
Friedrichs Wilhelmsbath
Gehringswalde
Geilnau
Geissliugen
Georgenbath
Geroldsgrün
Geroldstein
Giengen
Gleisweiler

GERMANY—continued.

Gleissen
Goczalkowitz
Godelheim
Godesberg
Goldbach
Goldberg
Göppingen
Görbersdorf
Goschwitz
Greifenberg
Greifswalde
Griesbach
Griesbath
Gross-Albertshofen
Grosskarben
Grüben
Grull
Grund
Grundhofen
Günthersbath
Hackelthal
Hafkreuz
Hall
Halle
Halsbrücke
Hambach
Hanau
Hardeck
Harzburg
Hechingen
Heckinghausen
Heidelberg
Heilbrunn
Heilstein
Heinrichbrunnen
Helmstädt
Heppingen
Heringsdorf
Hermannsbad
Heselwangen
Hinnewieder
Hofgeismar
Hohenberg
Höhenstadt
Hohenstein
Holzhausen
Homburg
Hubertusbrunnen
Hüttersbach
Ilmenau
Imnau
Inselbad
Jaxtfeld
Jobsbath
Johannisberg
Jordansbath
Jungbrunn
Kanitz
Karlshafen
Karlsruhe

Kaudenbach
Kellberg
Kiel
Kirchberg
Kirchbrunnen
Kirchheim
Kissingen
Kleinengstingen
Kleinern
Klein-Schirma
Klütz
Kochel
Kondrau
Königsborn
Königsbrunn
Königsdorff-Jastrzembs
Königstein
Kornwestheim
Kösen
Köstritz
Krähenbath
Kraukenheil
Kreuth
Kreuznach
Kronberg
Kronthal
Krumbach
Kunzendorf
Laer
Lamscheidt
Landeck
Landskron
Langenberg
Langenbrücken
Langensalza
Laubbach
Lauchstadt
Lausigk
Lauterberg
Lendershausen
Leutstetten
Lichtenthal
Liebenstein
Liebenzell
Limmer
Lindenholzhausen
Lippspringe
Löbenstein
Ludwigsbrunnen
Luisenbath
Lüneburg
Malmédy
Mariabrunnenbath
Marienfels
Markammer
Meinberg
Mergentheim
Michelstadt
Misdroy

Moching
Mondorf
Münchshofen
Münster
Munsterberg
Muritz
Muskau
Nammen
Nassau
Nauheim
Naumberg
Neundorf
Neuenheim
Neuhaus
Neuenahr
Neumarkt
Neu Ragóczi
Neuschwalheim
Neustadt (Haart)
Neustadt (Saale)
Neustadt-Eberswalde
Niederbronn
Niederhall
Nieder-Langenau
Niedernau
Niederselters
Nieratz
Nierstein
Nördlingen
Norderney
Northeim
Ober-Brambach
Ober-Herzern
Oberlahnstein
Obermendig
Oberselters
Obertiefenbach
Oberwinter
Ochsenhausen
Oeynhausen
Offenau
Offenstein
Okarben
Oldesloe
Olmenhausen
Oppenau
Osterspay
Osthofen
Ottenstein
Parchim
Partenkirchen
Peissenberg
Petersthal
Plaue
Polzin
Potsdam
Prenzlau
Pyrmont
Radeberg
Rappenau

GERMANY—continued.

Rastenberg
Recklinghausen
Rehburg
Rehme
Reiboldsgrün
Reichenhall
Reinerz
Reutlingen
Rietenau
Rippoldsau
Rodenberg
Rohnau
Roigheim
Roisdorf
Ronneburg
Rosenheim
Rosheim
Rosswein
Röthelbad
Rothenbach
Rothenburg
Rothenfelde
Rothenfels
Rothwell
Rudolstadt
Ruhla
Saargemünd
Sachsa
Säckingen
St. Blasien
St. Georgen
St. Ulrich
Salzbrunn
Salzhausen
Salzschlirf
Salzuffeln
Salzungen
Sassnitz
Schandau
Schieder
Schlackenbath
Schlangenbad

Schleusingen
Schmalkalden
Schmeckwitz
Schönbeck
Schwalbach
Schwalheim
Schwallungen
Schweizermühle
Schwelm
Sebastiansweiler
Seebruch
Seeon
Selters
Sennfeld
Sierk
Sinzig
Siromibath
Soden
Soest
Sohl
Staden
Stavenhagen
Steben
Steinabad
Steinfurt
Steinheyde
Streitberg
Suderode
Sulza
Sulzbad
Sulzbach (Alsatia)
Sulzbach (Baden)
Sulzmatt
Swinemünde
Sylt
Tambach
Tannenbrunnen
Tatenhausen
Tegernsee
Teinach
Tennstädt
Tharandt

Theusserbad
Tönnisstein
Traischorloff
Travemünde
Tuebingen
Ueberkingen
Ueberlingen
Uhlmühle
Valdorf
Veierbach
Vippach-Edelhausen
Wachenheim
Waldbad
Wangeroog
Warmbrunn
Warnicken
Warnemünde
Wassacherberg
Wattenweiler
Weilbach
Westerland-Sylt
Wiesan
Wiesbaden
Wiesenbad
Wiesloch
Wildbad
Wildungen
Wilhelmsbad (Silesia)
Wilhelmsbad (Saxony)
Wilhelmsbad (Hessen)
Wimpfen
Winterbach
Wipfeld
Wittekind
Wolfach
Wörth
Zaisenhausen
Zavelstein
Zerbst
Zinnowitz
Zittau
Zoppot

GREAT BRITAIN.

Aberayron
Aberbrothwick
Aberystwith
Aberdaron
Aghaloo
Airthrey
Aldborough
Alderney
Allonby
Alnmouth
Alton Towers
Amblesido
Amlwich
Amroth

Antrim Spa
Appledore
Ardrossan
Ashby-de-la-Zouche
Arnside
Athimonns
Athlone
Ballater
Ballycotton
Ballyspollan Spa
Bangor
Bantry
Barmouth
Bath

Beaumaris
Ben Rhydding
Berwick
Bexhill
Birchington
Bishops-Down Grove
Blackpool
Bognor
Bonnington
Bournemouth
Bray
Bridge of Allan
Bridge of Earn
Bridlington Quai

GREAT BRITAIN—continued

Bridport
Brighton
Brixham
Broadstairs
Broughty Ferry
Brownstown Spa
Bude
Budleigh-Salterton
Builth
Bundorran
Burnham
Bushey, the Hall
Butterby
Buxton
Campbelltown
Carnarvon
Castel Connell
Cheltenham
Clacton-on-Sea
Claremont
Cleethorpes
Clevedon
Clifton
Clynnog-Vawr
Colwyn Bay
Courtmacsherry
Cowes
Criccieth
Crieff
Cromer
Crosshaven
Cullercoats
Dale
Dalkey
Dartmouth
Dawlish
Deal
Derrindaff
Devonport
Dinsdale
Douglas
Dovedale
Dover
Drennau Springs
Drogheda
Droitwich
Drumsna
Drumgoon
Drumrastel
Dumlane
Dunbar
Dundrum
Dunmore
Dunoon
Durness
Eastbourne
Epsom
Exmouth
Falmouth
Felixstowe

Filey
Fleetwood
Flint
Folkestone
Fordel
Fowey
Frasersburgh
Freshwater
Galway Spa
Garryhill Spa
Gisland
Glanagarin
Glengariff
Gloucester
Golden Bridge
Grange-over-Sands
Gravesend
Great Yarmouth
Guernsey
Hampstead
Harrowgate
Hartlepool
Harwich
Hastings
Hayling Island
Helensburgh
Heligoland
Herne Bay
Holbeck
Holkham
Holyhead
Holywell
Holywood
Horley Green
Hornsea
Hovehampton
Hovingham
Howth
Hunstanton
Hythe
Ilfracombe
Ilkley
Instow
Islington
Johnstown
Keswick
Kilburn
Kilkee
Kilkenny
Killymard
Kilrush
Kinsale
Knaresborough
Krevenish
Lake District
Largs
Leamington
Lee
Lisdoonvarna
Littlehampton

Littlehaven
Llandrindod
Llandudno
Llandwrtyd
Llanfairfechan
Llangranog
Llanstephen
Looe
Lowestoft
Lucan
Lulworth West
Lymington
Lynmouth
Lytham
Mablethorpe
Malahide
Mallow
Malta
Malvern
Margate
Matlock
Melcombe Regis
Melksham
Millport
Minehead
Moffat
Morecambe
Mumbles
Mumby-cum-Chapel
Nairn
New Brighton
Newcastle
New Quai (Engl.)
New Quai (Wales)
Newtondale
Nottington
Oban
Paignton
Pannanich
Passage
Peebles
Pembrey
Penarth
Pendine
Penmaenmawr
Penzance
Pitcaithty
Pitlachy
Plymouth
Portland
Portobello
Port Rush
Port Steward
Pwllheli
Queenstown
Ramsgate
Ramsay
Redcar
Redruth
Rhyl

BB

GREAT BRITAIN—continued.

Rosstrevor
Ryde
Ryhope
St. Andrews
St. Anne's-on-Sea
St. Bee's
St. David's
St. Lawrence-on-Sea
St. Leonard's
St. Mary's
Saltcoaths
Salcombe
Saltburn
Saltfleet-Haven
Sandgate
Sandown
Sandrocks
Sark
Saundersfoot
Scarborough
Seaford
Seasenles
Seaton
Seaton Carew
Sea-View
Shanklin
Shap
Sheerness-on-Sea
Shotley
Sidmouth
Silloth
Skegness
Skinburness
Southampton
Southbourne
Southend
Southport
Southsea
South Shields
Southwold
Spital
Starbeck
Streatham
Strathpeffer
Sutton
Swanage
Swanlibar
Swansea
Temburg
Tenby
Thorpe Arch
Thurso Bay
Tissington
Torquay
Totland Bay
Towyn
Tramore
Troutbeck
Tunbridge Wells
Tynemouth
Ullswater
Upton
Ventnor
Vicarsbridge
Victoria Spa
Walton-on-the-Naze
Warrenspoint
Watchet
Wells-next-the-Sea
Wemyss Bay
Westgate-on-Sea
Weston-Super-Mare
Westward Ho
Weymouth
Whitburn
Whitby
Willoughby
Windermere
Windsor Forest
Withernsea
Worthing
Yarmouth

GREECE.

Adepsos
Aegina
Chios
Corfu
Cos
Dirce
Epidaurus
Gythium
Hellopia
Hermione
Hypate
Ikaria
Kaifa
Kaissariani
Kalauria
Karithena
Katharsion
Keuchreæ
Klemutzion
Kythmos
Langassa
Laurion
Lebetzoba
Lemnos
Lepanto
Lesbos
Letautus
Ligurio
Lintzi
Lucas
Lutraki
Melos
Milo
Methana
Moselli
Naxos
Nissyros
Páros
Peleikiton
Santorin
Skara-Chori
Tenos
Thera
Thermia
Thermopylae
Thesbis Spring
Vonitza
Zante
Zea

HOLLAND.

Hilversum
Katwyk
Scheveningen
Wyk aan Zee
Zandvoort
Zwolle.

ITALY.

Abano
Acireale
Acquacetosa
Acqua-Acidola
Acquae-Albulae
Acqua-Bolle
Acqua Puzzolente
Acqua Raineriana
Acqua Santa
(Terni)
Acqua Santa di Buy-
lato

CLASSIFICATION. 387

ITALY—*continued*.

Acqua-Santa (Genoa)
Acquae-Subreni-Homini
Acqui
Agnano
Aitora
Alassio
Albano
Alcama
Ali
Alica
Allegrezza
Allume
Amalfi
Ammoniac Gas Spring
Anacapri
Anagui
Ancona
Anguillara
Arcidosso
Arenzano
Armajolo
Asciano
Ascoli
Asinolunga
Aspio
Astroni
Baccanella
Bagnaccio
Bague-a-Baccanella
Bagnoli
Bagni-à-**Acqua**
Bagni-à-**Morba**
Bagni, Acqua dei
Bagno Bossalo
Bagno d'**Apollo**
Bagno-Fresco
Bagno dei Rhachitici
Bagno-in-Romagna
Baldini, Acqua dei
Balnea **d'Avignone**
Bari, Acqua dei
Battaglia, La
Bellaggio
Benetutti
Bergallo
Bivuto di Termini
Bobbio
Bocheggiano
Bordighera
Borgo-Maro
Borla
Bormio
Borra
Borrone
Bottaccio
Bresigala
Brindisi
Bron**ia**
Bruca

Buca dei Fori
Burrone
Caccio-Cotto
Cadena**bbia**
Caldanella
Caldieri
Caldini
Calliano
Calvello
Canal-grosso
Caprenne
Capri
Caprifico
Casale
Casamicciola
Casa Nuova
Casa Stronchino
Casciani
Casino
Casiola
Castel-Doria
Castellamare
Castelletto Mascagni
Castelletto d'Orba
Castel San Pietro
Caste**l**nuovo
Castiglione
Castro-Caro
Catania
Catena
Cattenaja
Cave
Caz-di-**Bagno**
Cecinella
Cefalu
Cerresoles
Cesalpino
Cetona
Chianciano
Chiatamone
Chiavari
Chitignano
Chiusa dei Monaci
Cinciano
Cipollo
Citára
Civilliano
Civita Vecchia
Colico
Como
Cornigliano-Ligure
Corticella
Cotone
Cotto
Courmayor
Craveggia
Doccio
Dofana
Dovadol,
Elba

Eufemia
Faënza
Falciano
Fano
Ficoncella
Filetta
Floren**ce**
Fonga
Fontaccia
Fordignano
Fossino
Fosso degli Onta**ni**
Gagliana
Galleraje
Gavorano
Gen**oa**
Gerace
Giglio
Giunco-Marino
Grotto **del** Cane
Imola
Irno (Val d')
Ischia
Isola Bona
Lama
La Salute
Leccia
Levana
Lipara
Livorno
Loreta
Lu
Lucca
Macerrato
Madonna à Papiano
Madonna di tre Fiume
Maria dell 'Aquila
Maria in Bagnos
Marsala
Masino
Massa
Mercantale
Messina
Meta
Miemo
Milan
Moggiona
Molla, Il
Monsummano
Montafia
Montagnone
Montalceto
Mont'Amata
Monte-Catini
Montegrotto
Monteflascone
Monte-Ortone
Monte-Rotondo
Montione
Morba

ITALY—continued.

Mortagone
Naples
Nave dell'Inferno
Nerone (Stufa di)
Nervi
Nitrolis
Nocera
Noceto
Palazonia
Palazzolo
Palermo
Palestrina
Pallanza
Pantano
Paterno
Pegli
Pejo
Pelagio
Pelago
Penna, La
Pergine
Pertino
Pesaro
Pescara
Petraglia
Petriolo
Pietra
Pigna, La
Pillo
Pisa
Pisciarelli
Pitigliano
Pizzo, Il
Pizzofalcone
Poggetti
Poggibonzi
Poggio Curatale
Poggio Pinci
Poggio Rosso
Pontano
Popoli
Porretta, La
Pozzuoli
Pré St. Didier
Pretiolo
Puzzola di Picuza
Querzola
Raddusa
Rapullano
Rapallo

Ravone-in-
  Casaglia
Recoara
Retorbido
Riando
Riguardo
Rimini
Rio
Riolo
Rio Meo
Rio Sordo
Rita
Rocca San Felice
Rocegno
Romagna
Rombole
Rome
Roselle
Rostona
Poggetti
Poggibonzi
St. Cassien
St. Genis
St. Vincent
Salceti
Salerno
Salice
Saló
San Casciano
San Daniele
San Fedele
San Filippo
San Genesio
San Germano
San Giuliano
San Ginseppe
San Leopoldo
San Martino
San Marziale
San Michele
San Remo
Santa Cesarea
Santa Gonda
Santa Margherita
Santa Restituta
Santa Vigone
Sasso di Maremma
Saturnia
Sciacca
Sclafani

Segesta
Seraglio
Serapis (Temple)
Seravelle
Sestri-Ponente
Sigliano
Sinigallia
Sorrento
Spezia, La
Sprofondo
Stresa
Sujo
Tabbiano
Talamanaccio
Taorozeina
Telese
Terme Luigiane
Termini
Termini Castroreale
Testa
Tolpa
Torre Anunziata
Torre del Grecco
Torretta, La
Trani
Trescore
Tritoli
Turin
Valdieri
Valenza
Vellebro
Venafro
Venice
Verrazano
Vescovo
Vialla
Viareggio
Vicarello
Victorie
Vignale
Vignone
Villa delle Caselle
Villa d'Este
Vinudio
Visone
Viterbo
Voltaggio
Volterra
Voltri
Zafarana

PORTUGAL.

Alcafache
Arogos
Bellas Agoas
Cabeça-da-Vida
Caldas do Geres
Caldas da Nossa
  Senhora
Caldas da Rainha

Chaves
Gafete
Gaiciras
Junqueiro
Lisbon
Madeira
Mirardella
Monsão

Oporto
Rede de Corvaçeira
Rio-Mayor
Rio-Real
Sau Pedro do Sul
Torres-Vedras
Vidago
Vizella

PRINCIPALITY OF MONACO.

Condamine | Monaco | Monte Carlo

RUSSIA.

Abas-Tuman
Alexander's Spring
Barbern
Beschtau Bath
Borschom
Busk
Bykowicz
Ciechocinek
Dubograedsk
Eisenberg
Essentuk
Gori
Helsingfors
Janischek
Kammietz-Podolsk
Kaschin
Kastanowka
Katharinbath

Kemmern
Kiszlawodsk
Klutschewsk
Kötschenowa
Kunda
Kungara
Kuppis
Langenberg
Lipezk
Neskutschnoie
Nowosselja
Odessa
Oesel
Orel
Orenburg
Oni
Pjätigorsk
Psekups Springs

Riga
Sarepta
Schelesna-**Wodsk**
Schmardau
Serdepol
Sergiewsk
Siekeriki
Sitka
**Slepzoff-Michael-
off Springs**
Spag
Stolypin
Toropetz
Twer
Uleaborg
Undary
Utzera
Zagwera

SPAIN.

Alange
Aláraz
Alcala
Alcantud
Alceda
Alhama de Aragon
**Alhama de
Granada**
Alhama de Murcia
Alicante
Alicum
Aliseda
Almeida
Almeria
Aramazona
Archena
Archevaleta
Arenosillo
Argentona
Arnedillo
Arranzarre
Artejo
Baños
Barambio
Baza
Bejar
Belascoin
Bellus
Benavente
Benimarfull
Bertua
Betelú

Bonar
Borines
Braque
Buyeres de Nava
Buzost
Cadiz
Caldas de Besaya
Caldas de Bohi
Caldas de Cuntis
Caldas de Estrac
Caldas de Malavella
Caldas de Mombuy
Caldas de Oviedo
Caldas de Reyes
Caldas de Tuy
Caldillas de San Miguel
Candio
Canena
Canillejas
Carbagnal
Carballo
Carballino
Carratraca
Casares
Castenar dei Bor
Cati
Cervera
Cestona-guezalaya
Chiclana
Chulilla
Concepcion de Peralta
Corcoles
Cortejada

Cuervo
Diezgo
Elche
Elorrio
Escoriaza
Esparaguera
Estadilla
Ferreira
Fitero
Fontarabbia
Fonté
Fortuna
Frailes y la Rivera
Fuen-Alamo
Fuen Caliente
Fuente Amargosa
Fuente de Piedra
Fuente del Rosal
Fuente del Toro
Fuente Sta. de Lorca
Fuente Sta. de
Gayangos
Galera
Gandesa
Garriga, La
Gava
Gaviria
Gigonza
Gijon
Grabálos
Graena
Grao

SPAIN—continued.

Gravalos
Guarda-Vieja
Guesalivar
Guitiriz
Hermida, La
Hervideros del
 Emperador
Hervideros de
 Fontillescа
Hervideros de
 Fuen Santa
Hervideros de Villar
 del Pozo
Horcajo de Lucena
Huelva
Humera
Ibero
Inchaurte
Jabalcuz
Jacintos
Jaen
Jaraba
La Garriga
La Salvadora
Landetta
Lanjarron
Ledesma
Lés
Lierganes
Lodova
Löeches
Lucarnena
Lujo
Madrid
Mala
Malaga
Marbella
Marmolejo
Martos
Mecina-Burbaron
Montanejos
Molar, El

Molina de Aragon
Molinar de Caranza
Monda
Mondariz
Mondon
Monegrillo
Montemayor
Mula
Murcia
Nanclares
Navajas
Navalpino
Novelda
Nuestra Señora de
 Abella
Nuestra Señora de las
 Mercedes
Olivera
Ontaneda
Orense
Ormaiztegui
Panticosa
Paracuellos
Paterna
Pixigueiro
Portovia
Portugos
Pozo-Amargo
Prelo
Puda, La
Puente Viesgo
Puertolano
Quinto
Rio-Tinto
Riva-Los-Boños
Rivera
Rubinat
Salinetas de Novelda
Sacedon
San Adrian
San Bartolomé de la
 Quadra
San Gregorio de Brozas

San Juan
San Juan de Campos
San Sebastian
San Vincens
Santa Agueda
Santa **Ana**
Santa Filomena de
 Gomillaz
Saudon
Seguara de Aragon
Segura
Sierra Albamilla
Sierra **Elvira**
Siete Aguas
Sobron
Solar de Cabras
Solares
Sousas y Caldellinas
Tardon
Teruel
Thermas, Las
Tiermas
Titus, Baños de
Torre de San Miguel
Trillo
Urberoaga de Alzola
Urberoaga de Ubilla
Vacia-Madrid
Valle de la Cueva
Valle de Rivas
Valdeganga
Valencia
Villacarillo
Villafafila
Villaharta
Villar del Pozo
Villaro
Villatoya
Villavieja de Nules
Vilo or Rosas
Zaldivar
Zujar

SWEDEN AND NORWAY.

Adolfsberg
Badstofuhver
Borås
Daneverd
Fällorne
Gustafsberg
Helenskilde
Hernösand

Lanaskede
Loka
Lund
Maestrand
Medewi
Modum
Porla
Ramlöza

Ronneby
Sandefjord
Stromstad
Torpa
Vaxholm
Wahlbergn
Wallby Baths

SWITZERLAND.

Aigle
Albisbrunn
Allerheiligen
Alveneu
Andeer
Appenzell

Arp
Arzilhe
Attisholz
Axenfels
Axenstein
Baden

Bagni di Craua
Bagno
Balzach
Beggenried
Belalp
Bellrière

CLASSIFICATION.

SWITZERLAND—*continued.*

Bex
Birmansdorf
Bleichbad
Blumenstein
Blumenthal
Bonn
Brigg
Brüttalen
Bubendorf
Bürgenstock
Campfer
Coire
Combe-Girard
Daetlingen
Davos
Dardagni
Dorfbad
Drise
Enatbühl
Engistein
Eppingen
Ermetschwyl
Evolena
Farnbühl
Fideris
Fläsch
Flims
Forstegg
Gais
Garmyswyl
Gempelenbath
Geneva
Gisi
Golaise
Gonten
Gränichenbad
Grindelwald
Gurnigel
Gyrenbath
Hübernbad
Haldenstein
Heiden
Heinrichsbath
Heustrichsbath
Horn
Ibenmoos
Iberg
Interlaken

Jenatz
Kirchleerau
Knutwyl
Laeunnoli
Lalliaz
Lausanne
Lavey
Leissigen
Leprese
Limpach
Locarno
Lochbachbad
Loëche-les-Bains
Losdorf
Lugano
Luxburg
Mahhá
Maloja
Martigny
Meltingen
Merligen
Montbarri
Montreux
Morgins
Mürren
Niederweil
Nydelbad
Ollon
Osterfingen
Paradies
Passugg
Peiden
Pfaeffers
Piguien
Pignol
Pontresina
Ponts, Les
Poschiavo
Ragatz
Ransbath
Rheinfelden
Righi
Riedbad
Rolle
Römerbath
Röslibath
Russwyl

St. Cergues
St. Moritz
Samaden
San Bernhardino
Saxon
Schinznach
Schimberg
Schmerikon
Schönbrunn
Schöneck
Schöngau
Schüls
Schüols
Schwarzenberg
Schwarzseebath
Schwefelbergerbath
Seewen
Stachelberg
Suot-Sass
Tarasp
Teufen
Thalgut
Thoune
Thusis
Trois-Torrens
Unterbath
Unterhallau
Unterseen
Urnäschen
Vals
Vevey
Veytaux
Villeneuve
Visibachbath
Waidbaldenbath
Waldstatt
Wattwyl
Weggis
Weissbad
Weissenburg
Weissenstein
Wildegg
Wildhausbath
Yverdon
Zermatt
Zürich
Zug

TURKEY (European and Asiatic).
Servia, Roumania, Bulgaria.

Ahioli
Aias
Aidos
Bobotsch
Brussa
Calliarhoë
Chios
Ecski-cherrer
Feredschik
Fineeschti

Gadara
Hassan-Pacha-Pallenka
Ildjak
Jallova
Kalimaneste
Kimpalangi
Kirkilissa
Kosia
Kostendil

Lidja
Metelin
Oloneschti
Otschin
Patmos
Salonichi
Serboueschte
Sibitschudi-Suz
Smokobo
Smyrna

SPECIALISTS NAMED IN THE BOOK.

Dr. Seegen	Diabetes	Karlsbad ... page 162
Dr. Debout d'Estrée	Gravel	Contrexéville ,, 89
Dr. Bennet	Lungs	Mentone ,, 266
Dr. Kaposi	Skin Diseases	Vienna ,, 339
Dr. Metzger	Contortions, Varicose Veins and Articular Pains	Amsterdam ,, 18
Dr. Schmitz	Diabetes	Nenenahr ,, 221
Dr. Cazaux	Throat	Eaux Bonnes ,, 102
Dr. Nothnagel	Balneology	Vienna ,, 339
Dr. Haase	Eyes and Ears	Homburg ,, 149
Dr. Veiel	Skin Diseases	Cannstadt ,, 69
Dr. Fischer	Nervous Complaints	Do. ,, 69
Dr. Averbeck	Morbus Brightii	Laubbach ,, 175
Dr. Von Langenbeck	Surgeon	Wiesbaden ,, 348

www.ingramcontent.com/pod-product-compliance
Lightning Source LLC
Chambersburg PA
CBHW051233300426
44114CB00011B/718